Animal Traditions
Behavioural Inheritance in Evolution

Animal Traditions maintains that the assumption that the selection of genes supplies both a sufficient explanation of the evolution of behaviour and a true description of its course is, despite its almost universal acclaim, wrong. Eytan Avital and Eva Jablonka contend that evolutionary explanations must take into account the well-established fact that, in mammals and birds, the transfer of learnt information across generations is both ubiquitous and indispensable. The introduction of the behavioural inheritance system into the Darwinian explanatory scheme enables the authors to offer new interpretations for common behaviours such as maternal behaviours, behavioural conflicts within families, adoption and helping. This approach offers a richer view of heredity and evolution, integrates developmental and evolutionary processes, suggests new lines for research and provides a constructive alternative to both the selfish gene and meme views of the world. It will make stimulating reading for all those interested in evolutionary biology, sociobiology, behavioural ecology and psychology.

EYTAN AVITAL is a lecturer in Zoology in the Department of Natural Sciences at David Yellin College of Education in Jerusalem. He is a highly experienced field biologist, and has written one zoology text and edited several others on zoology and evolution for the Israel Open university.

EVA JABLONKA is a senior lecturer in the Cohn Institute for the History and Philosophy of Science and Ideas, at Tel-Aiv University. She is the author of three books on heredity and evolution, most recently *Epigenetic Inheritance and Evolution* with Marion Lamb.

Animal Traditions

Behavioural Inheritance in Evolution

Eytan Avital
Department of Natural Sciences
David Yellin College of Education, Jerusalem

and

Eva Jablonka
The Cohn Institute for the History and Philosophy
Of Science and Ideas,
Tel-Aviv Univeristy

CAMBRIDGE
UNIVERSITY PRESS

PUBLISHED BY THE PRESS SYNDICATE OF THE UNIVERSITY OF CAMBRIDGE
The Pitt Building, Trumpington Street, Cambridge, United Kingdom

CAMBRIDGE UNIVERSITY PRESS
The Edinburgh Building, Cambridge CB2 2RU, UK
40 West 20th Street, New York, NY 10011-4211, USA
10 Stamford Road, Oakleigh, VIC 3166, Australia
Ruiz de Alarcón 13, 28014 Madrid, Spain
Dock House, The Waterfront, Cape Town 8001, South Africa

http://www.cambridge.org

First published 2000

Printed in the United Kingdom at the University Press, Cambridge

Typeface in Swift 9.5/13pt, in QuarkXPress™[wv]

A catalogue record for this book is available from the British Library

Library of Congress Cataloguing in Publication data
Avital, Eytan, 1951–
    Animal traditions: behavioural inheritance in evoloution / Eytan Avital and
    Eva Jablonka.
        p. cm.
    Includes bibliographical reference.
    ISBN 0 521 66273 7
    1. Animal behavior. 2. Behavior evolution. 3. Behavior genetics I. Jablonka, Eva II.
    Title

QL751 .A94 2000
591.5–dc21   00-037819

ISBN 0 521 66273 7 hardback

To Marion and Silvi with love and gratitude

Contents

Preface

This book is about the way in which the evolution of birds and mammals is affected by social learning and by the traditions formed by social learning. From observation and experiment, we know that higher animals can acquire information from or through the behaviour of others, and through their own behaviour they can transmit this information to the next generation. Variations in such socially acquired and transmitted behaviour-influencing information cannot be under direct genetic control, since animals with very similar genes can have, and pass on, very different behaviours and traditions. There is clearly another inheritance system, a behavioural system of information transmission, which is superimposed on the genetic system. Some years ago we decided that the evolutionary consequences of this additional tier of variation and inheritance were worth exploring, and set out to see how our view of the evolution of higher animals is altered by incorporating non-genetic behavioural inheritance and the traditions that it produces. This book is the outcome of that endeavour.

We found that adding the behavioural system of information transmission has some radical implications for the current gene-centred view of evolution. For example, the classical distinctions between development and evolution become very blurred. An animal tradition is the product of a historical, evolutionary process, yet it can be formed and transmitted only if it is actively constructed during the behavioural development of individuals and groups. Unlike genetic information, behavioural information must be used and displayed for it to be transmitted. The generation and transmission of learnt behaviours are therefore not independent of their development, since any change acquired during the development of a behaviour can be passed on. Consequently, evolutionary adaptations are not shaped exclusively by selective processes; the evolution of behavioural adaptations involves the inheritance of acquired characters.

In acknowledging the importance of behavioural inheritance, we reject the rigid gene-centred sociobiological view of evolution, which ignores the influence of habits and traditions. As we see it, any evolutionary interpretation of social behaviour requires a consideration of both genetic and 'cultural' factors. Variations in socially transmitted behaviours affect evolution in two ways: first, the variant behaviours are an additional source of raw material for selection; second, social behaviour forms part of the selective regime in which individuals live, learn and reproduce. Therefore, habits and traditions are not merely the products of evolution – they are also one of its major constructing agents.

In presenting our ideas, we have not attempted to provide a comprehensive review of animal social learning and cultural evolution, since they are subjects which, much to our delight, are reaching vast, textbook dimensions. Instead, we have explored the consequences of the two-tier thinking that we believe is necessary for the evolutionary interpretation of various types of behaviour – thinking that includes both the behavioural and the genetic inheritance systems. In most chapters, we start our discussion by presenting a description of real-life behaviour based on our own observations of animals, because we feel that such descriptions make the reading more enjoyable and introduce a social and ecological framework that makes the subsequent discussion easier to follow. To avoid the inevitable stumbling over unfamiliar Latin names, we use common species names throughout the book, but provide a detailed index of species at the end. We present many examples that portray the scope and breadth of animal traditions, and explain our reasons for believing that this is merely the tip of a large iceberg. We show that when we consider the exchange and the sharing of learnt information in an evolutionary framework, we are often compelled to change previous assumptions about the costs and benefits of the behaviour for the interacting partners, and hence change established evolutionary interpretations. We also illustrate how the different ways that information can be transmitted by the behavioural inheritance system – through two parents, through a single parent or through non-parents – may alter the manner and the direction in which evolution proceeds. We go on to argue that social learning and the establishment of traditions may lead to a shift in the level at which selection occurs, and sometimes to rapid and effective speciation. Finally, we examine the intricate and subtle ways in which learning and traditions affect the evolution of the genetic basis of learning and of morphological characters. The overall result is a picture of the evolution of animal behaviour that is driven and shaped by habits and traditions.

Although this book is mainly about birds and mammals, some aspects of the developmental approach we advocate are also relevant to invertebrates. The implications of the view we present also bear on how one sees the structure of the human mind, and how the developmental and evolutionary relationships between genes and behaviour are understood. These two problems are related, and are at present hotly debated. Our position, which we explain in the introductory chapters and return to in the final chapter, is that the facile assumption that many animal and human behaviours are underlain by pre-existing, specifically selected mental modules is often unfounded. The relationship between mind, behaviour and genes is usually a lot more subtle and circuitous than is assumed by most sociobiologists and evolutionary psychologists. The structure, the limits and the possibilities of the wide behavioural plasticity of birds and mammals have hardly been explored, and, although all behaviour has a genetic basis, the attempt to derive behaviours or mental states solely from pre-existing genetic programs is at best problematical, and at worse absurd.

Writing this book has been a very gratifying experience. We were encouraged by colleagues from our home institutions and from elsewhere. We are very grateful to the Wissenschaftskolleg in Berlin, which enabled E. J. to spend a year in a stimulating and challenging environment, and provided ideal working conditions. The discussions in the biology group, the excellent library services and the encouragement provided by the fellows and the staff of the Kolleg have been invaluable.

We were, in fact, extremely fortunate to encounter the scientific community at its best. Almost all of the very many individuals to whom we turned for material or advice have been enormously generous, often without knowing us previously. They devoted much of their time to our questions and requests, sent us papers and even books that we could not obtain, corresponded with us and helped us in every conceivable way. In this cynical age, it was a heart-warming experience. We have learnt and benefited enormously from the flood of material we received, although because of lack of space we could not incorporate all of it into the final manuscript. We are very grateful to all of them. In particular, we would like to thank the following friends, colleagues, students and former teachers for various forms of help, advice and encouragement: Israel Avital, Orit Avital, Zvi Atzmon, Carol Berman, Sharmila Choudhury, Yehuda Elkana, Rachel Galun, Lilach Gang, Dani Golani, J. Lee Kavanau, Mikhal Lederer, Yaron Lehavi, Alicia F. Lieberman, Shlomit Magen, Yoram Okhanuna, Laor Orshan, Craig Packer, Meir P. Pener,

Frederick D. Provenza, Alexander F. Skutch, Iddo Tavory, Timothy H. Tear and Amotz Zahavi.

Special thanks are due to our colleagues and friends Evelyn Fox Keller, Simona Ginzburg, Claudia Goebel, Rainer Goebel, James Griesemer, Peter Hammerstein, Dan-Eric Nilsson, Ekkehart Schlicht and Eric Warrant, who each read one or more chapters of the book and gave us invaluable criticism, penetrating comments and a lot of encouragement. We thank our editor, Tracey Sanderson, for her help and patience through the gestation and delivery periods of this book. We also thank our mothers for the very valuable and long-lasting legacies they have transmitted to us. Silvi Fridman-Avital deserves special thanks for her competent editorial assistance, useful criticism of some of the chapters, and for the construction (along with Dror and Shakhaf) of a long-term, loving writing-niche for E. A. The greatest thanks go to our friend and colleague, Marion Lamb, without whom this book would never have seen the light of day. She read the whole book, scrutinised and sharpened our arguments, pointed out numerous inconsistencies and contradictions, and improved the English. It is to Marion and to Silvi that we dedicate this book.

Eytan Avital
Eva Jablonka

Acknowledgements

We are grateful to the following publishers, institutions and authors for permission to quote from their books or papers: Academic Press, for the quotation (p. 94) from C. M. Heyes' paper in *Animal Behaviour*, 1993; Bloomsbury for the quotation (p. 96) from R. Wrangham & D. Peterson's 1995 book, *Demonic Males*; Cambridge University Press for the quotation (pp. 320–1) from *Darwin's Biological Work* (ed. P. R. Bell, J. B. S. Haldane, P. Marler, H. L. K. Whitehouse & J. S. Wilkie, 1959); Sigmund Freud © Copyrights, the Institute of Psychoanalysis and the Hogarth Press for the quotation (p. 299) from the Standard Edition of *The Complete Psychological Works of Sigmund Freud*, 1952; Houghton Mifflin Company for the definition (p. 22) taken from *The American Heritage® Dictionary*, 3rd Edition, 1992; Oxford University Press for the quotations (pp. 248 and 369) from S. Asch's *Social Psychology* (1952), and the quotation (p. 181) from D. W. Mock & G. A. Parker's *The Evolution of Sibling Rivalry* (1997); Princeton University Press for the quotation (p. 227) from H. Kummer's (1995) book *In Quest of the Sacred Baboon*; Routledge for the quotation (p. 61) from H. Hendrichs's paper in *Evolution, Order and Complexity* (ed. E. L. Khalil & K. E. Boulding, 1996); the Royal Society of Medicine for the quotation (pp. 264–5) taken from J. B. Calhoun's paper in *Proceedings of the Royal Society of Medicine*, (1973); Robert Trivers for the quotation (p. 183), from his book *Social Evolution* (1985, Benjamin Cummings: Menlo Park, California); W. H. Freeman for the quotations (p. 25 and p. 358) from R. Dawkins' (1982) *The Extended Phenotype*; W. W. Norton & Co. Inc. for the quotation (p. 48) from J. B. Watson's 1924 book, *Behaviorism*.

1   New rules for old games

If you ask a biologist to explain the evolution of the elaborate morning song of a great tit, the subtle food preferences of a domestic mouse, or the efficient hunting techniques of a pack of wolves, what sort of explanation will you get? The chances are you will be told that this type of behaviour can readily be explained by the conventional theory of natural selection acting on genetic differences between individuals. Ever since Darwin, the theory of natural selection has been applied to all sorts of biological problems, from the origin of life to the origin of language, and for most of this century it has been assumed that genetic differences between individuals underlie the variation on which natural selection acts. It is not surprising, therefore, that behavioural evolution is also seen as the outcome of the selection of genetic variations. But is this view correct? In this book we are going to argue that when applied to the behaviour of higher animals, conventional evolutionary theory is rarely adequate and is often misleading. Natural selection acting on genetic differences between individuals is not a sufficient explanation for the evolution of the behaviour of the great tit, the mouse or the wolf.

To understand why we are not satisfied with the current application of Darwin's theory to behaviour, we need to go back to basics. Darwin's theory depends on some fundamental properties of biological entities: on their ability to reproduce, on the differences between individuals and on the heritable nature of some of these differences. In situations in which resources are limited, the interaction of these properties leads to natural selection: heritable variations that increase the chances that the individuals carrying them survive and reproduce will, in time, become more frequent. Eventually, the cumulative effects of selection lead to evolutionary adaptations – to the wing of the swallow, the song of the nightingale, the dam of the beaver. In this general formulation, the theory is comprehensive and powerful, and can bear upon evolutionary processes of all kinds and at all levels. Like most biologists, we accept that Darwinian natural selection is of central importance in the evolution of behaviour. What we are dissatisfied with is not Darwinism, but the currently fashionable version of Darwinism, which we will refer to as 'genic' Darwinism. Many of the assumptions made by the proponents of the genic version of Darwinism seem to us to be oversimplified and restrictive. We are therefore going to look again at some basic questions that are relevant to the application of Darwinian evolutionary theory to behaviour. We want to ask: what is the nature of the raw material of behavioural evolution? What is the origin of heritable variation? How are variations transmitted? How does behavioural evolution by natural selection work?

These questions may sound strange, even if not downright silly and unnecessary. After a century of genetics and over half a century of molecular biology, many people feel that they know the answers: the hereditary variations are variations in genes, in DNA base sequences. New variants arise through random changes in these DNA sequences, and are transmitted when DNA is replicated. The processes that lead to changes in genes are 'blind', so the new variants are not adaptive responses to the life experiences of the organisms that produce them, and do not anticipate the needs of the offspring that inherit them. The effects that these random changes in DNA have on the characteristics of organisms lead to differences in their ability to survive and their success in producing offspring. Over time, genes with effects that improve an individual's chances of leaving descendants – that increase fitness – become more frequent in the population.[1] Natural selection is basically gene selection.

What is wrong with these gene-centred answers to our questions? We are certainly not going to deny the fundamental importance of genetic variation in the evolution of behaviour. What we are going to maintain, however, is that explaining the evolution of animal behaviour in terms of gene selection alone is a mistake. Gene selection alone cannot account for a lot of the behaviour seen in higher animals, including the song of the great tit, the behaviour of the wolf pack and the food preferences of the mouse. These three examples were not chosen at random. What they have in common is that they all involve a special type of learning – social learning. With social learning, animals learn from others how to behave. Generally, in discussions of the evolution of behaviour, social learning is treated merely as a product of gene selection, but social learning is more than this: social learning can be an important agent of evolutionary change. We therefore think that it should be given a more prominent place in evolutionary theory. Darwinian evolution depends on heritable differences between individuals, but not all heritable differences stem from genetic differences. The behavioural differences that are transmitted through social learning also provide the raw material on which natural selection acts.

To illustrate our point we want to carry out a thought experiment that will enable us to think about the evolution of behaviour without resorting to the selection of genes. Imagine a large population of small, brownish, omnivorous, rodent-like mammals, living in small family groups in a species-rich, semi-desert habitat. Call them 'tarbutniks'.[2] Each family consists of a pair of parents and young of various ages. All individuals in the population, indeed in the whole species, are genetically identical. Furthermore, not only are all the tarbutniks genetically identical, but their genes never mutate, so there is not even the possibility of genetic differences between them. However, they are not all identical in appearance and behaviour. Some are larger than others, there are slight differences in their coat colour, their calls are not identical, they produce different numbers of offspring, and there are various other small differences in their anatomy and the way they behave. But there is no correlation between parents and offspring in either appearance or behaviour: the tarbutnik-pups are no more similar to their parents than to any other individual in the population. The differences between individuals are the result of accidental events during their development, and these variations are not heritable. Consequently, although the population may increase or decrease in size, may fill the earth or go extinct, since the variations are not inherited, it does not evolve.

Our tarbutniks start their lives as helpless young, sucking their mother's milk; they grow rapidly, and are soon foraging with their parents for anything that is edible. They are extremely curious, and can learn about their environment through individual trial and error. By trying again and again, they eventually discover a good way of opening nuts and getting at the seeds. After some bitter experiences, they learn that black-and-red striped bugs are best avoided. This ability to learn is important: they possess an excellent memory, so they usually benefit greatly from their past experiences. But they cannot learn from the experience of other individuals, and can never be influenced by anyone else's behaviour. Whatever experience an individual has accumulated, whatever useful information it has acquired about its surroundings, this knowledge is never shared. Each young tarbutnik has to find out about the world through his or her own trial-and-error learning.

Now let us change just one single factor in our imaginary world: let us add to tarbutnik life social learning. By social learning we mean that individuals can learn not just from their own experience, but also from the experience of others. Since age groups overlap, information is transmitted between, as well as within, the generations. A mother can transmit information to her young, young can learn from their fathers and from neighbours, peers can learn from each other. Gradually, patterns of behaviour spread among individuals. What is more, the socially transmitted behaviour patterns can change progressively. An individual tarbutnik that somehow discovers or learns by trial and error something new and useful, such as an additional type of food, can transmit this knowledge to its offspring. Thanks to its new food source, this tarbutnik may be more successful than others in producing and rearing pups. Its lineage will thrive. Even if the better-informed individual does not have more biological offspring, it may have more 'students' ('cultural offspring') who learn its new and useful pattern of behaviour. The new behaviour may thus spread in the population. The addition of social learning to a social organisation in which young and adult individuals regularly interact has introduced the possibility that behaviour patterns can be transferred from one generation to the next. Since some variations in behaviour are now heritable, Darwinian evolution is possible!

It is easy to imagine how new and useful learnt behaviours in our tarbutnik population can accumulate and become perfected by natural selection, so that a complex behavioural adaptation, such as constructing and using a burrow, can evolve. First, a tarbutnik may discover by chance, or through individual trial-and-error learning, or perhaps even by observing individuals of other species, that by occupying a simple hole in the ground they and their offspring are better hidden from predators. The offspring do not have to reinvent or rediscover this: they, as well as other individuals in the group, learn this useful habit from experienced parents, and some may even elaborate on it. They may start extending existing holes by digging, and produce something resembling a short tunnel, which gives them even better protection, not only from predators but also from the extremities of the weather in their semi-desert habitat. By chance, some may dig a tunnel with an entrance and an exit. The tarbutniks who do this evade snake attacks and survive better than others, so the habit spreads. Some tarbutnik mothers produce their young in the burrow they dig, and this habit, which protects both mother and young, also spreads. The individually acquired inventions may be combined and accumulate, producing traditions that change the life style of the animals.

The evolution of traditions, which involves the modification and selection of behaviours learnt from family and neighbours, can lead to more than artefacts like burrows. Foraging traditions, traditions of parental care or traditions of mate choice may also evolve through the selective accumulation of individual variations in behaviour. The way tarbutniks communicate with each other may also be influenced by such evolved traditions. Imagine that a parent discovers that in dense cover, but not in the open, its young respond more readily to an alarm call of a particular frequency. The use of this dense-cover call will probably spread, because the young are less likely to get lost or be eaten by predators, and when they themselves become parents they will use, and hence transmit, the alarm call they learnt. Similarly, think of what might happen if a male discovers that females who are given their favourite food, red berries, are more willing to accept his advances. Thanks to this discovery, he fathers more young than his rivals. His observant sons and their young male friends soon learn and repeat this behaviour. The habit spreads.

But we can go even further. Imagine that the original large tarbutnik population becomes fragmented – massive flooding makes a river change its course and splits the original population into two groups, unable to contact each other. The individuals in one group may, in time, become so different in habits and preferences from members of the other group that, even if they had the chance, they would never, or seldom, communicate with, mate with or learn from members of the other group. One group's courtship offering is red berries, but the other uses nuts, which berry-preferrers have no idea how to deal with. Males offering nuts to berry-preferring females are rejected, and nut-preferring females do not accept berry-offering males. An effective reproductive barrier has been established. Behavioural speciation has occurred,[3] and may lead to the groups diverging even more. Remember that no genetic change is possible in our tarbutniks, so all of their evolution is through the transmission of behaviours. What we see is cultural evolution.

Now let us return to the real world. Unlike our tarbutniks, real organisms are not immune to genetic changes. There is an almost unlimited supply of genetic variation in real animals, which makes it impossible to focus exclusively on cultural evolution. But this is not a good reason for ignoring the role of the cultural inheritance of habits. To do so, leaves too much unexplained. For example, how can we explain differences, such as the different song dialects of family groups of sperm whales, which cannot be attributed to differences in genes? It seems that these dialects are not related to gene differences, but are determined by evolving local traditions, passed on by vocal imitation. In a case like this, we can focus on the transmission of behavioural variations through social learning while ignoring, for the time being, the effects of any gene differences. Of course this does not mean that genes are unnecessary and dispensable. What it does mean is that differences in genes may be irrelevant for some variation in heritable behaviour, at least for a while. So, when we talk about behavioural transmission, we mean that the transmitted differences in behaviour do not depend on genetic differences, but we do not mean that behaviour is devoid of a genetic basis, that it is gene-free!

It can be argued, of course, that, although cultural evolution can, in theory, lead to staggering diversity and spectacular adaptations, it is really a relatively minor and unimportant process, of no significance in the evolution of the basic patterns of behaviour in animals, or even in man. According to this line of argument, all the significant questions about the song of the great tit, the hunting of the wolves or the food preferences of mice, can be answered in terms of gene selection alone, without recourse to non-genetic transmission of behaviour. This gene-centred view is the prevalent view today, so we need to look at it more closely.

Why genes are not enough

The gene-centred view of behavioural evolution is the one offered by classical sociobiology theory. Through the publication of E. O. Wilson's milestone book *Sociobiology*, the grand ambition of sociobiology was clearly spelled out: to understand the social behaviour of animals, and even of man, in terms of gene selection. According to the sociobiologists, variations in genes determine heritable variations in social behaviour; some behaviours result in the production and survival of more offspring than others, so the genes responsible increase in frequency and the social behaviour of the population evolves. Psychology and sociology were to be incorporated into biology, since explanations of human behaviour would be found in the genes that have been selected during evolutionary history. This idea, and in particular its supposed implications for human freedom of action, was, and still is, hotly debated, and split the scientific community into excited supporters and scornful dismissers.

The case for a gene-centred view of evolution in general, and of the evolution of non-human behaviour in particular, was persuasively advocated by Richard Dawkins in books such as *The Selfish Gene* and *The Extended Phenotype*. In time, this once controversial view became the standard evolutionary wisdom. Dawkins argued that the most fruitful and economical way of interpreting adaptive evolution is to look at it through the lens of the gene: to consider the gene as the unit of variation and selection. The catch-phrase Dawkins coined, 'the selfish gene', in fact denotes the way copies of a gene spread through a population at the expense of other variants of the same gene. It is a different way of formulating the old view that evolution is a change in gene frequencies. Using ideas developed by William Hamilton and George Williams in the 1960s, Dawkins showed how many of the long-standing problems in evolutionary biology disappeared if the gene, rather than the individual, was made the principal level of analysis. In particular, the unselfish, altruistic acts of social animals made evolutionary sense when looked at from the selfish gene's point of view.

The selfish gene idea generated a lot of controversy. Some critics attacked it for being a restrictive view of evolution which, because it ignores other levels of selection and variation, leads to more or less (usually less) sophisticated versions of genetic determinism, of the notion that genes govern everything animals are and do.[4] However, most of the critics were less concerned about general issues, and far more worried about the implication of the gene-centred approach for interpreting human social behaviour. They ignored, or uncritically accepted, its implications for animal social evolution, but attacked its application to humans. These critics felt that something rather important – culture – had been left out. However, even in his first book, *The Selfish Gene*, Dawkins had suggested that something extra was involved in human evolution: he argued that cultural evolution proceeded through the selection of 'memes'. He defined memes as units of information (such as ideas) which reside in the brain and are transmitted from one person to another by behavioural means. He envisaged human cultural evolution as being dominated by the replication and selection of memes rather than genes.[5] Nevertheless, in spite of the meme idea, the majority of sociobiologists, who endorsed Dawkins' view of evolution, regarded human culture as an adaptive by-product of the selection of genes. The transmission of memes did not alter the basic rules of the evolutionary game. It was assumed that, since the ability to pass on ideas and behaviours is itself a result of gene-based selection, it is only the genetically determined ability to produce culture that is evolution-arily interesting. Culture is, in fact, still considered as a kind of 'icing on the cake', even when thinking about human evolution. It is usually excluded from the interpretations of the evolution of those funda-mental species-specific human behaviours that have a significant 'innate' component. So cultural inheritance is deemed irrelevant to the evolution of the *ability* to acquire language, the *ability* to have complex and multiple social interactions, the *ability* to control muscles and emotions, and so on. Gene differences are so obviously involved in the evolution of 'innate' behaviours, that most evolutionary biologists automatically exclude any role for culture in their evolution.

It is important to clarify at this early point what we mean by 'innate'. 'Innate behaviour' is the term used for a pattern of behaviour whose development is not dependent (or is only slightly dependent) on a process of learning, and is not altered by variations in the environmental conditions that the animal experiences. This does not mean that envi-ronmental conditions and experience are unimportant; like any other trait, a pattern of behaviour is always the result of interactions between the animal and its environment. What it means is that most of the differences in individual experiences and conditions make no difference to the development of the mature, species-specific, behaviour. 'Innate' behaviour is relatively independent of learning. Most people think of 'innate' behaviour as 'genetically determined' behaviour, but, as we shall see in this and later chapters, there are problems with this view.

The relative contribution of culture and genes to the development of social behaviour is a complex issue and one that is often misunderstood. No biologist in his or her right mind would deny that there is a genet-ic basis for the ability to transmit cultural practices. Equally, even the most fanatical sociobiologist would happily admit that many behaviours are the result of the way genes are expressed in a particular environ-ment, and that genetically identical organisms, such as identical twins, can display different behaviours as a consequence of differences in diet, education and family relationships and for other complex reasons. However, the sociobiologists argue that, since the *range* of cultural practices depends on genes, the genetic level is the preferred level of explanation. Thus they argue that what needs to be explained is not the evolution of a particular 'cultural' practice, such as Christmas dinner or the Jewish Seder, but rather the evolution of the genetically determined psychological mechanism, the genetic strategy, that leads to food-sharing. It should be noted, however, that not only is it assumed that a defined strategy is inscribed in the genetic material, but it is also often assumed that the regulation of this strategy by the environment is genetically determined. Robert Wright, one of the spokesmen for modern human sociobiology, asserts that not only are the 'knobs of human nature' (for example food-sharing) genetically determined, but so also are the ways in which the 'knobs' can be calibrated (where, when and how to share food). The calibration is accomplished 'by a generic, species-wide developmental program that absorbs information from the social environment and adjusts the maturing mind accordingly'.[6] According to such sociobiologists, it is possible to explain not only general cognitive, emotional and social patterns of behaviour in terms of genes, but also more specific ones – self-deception and a sense of duty, humour and a hatred of strangers.[7]

This way of thinking has led most human sociobiologists to argue that the genetic strategies that have evolved are embodied in the mind of man as highly specialised semi-autonomous cognitive units, which they refer to as 'modules'. A neural module is a dedicated neural circuit in the brain that processes only a certain type of incoming information (e.g. information about potential mates) rapidly and in an unconscious way.[8] These genetically determined modules, which underlie our allegedly very definite human nature, are the consequence of past selection in 'the environment of evolutionary adaptation' or 'the ancestral environment'. This environment is that imagined for our hominid ancestors, starting about two million years ago, when *Homo erectus* first roamed the plains of Africa. By making fitting assumptions about what the 'ancestral environment' was like, the past function of each and every behaviour is inferred. A specific psychological mechanism is then assumed to underlie each observed type of behaviour. It is assumed that genes for each mechanism have been selected, so that it is embodied in the brain as an independent cognitive module. The same explanatory strategy is used to provide explanations for all social behaviour patterns, however esoteric. Since this type of argument can readily explain every conceivable behaviour, why do we maintain that evolutionary biologists need to incorporate an additional inheritance system into their explanations? Why are genetic strategies not enough? What is wrong with the assumption that the mind is an assembly of separately selected semi-autonomous cognitive modules?

There are several reasons why something is wrong. As we shall discuss in more detail in the next chapters, for some traits in animals and man there is little evidence for substantial genetic determination. In fact, even seemingly 'fundamental' and 'innate' patterns of behaviour, such as whether or not a relationship is monogamous, or how the young are cared for and by whom, differ between populations of the same species.[9] It is often impossible to predict the mating system or the type of parental care that will be found without knowing the ecology and history of the population. Moreover, not only are there many ecological and historical variations in patterns of behaviour, but we also know that some of them are passed on from one generation to the next. They are cultural and heritable. Many people argue that using the term 'culture' for animal traditions is inappropriate, and we shall discuss these difficulties in a later section. For the time being we will use the term 'culture' in a diffuse and intuitive manner to mean social traditions and sets of social traditions. One example of what we regard as animal culture is the well-studied food-handling behaviour of the group of Japanese macaques living on the small, wooded island of Koshima. These monkeys used to live and forage in the forests, but Japanese primatologists started to feed them by scattering sweet potatoes on the sandy beach. Soon, the monkey troop began to leave the forest and feed on the beach. About a year after the feeding started, a young female monkey was observed to wash the potatoes in a nearby brook, actively removing the adhering sand. Within the next few years, potato-washing spread through the troop, and the practice was transferred from the brook to the sea. As well as potato-washing in the sea, several other habits associated with feeding on the sandy beach are now well established in the group of macaques on Koshima. The habits are transmitted from mothers and other group members to the infants.[10]

Japanese macaques are not the only animals to have changed their behaviour in recent times. In many cities and towns, European red foxes have successfully adapted to their new and complex urban habitat over a period that has been far too short to allow adaptation through the selection of genes. The same is true of common racoons in America and Palestine sunbirds in Israel. These facts do not disappear into thin air just because they do not fit gene-based selection theories. In birds and mammals, the inheritance of habits, of information transmitted through social learning, is both ubiquitous and indispensable. Given the existence of patterns of behaviour that are reliably transmitted from one generation to the next and are selected at the 'cultural' not the genetic level, it is illogical to base theories about the evolution of behaviour solely on specific brain modules that were constructed via the selection of genes. The course of the evolution of behaviour cannot be adequately described and understood without incorporating 'culture' as an active and interacting evolutionary agent that affects the selection of genes. Genes are not enough.

Why culture is not enough

With respect to humans, the opposite view to that of the sociobiologists is also common, particularly among social scientists, who deny a role for genes in human cultural evolution. These social scientists argue that the range of behaviours that an individual human being with a particular set of genes can exhibit (what is known in biological jargon as 'behavioural plasticity') is very wide, practically indeterminable. What is more, they argue, people with different genes can show very similar behaviours. Gene differences are therefore deemed to be irrelevant to the behaviour seen in society, because they do not underlie differences in behaviour. Although genetic evolution may have led to the ability to produce habits and traditions, once this ability is in place, genes only limit the range of possible behaviours, and these limits are so wide that gene differences are, in effect, negligible. According to this view, the explanation of cultural differences and cultural change lies in the realm of the social sciences, not biology. Purely cultural evolution, such as that which we described in our imaginary tarbutniks, is sufficient.

We agree that genes limit rather than determine traditional or cultural differences, and in many cases variations in genes can safely be ignored. However, even when there is a lot of plasticity, variation is often constrained and organised so that among the many things that can be learnt and transmitted, some are learnt and transmitted more easily than others. For example, it is easier for a rat to associate gastric discomfort with taste than with sound, and it is easier for humans to memorise rhymes than to memorise part of the telephone directory. The need to understand the structure of plasticity compels us to think about the structuring factors, and genes are among them. Even when genetic differences do not underlie differences in traditions, we need to know how the genetic constitution influences the evolution of traditions, and how 'cultural' evolution shapes the way genes function and interact with habits. Our point will be clearer if we return to our thought experiment with the genetically identical tarbutniks.

Imagine that two populations of genetically identical tarbutniks are founded in two different habitats. The food sources in the two habitats are not the same, so the nutritional problems the tarbutniks face are different. In one habitat, (A), fruit is particularly abundant. It is tasty and easy to handle, but tends to be acidic and give the tarbutniks digestive problems. Fortunately, some tarbutniks discover by chance that eating mud after ingesting the local fruits helps their digestion. Through social learning, the mud-eating habit spreads. Tarbutniks then find that eating mud from one particular area has a dramatic effect not only on the digestion of fruits, but also on the digestion of other types of plant material. (It is because some microorganisms in this soil have enzymes that degrade cellulose.) The habit of eating the special mud also spreads. Not surprisingly, the pre-existing tendency of young tarbutniks to taste the faeces of their parents is reinforced: by eating faeces, the young acquire some of the beneficial microorganisms from their parents. The outcome of this set of feeding adaptations is that the food available in habitat A can be digested very efficiently.

Now look at what happens in the other habitat, (B). This habitat is very rich in nutritious nuts, so the tarbutniks have different problems. The shells of the nuts are hard, and rarely break when the nuts fall. However, some tarbutniks find ways of cracking the nuts open, and, through parental example, the young learn nut-opening techniques. Gradually, better techniques develop and spread through the population by social learning. Some of the materials in the nuts enhance the transcriptional activity of a gene coding for a major digestive enzyme, so the tarbutniks in habitat B digest the nuts very efficiently. Thus, in both habitats A and B, effective ways of dealing with the local food sources evolve. An unsuspecting biologist, seeing the behavioural and physiological differences between the two populations, and being unaware of the genetically identical and immutable nature of tarbutniks, might start looking for the underlying genetic differences. She would find none, of course. What has happened is that 'cultural' evolution has produced two different, but effective ways of using exactly the same set of genes to satisfy the tarbutniks' nutritional requirements. The differences in the digestive physiology of tarbutniks in populations A and B involve differences in gene expression, not differences in genes.

In this imaginary case, traditions evolved and shaped gene activity. This evolution led to an excellent fit between habits and genes, without the latter changing at all. Although in real organisms genetic variation is abundant, and genetic variation will be mobilised by natural selection, we believe that this imaginary case highlights two important points: first, behavioural adaptations are often primary, and can lead to complex sets of physiological adaptations on the basis of the set of genes already present. Second, since traditions usually evolve much faster than genes, it is much more plausible that traditions evolve to fit and utilise the existing genes, than that genetic evolution drives the evolution of traditions. Of course, in the study of long-term evolution, genetic variation is of great importance. Sooner or later the two inheritance systems, the genetic and the social–cultural, will interact and genes and culture will co-evolve.

The American anthropologist William Durham has given an excellent example of one type of co-evolution of genes and culture: it is concerned with the ability of adult humans to make use of the milk sugar lactose, and with the role of cows and bulls in human societies.[11] Fresh milk contains the sugar lactose, which can be broken down into its useful components (glucose and galactose) by an enzyme, lactase-I, which all mammals are able to synthesise. The level of this enzyme is normally very high in the young mammal just after birth, but decreases dramatically during weaning. Normally, therefore, fresh milk is digestible only during the suckling period. Adults outgrow their ability to digest the milk-sugar because their lactase-I level goes down. Consequently, when adults drink fresh milk, it does not yield much energy, and often gives them mild indigestion and sometimes diarrhoea. This situation is characteristic not only of non-human mammals, but also of most human populations. But there are some illuminating exceptions. There are adults who have genes that enable them to break down lactose ('lactose absorbers') and hence benefit from drinking fresh milk.[12] A high proportion of these people is found in the dairying populations of northern Europe, and among wandering pastoralists, such as the Tussi population in the Congo basin. In other populations, including many dairying populations, lactose absorbers are much less common. How can we explain this rather odd distribution?

History and ecology provide the clues. The domestication of cattle led not only to an increase in beef eating, but also, about 4000–6000 years ago, to the use of fresh milk and processed milk products such as cheese. With processed milk, there are no problems with lactose absorption, because processing removes most of the lactose. However, the life style of the wandering pastoralists of the Congo basin probably made it difficult to process milk, and so they came to depend heavily on fresh milk as a ready food source. Adults needed the ability to digest the lactose in this milk. Those with the genetic ability to do so thrived and reproduced, and the gene or genes responsible spread through the population.

What about the other, non-wandering, dairying populations with a high frequency of adults who can absorb lactose? They can and do process milk, so is there any reason why they should drink fresh milk? Is the ability to absorb lactose of any particular benefit to them? It is: lactose, the sugar found in fresh milk, is not only an excellent energy source, it also acts like a vitamin D supplement, facilitating the absorption of calcium. This is of great importance in environments where there is a deficiency of vitamin D. People living in sunny areas have a constant supply of vitamin D, because solar radiation converts precursor steroids to the vitamin. But, in regions that receive little sunlight, vitamin D may be in short supply. If so, the ability to drink milk after weaning has a great advantage, because it both supplies calcium and facilitates its absorption, thus preventing rickets, the crippling softening of the bones that results from calcium deficiency. Consequently, in populations that use cattle as a source of food and live in regions with limited sunlight, individuals who are able to absorb lactose as adults have an advantage over non-absorbers. We therefore expect such individuals to leave more descendants, and in time to become the majority in the population. The distribution of lactose absorbers fits this expectation – their frequency is particularly high in populations living in northern latitudes where there are periods of the year with little sunlight.

The increase in the frequency of the genes enabling adults to make good use of fresh milk is therefore the result of a cultural change, the domestication of cattle. Domestication was beneficial for individuals in all communities in which it was practised, because beef and milk products are energy-rich foods, but in some populations, such as those of the wandering pastoralists and the populations of northern Europe, fresh milk became particularly important. What is fascinating is that in these populations we see not only a high frequency of lactose absorbers, but also a high cultural regard for fresh milk. According to Durham, in the creation myths of the Indo-European people, the importance of the cow, the source of fresh milk, becomes greater the higher the latitude. In the myths of the most northern populations, the first animals or the first bovines to be created were female cows, who produced a lot of milk. Their milk was drunk fresh by giants and gods, and is considered to be the source of their great size and strength. The first bovine of creation was not used for food or sacrifice, but continues to nurture the world. These myths thus reflect the importance of fresh milk and at the same time reinforce and encourage its consumption, leading to even stronger selection for lactose absorbers; the increase in the frequency of adult lactose absorbers further enhanced the 'educational' value of the myth. A positive, multigeneration, feedback-loop between genetic and cultural evolution was thus formed. Culture and genes co-evolved, affecting one another. Culture alone was not enough, although it became the guiding selective force, opening up new possibilities of genetic evolution.

The evolution of lactose absorption is an example of harmonious and simple co-evolution between genes and culture. Other, less straightforward types of co-evolution are possible, but we will leave these for later chapters. Social learning is usually the driving and directing force of this co-evolution, leading organisms to construct, regulate and stabilise their biotic and social environments, and consequently to influence the selection pressures the environment imposes on them. Acquired and socially transmitted behaviours occupy the driver's seat because adaptations to local changes occur more quickly through behaviour than through genes.

Genes and culture: new studies and new problems

Human cultural evolution, and the interplay of genetic and cultural factors in the evolution of cultural practices, have been the subjects of some important theoretical work during the last twenty-five years. Geneticists Cavalli-Sforza and Feldman pioneered this new approach, which was soon taken up and developed by others, notably the anthropologist Robert Boyd and the ecologist Peter Richerson. All borrowed the mathematical tools of theoretical population genetics, quantitative genetics and epidemiology, and applied them to culture. To follow the changes in frequency and the spread of new cultural practices in populations, they treated cultural practices as if they were discrete transmissible entities, much the same as infectious viruses.[13] New fashions in music or clothes, for example, were assumed to be transmitted, and to spread like measles or chicken-pox.

This genetic–epidemiological viewpoint has highlighted several major differences between genetic and cultural transmission. First, cultural offspring need not be genetic offspring: genes are passed on only to one's children, but behaviour can be learnt by both kin and non-kin 'students'. Second, cultural variations are not 'random': they are acquired by a process of learning, and learning is not a 'blind' process, even when there is an element of trial and error in it. It is guided by goals, and organised by rules that allow effective categorisations and generalisations. Third, while, with very few exceptions, genes are all transmitted according to the stable laws of genetic segregation, which results in each gene having the same transmissibility (present in 100% of the offspring of an asexual organism and 50% of those of a sexual organism), patterns of behaviour have variable transmissibilities. If one particular pattern of behaviour is more easily perceived, learnt or memorised than others, it becomes more common in the next generation even if it does not have a particularly beneficial effect on the animal. Imagine, for example, a monkey population in which the young learn from their mothers what is good to eat. Two new types of food are introduced into the environment of such monkeys. The animals eat both foods, which are equally abundant and energy-rich, but one food type has features that make it more tasty, or slightly addictive. Naturally, this is the one that mothers are soon eating most frequently, and from their mothers the young learn to eat it too. The less tasty but equally nutritious food is ignored by the youngsters, at least for the time being. Such 'biased transmission' is rare in the genetic system, but it is probably the rule rather than the exception in behavioural evolution. One behaviour may be more easily learnt and remembered than another because of its inherent qualities, as with the monkeys' food, or because of the way it is acquired. For example, if the behaviour can be learnt from multiple 'teachers' (through the influence of parents, neighbours and other individuals), it is more likely to be both acquired and passed on.

Recognising that cultural inheritance is not only possible, but also can be profitably modelled, challenged the sociobiological view of human social evolution. Yet the cultural inheritance approach did not become mainstream, either in anthropology or in behavioural ecology. On the one hand, the anthropologists argue that the 'units' of cultural inheritance on which the models are based are too artificial: cultural practices cannot be treated like 'atoms' because they are part of a practically indivisible whole, a cultural 'package'. Even if one traces the evolution of a relatively simple human cultural product, for example, the pre-fabricated house, one has to take into account changes in family size and standards, immigration, supply and demand, political consequences, the strength of trade unions and so on. All of these affect the production and marketing of the product, and may also introduce modifications into the design of new models. Hence, social organisation not only affects the selection of the product, but often introduces alterations into the actual technological innovation itself, as well as influencing its rate and mode of transmission. The anthropologists stress that culture is a system of practices and institutions that is very difficult to tease apart, and information is transferred and reconstructed at several levels of social organisation. Moreover, every change in behaviour, practices or ideas can have direct modifying effects on a whole repertoire of behaviours, and reverberate through the whole system.

Most behavioural ecologists, on the other hand, tend to ignore cultural evolution theory. Often it has been presented in a rather inaccessible mathematical form, and it was not clear that it could offer them any interesting new insights into animal behaviour. Since culture is generally assumed to be of major significance only in humans, and to have only very minor influences on animal behaviour, the cultural–evolution approach seemed to be of little importance for understanding the evolution of the social behaviour of mice, rats and bee-eaters. Only rarely have the analytical tools offered by the cultural evolutionists been applied to the social behaviour of birds and mammals.

The general framework used in the theoretical studies of the cultural-evolution school is the one that we are using in this book. Our approach, however, is different in two principal ways. First, we are focusing on transmitted behaviour in birds and mammals, rather than in humans. We shall use the rich field data on social organisation and interactions to re-examine the evolution of behaviour from a perspective that incorporates social learning and traditions as agents of evolution. This approach provides some simple (and testable) evolutionary interpretations of patterns of behaviour that have not been explained satisfactorily within the conventional framework. It also provides additional or alternative explanations for some behaviours that do have orthodox explanations. Second, we are going to stress the networks of ecological and developmental interactions in which particular patterns of behaviour are embedded. We shall be looking at the ways in which these networks are constructed and reproduced anew every generation, at the conditions that make them stable, and at the kind of heritable variations that they can support. In other words, we shall not treat patterns of behaviour in isolation, but rather as dynamic packages, parts of a developmental system that, in most cases, is transmitted and evolves as a whole.

The way non-human animals pass on learnt patterns of behaviour has been studied both by experimental psychologists, in the controlled and unnatural conditions of their laboratory, and by ethologists, under natural field conditions. Unfortunately, until about twelve years ago, many of the field observations were anecdotal, and were rarely integrated with laboratory studies and brought under a common theoretical roof. Although in 1980 John Bonner and Paul Mundinger published pioneering reviews on the evolution of culture in animals, and these reviews increased interest in the subject, the study of animal cultures has remained marginal. Mundinger's comprehensive review had little direct impact, while Bonner and the few evolutionary biologists who followed him were concerned mainly with the genetic basis of the ability to produce culture, not with the evolution of animal culture itself or its effects on genetic evolution. They regarded cultural evolution in animals as limited in scope, as a product rather than an agent of the social evolution of behaviour. However, this view of animal cultural evolution has been changing.[14] Several excellent symposium volumes that document and analyse social learning and 'cultural' practices in animals have been published during the last twelve years; collaboration between experimental psychologists and behavioural ecologists is growing. Increasingly it is recognised that, if we are to understand how animal psychology develops and how it has been evolving, social learning has to be considered. Today there is hardly an issue of an evolution-oriented behavioural journal that does not publish at least one article on social learning or the local traditions of animal populations. This interest in animal traditions should lead to a better understanding of the relationship between genes and culture during evolutionary time. At a more general level, it will alter the way in which we think about the processes that drive evolution.

Selective and instructive processes in evolution – Darwinian Lamarckism

Cultural evolution involves natural selection between alternative patterns of behaviour. It is therefore Darwinian. However, the origin and transmission of some of the behavioural variations on which natural selection acts depend on learning. Since animals can adapt by learning, and through social interactions can pass on some of their new adaptations to their progeny, changes in heritable behaviour can occur in direct response to changes in conditions. Such evolutionary change is said to be 'Lamarckian', by which most people mean that it involves the inheritance of acquired characters. In modern biology Lamarckism has usually either been ignored or ridiculed. In recent books on evolution, it has been almost a rite to point to the weakness of Lamarckism in order to illustrate the strength of Darwinism. The problem has been that, for as long as there seemed to be no evidence of a mechanism through which newly acquired adaptive characters could be transmitted while non-adaptive ones were not, Lamarckism seemed to introduce some kind of mystical goal into evolution. However, now that it has been recognised that there are inheritance systems (of which social learning is but one) that make it possible for adaptive characters to be acquired and transmitted, the ghost of teleology can be exorcised. Some acquired characters can be transmitted because there are inheritance systems that have evolved to do exactly that.

There is nothing surprising or unusual about the evolution of the ability to transmit some acquired characters. It is the outcome of natural selection: those individuals who could transmit to their progeny the beneficial adjustments that they had made during their lifetime were reproductively more successful than others, so the genes that made this possible spread. Lamarckian mechanisms thus evolved under the auspices of Darwinian ones, through the natural selection of random genetic variations.[15] The Darwinian origin of the mechanisms that transmit acquired characters does not diminish the evolutionary importance of these mechanisms once they are in place. Darwinian evolution, based on the selection of largely random genetic variations, has constructed new (Lamarckian) rules for the evolutionary game. To interpret evolutionary phenomena without incorporating these new rules is unreasonable. Lamarckism is not an alternative to Darwinism; it complements it to form a broader theory of Darwinian evolution.

So why is it so commonly assumed that Darwinism and Lamarckism are irreconcilable alternatives? The answer seems to lie in the fascinating sociological and historical developments that led to simplifying assumptions about each of the two theories.[16] A fundamental (and artificial) dichotomy was created between them. Darwinian theory focuses on selective processes – on choice between alternatives, while Lamarckian theories focus on instructive processes, on the acquisition of transmissible information. When selection is considered to be the major (or exclusive) cause of adaptation, the origin of variation is deemed unimportant; the most extreme and simplest selectionist model would, in fact, assume that the origin of variation is totally random with respect to selection. On the other hand, when variation is supposed to be acquired (by learning or any other process), the simplest instructionist model assumes that instruction results in a single typical result, and therefore in a population of similar individuals in which selection could play no role. If all mother hens acquire a preference for red grains, and this preference is passed on to the chicks as they observe their mothers eating red grains, selection seems irrelevant to the ultimate fate of the acquired variant, since all chicks are assumed to prefer red grains.

Such an extreme instructionist example is, of course, only a convenient straw man. Some of the chicks in our example may not learn at all and, among those who do, there may be differences in the strength of the preference (some of the mothers from whom the chicks learn may be very choosy, others less so), or in the time it takes to learn the preference, and so on. Some of these variations, if heritable, can be transmitted and provide the raw material of natural selection. A brief look at our own experiences reveals more clearly the fallacy of the assumption that behaviourally acquired characters lead to uniformity. When we learn a new skill, for example how to paint a room, we often modify somewhat the method that we have learnt from our teacher. We may use the various brushes in somewhat different ways, or paint doors, windows, walls and ceilings in a different sequence. A maladaptive variant, such as omitting to take down the lamp-shades before starting, is usually selected out. Only rarely will we observe a single technique in the population; as the techniques are passed on, we see clusters of variant methods. Thus, learning defines the direction and the general nature of change, while selection fine-tunes it. By directing variation, learning decreases the load of completely blind individual mistakes. It rarely eliminates all variation. The combination of learning and selection enables the organism to respond to the environment in a cheaper, quicker and generally more efficient way than would be possible through the selection of randomly generated variants. Selection and instruction are perceived by many evolutionary biologists as mutually exclusive, not because this is logically necessary, but because for each approach the exclusion of the other results in the simplest model. The artificial dichotomy between Darwinism and Lamarckism that has been created by modern Darwinians has done little except encourage the spread of the dogmatic, genic version of Darwinism.

Are there animal cultures?

No one denies that instructive processes involving the transfer of patterns of behaviour through social learning are an essential part of animal life, and that some of the habits observed in animal groups represent learnt traditions. But are we justified in calling these traditions 'culture'? In human societies, small variations in behaviour that are passed on from one generation to the next can accumulate, and cultural evolution leading to complex cultural adaptations is widely accepted as a major component of human social history. We can see this clearly with the evolution of local dialects, or with technological evolution. The question is whether such cumulative 'cultural' evolution also happens in animals. If small idiosyncrasies in non-human animals are not inherited, evolutionary improvement of a pattern of behaviour by the progressive accumulation of small selected changes is impossible. As we have already indicated, some investigators argue that using the term 'culture' for the socially transmitted habits of animals is inappropriate because the evolution that occurs is not cumulative in the way that it is with human cultural evolution. A second objection is that culture depends on something that animals do not have – representation and communication through a symbolic system, a system that uses arbitrary signs (symbols) that refer not only to things in the outside world, but also to each other. Since using the term 'culture' or 'cultural evolution' in the non-human context is clearly not without problems, we have to evaluate the objections raised to their use for animals.

Human culture is certainly permeated by symbols, and is defined as involving symbolic representation and communication. The dictionary definition of culture is clearly symbol and human-oriented. The *American Heritage Dictionary* defines culture as:

> a. The totality of socially transmitted behaviour patterns, arts, beliefs, institutions and all other products of human work and thought. b. These patterns, traits, and products considered as the expression of a particular period, class, community, or population: Edwardian culture; Japanese culture; the culture of poverty; c. These patterns, traits, and products considered with respect to a particular category, such as field, subject or mode of expression: religious culture in the Middle Ages; musical culture; oral culture.[17]

Although symbols (as we defined them) are part of human culture and are not found in animal societies, a lot of information in human societies is not represented and transmitted symbolically, and we feel that the lack of symbols is not a sufficient reason to deny culture to animals. There is a similar, though more general definition that can readily be applied to non-human animals. We are going to us the term 'culture' for 'a set of behaviour patterns or products of animal activities that are socially transmitted in an animal lineage, group or population'. When referring to the symbolic aspects of human culture, we will say explicitly 'symbolic culture'. Essentially, what we are suggesting in this definition is that, even when symbols are not involved, if there is a socially transmitted package of behaviours or products of animal activities, it is legitimate to talk about it as 'culture'. Even when the transmitted elements are not independent of the actual conditions that induced or constructed them, and would probably not be maintained if conditions changed, they still constitute a culture. Consider, for example, an event that unfortunately has occurred many times in human history – the enslavement of people following wars or colonisation. The enslaved people have to alter their life style as a direct result of the ongoing conditions of slavery, but at the same time they maintain some old habits, some old modes of behaviour, from their former societies. Their overall life style is very different from what it was when they were free. Does the fact that many of the behaviour patterns of the slaves are contingent upon their conditions of slavery and will disappear when freedom comes preclude the use of the term 'slave culture'? Surely this cannot be the case. It is the total life style that constitutes the culture, and this includes those socially transmitted elements that are partially dependent for their expression upon the contingent conditions. The same kind of argument can be used to justify calling the socially learnt and transmissible life style of red foxes that have colonised urban areas 'urban-fox culture'. Of course, it does not make sense to call a change in behaviour that is totally dependent on local conditions a 'cultural' change. To say that differences in the food handling techniques of two populations of finches are cultural when these differences depend solely on the seeds available is nonsense. On the other hand, differences in dietary habits that depend on the availability of the food items *and* are transmitted across generations through social learning, such as the differences in food manipulation observed in chimpanzee groups, can be called cultural. We therefore use the term culture when we have a good reason to believe that the behaviour patterns that constitute the life style of the community are socially learnt and are transmitted to subsequent generations. We do not demand that they would always be transmitted to the next generation if the conditions of life change.

What about the objection that the socially transmitted habits of animals are not truly cultural because culture implies a progressive accumulation of modifications over time? Some investigators have argued that the socially learnt habits of monkeys, birds or rats are much less complex than those of man, and that there is no evidence that the habits change progressively and cumulatively. Consequently, they say, these habits should be referred to as 'traditions' rather than 'culture'.[18] For example, they would claim that, whereas we can follow the cumulative cultural evolution of a human behaviour such as a religious ritual, or an artefact such as a boat or a bicycle, this is not the case with animal products like nests, or with behaviours such as sexual displays. This claim, however, is based on the usually untested assumption that most aspects of the observed behaviour are innate. It may be ill-founded. As we will show in the following chapters, there are patterns of behaviour and products of animal activities that strongly suggest that there has been cumulative cultural evolution rather similar to that found in humans. Nevertheless, we agree that progressive and relatively autonomous change in a single behaviour, or a single behavioural product, is much more likely in humans, who are endowed with symbolic representation, than in animals. But, even in animals, cumulative changes can occur, albeit in a less straightforward way. They can occur through the effects of one socially transmitted pattern of behaviour on another related behaviour. Think about foxes that move into towns: they acquire a new way of life that is socially transmitted to their offspring. They learn how, when and where to forage in the town's rubbish bins and gardens, how to avoid humans, where to dig a den and so on. Similarly, as towns sprawl into the countryside, birds learn that some humans in suburban gardens are keen on feeding them; they learn when and where to come to feed, and transmit this information to their offspring; a new tolerance towards humans may develop, and as a consequence the birds may alter their nesting sites and nesting materials, preferring to nest close to human habitation. What seems to happen is that, as new behaviours are learnt and transmitted to others, new behavioural variability is created and new evolutionary possibilities are opened up. The newly acquired and transmitted habits affect the selective value of the established behaviours. Existing patterns of behaviour become modified or are abandoned, and gradually a new life style develops. Although individual patterns of behaviour do not evolve autonomously, the whole changing life style affects the selection and evolution of the behaviours that contribute to it; selective changes in one individual pattern of behaviour are integrally linked to the evolutionary transformations occurring in the whole web of related behaviours. Thus, cumulative cultural evolution in non-human animals is usually net-like, and comes in more diffuse and broader behavioural 'packages' than in the human case, but the accumulation of modifications over time does occur. We feel, therefore, that applying the term 'culture' to a set of socially transmitted habits in higher animals is justifiable.

Memes and reproducers: the 'unit' problem

If the evolution of culture involves a complicated web of interrelated behaviours that are affected by both instructive and selective processes, how can we study it? What is the 'unit' of evolution? We have some idea of what we are talking about when we discuss genes as units of evolution: it is possible to develop models and follow the rise and fall of the unit, of a particular gene such as the lactase-I gene. But can we define a transmissible unit of variation and selection for a tradition or culture? Certainly, most anthropologists would argue that 'units' of culture are impossible to define. Yet there are situations, especially in technological evolution, where it is feasible to define a cultural 'unit-trait' and follow its transformations through time. The 'unit-traits' of a bicycle would be its parts – the brakes, the pedals, the wheels, the gears and so on. Tracing the rise and fall of variants of these parts would provide interesting information about relative rates and patterns of evolution, divergence and convergence, and the steadiness of trends. It could tell us how relative wheel size changed, when and where back-wheel drive and front-wheel drive were used and so on. However, even in this extreme and apparently clear-cut case, once we wish to understand *how* innovations arise and spread, the 'unit' problem returns. The 'unit-trait' whose change has been followed is a result of a process that needs to be understood. We need to know what psychological and social factors lead to innovatory patterns of behaviour or innovatory ideas in the first place, and how and why new ideas are implemented and new patterns of behaviour established. In other words, we need to know how new ideas and behaviours are produced and how they spread. Looking at the unit-trait, which is the product of these processes, is not enough.

At present, Dawkins' solution to the analysis of cultural evolution – the shift to the meme – is widely accepted and used.[19] However, we find this term very problematical, and impossible to use in the context of animal cultural evolution. If 'meme' was simply a useful shorthand for a behaviour that is transmitted by non-genetic means, it would be very useful. But it is not. The meme as defined by Dawkins is a representation of a behavioural act stored in the nervous system and transmitted by social learning through generations of communicating individuals. A meme can be the representation in a chick's brain of the kind of grain to eat, or of a green-eyed mate. The effects of the meme are to produce, under the right conditions, preferential pecking of red grains, or pursuit of a green-eyed mate. The outward effects of the memes:

> may be perceived by the sense organs of other individuals, and they may so imprint themselves on the brains of the receiving individuals that a copy (not necessarily exact) of the original meme is graven in the receiving brain. The new copy of the meme is then in a position to broadcast its phenotypic effects, with the result that further copies of itself may be made in yet other brains.
>
> (Dawkins, 1982, p. 109)

In the special case of humans, the meme can also be an idea of some kind that can be represented in a symbolical form. For example, the plan for building a bicycle, or a subplan for building the frame or a wheel, can be considered a 'meme'.

Because a meme is defined as a mental representation of a pattern of behaviour, as something tangible located in the brain, it may mislead us into thinking that memes are independently varying transmissible units, similar to genes. They are not. There is a fundamental difference. Whereas genes can be replicated and transmitted whether or not they are expressed, a meme cannot be transmitted unless the behaviour it represents is displayed. The only exception is when human beings use symbols (such as words and numbers) which can be replicated and transmitted without reference to their content and without the implementation of the ideas they contain. Apart from this, behaviours are not transmitted unless they are displayed. The way that a chick learns to preferentially peck at red grains depends on the mother pecking red grains in the chick's presence. Her red-grain-pecking meme will not be transferred unless she actually does this. The behaviour and its transmission are tied together; they are a result of the interaction of the animal with the environment, and it is impossible to separate them. It is, therefore, misleading to speak about the 'replication' of memes, although it is not misleading to talk about the reconstitution or re-generation of patterns of behaviour.

Another fallacy stemming from the use of the meme concept is that there is a simple correspondence between a particular transmissible behaviour and a particular circuit in the brain. An example may help to illustrate why this is wrong. It is known that some psychopathic human mothers avoid touching their babies. This early deprivation results in the development of the same psychopathic disorder in their children. When the deprived, psychopathic daughters become mothers, they, too, avoid touching their babies, who grow up with the same type of psychopathic disorder. Hence the psychopathic disorder is transmitted in families.[20] The heritability of the disorder is due to the similar social and psychological conditions experienced by mothers and daughters. It is quite clear here that no 'meme-for-not-touching-babies' was transferred from generation to generation or from brain to brain. What was re-generated in the daughters was the pathological maternal behaviour that was brought about by recurring pathological interactions with the mother. Although a behaviour that we could call 'not-touching-babies' can be named and followed, it cannot be isolated; 'not-touch-ing-babies' both causes the disorder to be passed on, and at the same time is a symptom of it. This pathological set of behaviours cannot be fragmented into separate parts, into autonomously transmissible 'memes'.

There are many similar examples of heritable behaviour that results from complex self-reproducing interactions between individuals and the environment in which they are raised. The term 'meme' does not cap-ture this type of re-production of acquired behaviour patterns, and erroneously implies the existence of independently varying and trans-missible parts of the system.

This is not the place for a detailed critical review of the usefulness or otherwise of the meme concept, but it is worthwhile highlighting some of the difficulties in applying it to cultural evolution.[21] To use the term 'meme' in the original sense of a defined representation of a behaviour pattern that can be transferred from brain to brain is to neglect the developmental system that underlies the re-generation of behavioural practices. Because the transmission and 'expression' of a mental repre-sentation cannot be separated, the accuracy of transmission depends on the conditions in which the behaviour is produced. This is not so with genes, where the origin of genetic variation is usually assumed to be unrelated to environmental conditions. There is a single mechanism of DNA replication, which is usually unaffected by the function of the genes it replicates,[22] but the mechanisms underlying the re-generation of behaviour are numerous, and not separate from the behaviour itself. It is impossible to ignore these developmental mechanisms: they are part of the reproduced behaviour. The ease with which we learn and pass on a new song depends on the content of the song, its melody, its rhythm, when we learnt it, the other songs we know and the type of music we are familiar with. Similarly, song-learning by a young male songbird depends on the songs it hears, when it hears them, how its 'tutors' behave when they sing and so on. The transmission process typ-ical of the behavioural inheritance system depends on the form and con-tent of the information, as well as on the circumstances in which learning occurs.

For all the reasons just outlined, we do not find the meme concept helpful in thinking about the role of socially learnt, transmissible pat-terns of behaviour in evolution. We believe that the most effective and useful way to approach the whole problem of the 'unit of evolution' is to use the developmental framework that has been developed by people such as the biologist Patrick Bateson, psychologist Susan Oyama and philosophers James Griesemer, Paul Griffiths and Russell Gray.[23] Unlike the gene-centred approach of Dawkins, for whom genes or memes are the only legitimate units of heritable variation, the developmental approach incorporates all the different ways through which information can be re-produced. Everything that enables an individual to re-produce its own development is part of what James Griesemer calls the 'reproducer'. This concept inevitably leads to concurrent consideration of selection at different levels of organisation – the gene level, the cell level, the organism level and so on. It also requires consideration of all the inheritance systems involved (the genetic, the cellular and the cultural), and the way in which selection acting on variation transmitted by one system affects the selection of variation transmitted by another.

We have already looked at one example of the complex interplay between the selection of variations transmitted by the cultural system and those transmitted by the genetic system. The domestication of cattle, a cultural change, created an environment in which for some populations it was an advantage for adults to have the gene enabling lactose absorption. It is an example of what today is called 'niche construction': the people themselves created the environment in which the gene was selected. Many years ago, Conrad Waddington and Richard Lewontin both stressed that living organisms are not passive entities, but that they actively determine and construct their environment and, hence, the selective regimes in which they live and breed.[24] If genetic variation was possible in tarbutniks, we would find that the traditions that evolve in different populations would, in time, lead to selection of genetic variants appropriate for the environment the tarbutniks construct. Perhaps genes that gave a spade-like foot would be selected in the population of burrowers; in the population in which males seduce the females with red berries, genes producing small red noses might benefit males! The selective environment constructed by populations with different traditions would lead to genetic divergence, which would, in turn, stabilise and reinforce the traditions. The dynamic, cyclical and multigenerational nature of the interaction between the organism and its environment was recognised in the 'niche construction' concept developed by Odling-Smee.[25] He argued that the transmission to descendants of ecological legacies, formed through the activities of organisms, is an integral part of evolution. Recently, the evolutionary consequences of niche construction have been explored using mathematical models, and it has been shown that niche construction fundamentally affects the rate and the direction of genetic changes.[26]

Clearly, the developing individual is not a passive vehicle for the acquisition and transfer of discrete 'units' of behavioural information, nor is it a passive target of selection. It actively stabilises, regulates and constructs its biotic and social environment, and is to some extent able to control the type of selection that will affect it. It also transmits to its descendants the selecting environment that it has constructed by its own activity. It transmits its learnt information as a co-ordinated, updated and coherent package. With social learning the efficiency of transmission is often linked to the benefit that the transmitted information confers. Consider, for example, animals that learn to relieve their indigestion by eating plants with medicinal effects: the beneficial effects of eating the plant reinforce the habit, and, because it is repeated, the habit will be more readily transmitted by social learning. Moreover, with social learning, the behavioural information that the individual acquires, controls and transmits depends on its status in the social group, its sex and so on. Because of the complexity of the relationship between a particular behaviour, the probability of its transmission, and the effect the animal has over its social environment, it is difficult to think about behavioural evolution in terms of discrete behavioural units. Understanding the evolution of behaviour requires a developmental perspective which recognises this complexity and integrates the different ways in which behavioural information is generated, transmitted and selected.

Back to Darwin

Where does the focus on social habits, on social learning and on traditions lead us when we attempt to understand the evolution of behaviour? Curiously, it leads us right back to Darwin, to the ideas that Darwin developed in two well-known but rarely read books: *The Descent of Man* and *The Expression of the Emotions in Man and Animals*. Darwin's ideas about the evolution of instincts have been discredited because of their strong 'Lamarckian' flavour: Darwin suggested that learnt behaviour that becomes habitual eventually becomes hereditary. First, he said, an individual acquires a habit, and then the habit becomes inherited in the lineage. In another book, *The Variations of Animals and Plants under Domestication*, Darwin suggested a mechanism that could account for the inheritance of acquired characters and the features modified by use or disuse. This heredity theory suggested that representative particles are secreted from all parts of the body during development, accumulate in the sex cells and become the hereditary material for the next generation. The theory turned out to be completely wrong, and the notion that acquired habits can become hereditary was rejected along with his mechanism of inheritance.

Yet it is to Darwin, we believe, that we should turn again. Not, of course, to his theory of inheritance, but to his belief that individual and social habits drive much of the evolution of behaviour. As we have seen, there is no necessary contradiction between selection and instruction in evolution, and the two often go hand in hand. The way we view the evolution of behaviour is very close to that advocated by Darwin, although we put more emphasis on social learning than he did, and we employ conventional genetics and natural selection to explain the transition from a behaviour that was dependent on many learning trials (learnt behaviour) to one that is dependent on fewer trials (a more 'innate' behaviour). Emphasising social learning makes us focus on the ways in which changes in the nature and means of transfer of information can affect the evolution of behaviour.

The present tendency to explain all adaptive evolution in terms of gene selection is unfortunate. It is clear that during evolutionary history, the 'currency' of evolution – heritable, selectable variations – has itself evolved. The ways in which information is stored and transmitted have changed, and the changes have affected subsequent evolutionary history. The evolutionary game has acquired additional dimensions – the old game has evolved new rules! Dawkins compared evolution by natural selection to a blind watchmaker, evolving complex, watch-like adaptations by the natural selection of random variations. But the blind watchmaker is no longer blind, it is merely short-sighted. It has evolved some light-sensitive cells, and even rudimentary eyes!

Summary

Just like the anatomical and physiological features of animals and man, complex behaviours are the products of evolution. The common assumption is that the evolution of all adaptations, including behavioural adaptations, occurs through natural selection of genetic variants. We question this assumption, not because we doubt the efficacy of Darwinian natural selection, but because, in higher animals, many heritable differences in behaviour are not the result of genetic differences. Some behavioural variations are inherited because they are transmitted to others through social learning – individuals learn from others how to behave. Natural selection of such socially transmitted behaviour can lead to the formation of intricate animal traditions. In theory, behavioural evolution could occur through the selection of socially learnt variations, without any genetic change at all. In reality, however, the selection of socially transmitted behavioural variations usually interacts with the selection of genetic variations, leading to the co-evolution of genes and traditions. New behaviours that are generated in response to changes in environmental conditions are often both specific and adaptive, so they are 'acquired characters', in the classical sense. The evolution of behavioural traditions is thus both Lamarckian and Darwinian.

Recognising that information is transmitted through social learning leads to a very different view of heredity, variation and evolution. The selection of socially acquired and transmitted behavioural variations usually interacts with the selection of genetic variations, so models of evolutionary change have to be more complex. Defining the unit of evolution also becomes more difficult. Nevertheless, even though incorporating behavioural inheritance complicates some aspects of evolutionary theory, this approach also enriches it, because it recognises that evolution itself is an evolving and increasingly more discriminating process.

Notes

[1] 'Fitness' is a measurement of survival and reproduction. Formally, it is the product of the probability of an individual surviving to reproductive age and the expected number of viable and fertile offspring that it is likely to produce given that it does survive. Evolutionary biologists are usually more interested in 'relative fitness', which is the lifelong potential of one type of individual to survive and reproduce relative to other types of individual in the same population.

[2] Tarbutniks take their name from 'tarbut', the Hebrew word for 'culture'.

[3] There are many problems associated with the terms 'species' and 'speciation', but a commonly accepted view is that speciation has occurred when two populations become reproductively isolated. If individuals from two populations

meet, but for ecological, physical, behavioural or other reasons they do not mate or do not produce viable and fertile offspring, the populations are usually seen as being of different species.

[4] See, for example, Lewontin, 1993; Rose, Lewontin & Kamin, 1984.

[5] Dawkins, 1976.

[6] Wright, 1994, p. 9.

[7] In order to avoid the political stigma of the previous (1970–80) generation of sociobiologists, the human sociobiologists of the late 1990s do not call themselves by that name. Instead they are called 'evolutionary psychologists' or 'evolutionary anthropologists'. They present human sociobiology in a more sophisticated way than their predecessors, and provide empirical data, which they claim support their arguments. However, they share identical assumptions and research agendas with the older generation.

[8] A neural module works automatically and rapidly, and its function is to some extent autonomous from the function of other neural circuits. A module is sometimes assumed to have a more or less fixed neural architecture and ontogeny, and a typical pattern of breakdown, so that, when the module is damaged, the damage leads to typical abnormal behaviour. This characterisation follows the definition of an 'input module' in Fodor's sense. (See Fodor, 1983.)

[9] Lott, 1984. Evidence of the dramatic effects of habitat on primate social systems can be found in Normile, 1998, which summarises the work that was presented at a conference on *Recent Trends in Primate Socioecology* held in January 1998.

[10] Kawai, 1965; Kawamura, 1959; Kummer, 1971; Watanabe, 1989.

[11] Durham, 1991, pp. 226–85.

[12] Lactose absorption seems to be inherited as an autosomal dominant trait.

[13] Boyd & Richerson, 1985; Cavalli-Sforza & Feldman, 1981; Durham, 1991.

[14] For early reviews see Bonner, 1980; Mundinger, 1980. For recent documentation of social learning and cultural inheritance in animals, see Box & Gibson, 1999; Heyes & Galef, 1996; Wrangham *et al.*, 1994; Zentall & Galef, 1988.

[15] Jablonka, Lamb & Avital, 1998.

[16] Discussed more fully in Jablonka & Lamb, 1995, chapter 1.

[17] Soukhanov, 1992.

[18] Galef, 1992; Tomasello, Kruger & Ratner, 1993. Although we agree with Tomasello and his colleagues that cultural learning in humans involves more sophisticated cognitive mechanisms than in other animals, in both cases the resulting cultural product may involve the accumulation of learnt variations.

[19] See for example: Blackmore, 1999; Dennett, 1995.

[20] Discussed by Peter Molnar of the Semmelweis Medical School, Budapest, Hungary, in a lecture given at an interdisciplinary symposium in Bielefeld University, Germany, 1991.

[21] The many differences between memes and genes made Dawkins himself sceptical about the usefulness of the gene analogy for understanding culture (Dawkins, 1982, p. 112). See also Lack, 1998.

[22] There are some exceptions to this rule: genes that are actively transcribed, or have a chromatin conformation that leads to ready activation, seem to be more prone to mutation, recombination and transposition (reviewed by Jablonka & Lamb, 1995, chapter 7).

[23] Bateson, 1976, 1978, 1988, 1991; Oyama, 1985; Griffiths & Gray, 1994; Griesemer, *in prep*. Evelyn Fox Keller (Fox Keller, 1992, pp. 128–43) has noted the problems associated with the use of the term 'reproduction', which may be misleading when sexual reproduction is involved. We use this term to denote the generation of recurring patterns of behaviour.

[24] Lewontin, 1978; Waddington, 1959.

[25] Odling-Smee, 1988, 1995.

[26] Laland, Odling-Smee & Feldman, 1996; Odling-Smee, Laland & Feldman, 1996. The same idea, although from a somewhat different perspective, has been discussed and modelled by Wolf and others: see Wolf *et al.*, 1998, and references therein.

2 What is pulling the strings of behaviour?

Any discussion of evolution must assume something about heredity, so ideas about evolution and notions of heredity are intimately linked. From the outline of our views given in the previous chapter, it will be clear that we believe that something is wrong with the assumptions about heredity that underlie a lot of present-day evolutionary thinking. In this chapter, therefore, we are going to take a closer look at the hereditary basis of behaviour, focusing on its genetic basis. What does it mean to say that genes determine behaviour? What is the difference between this assertion and the claim that patterns of behaviour have a genetic basis? To what extent do heritable differences in behaviour reflect genetic differences?

Often the easiest and most fruitful way of thinking about the evolution of behaviour is to have some actual animal behaviour in mind, so in this and most subsequent chapters we are going to ground our discussion on some observations of real animals in their natural habitat. This time we take ourselves at sunrise to an old olive orchard in the Judaean hills near Jerusalem.[1]

It is April, and the ground is covered by a dense multicoloured carpet of flowers. A small, black, white and yellow bird hops silently and effortlessly from branch to branch, eventually reaching the top of an olive tree. The soft light is brightening rapidly, and the first burst of song from the treetop does not leave any doubt: it is a fine resident male great tit. The great tit is a common, conspicuous and well-researched denizen of woodlands, plantations, parks and gardens in Britain and Europe, as well as here, in Israel.

For such a small bird, our male sings his distinctive song loudly, and the clear morning air carries the song a long way. We can still hear his song, a recurring pattern of a short burst of singing followed by a slightly longer period of silence, a couple of hundred metres away, at the edge of the orchard. A second male starts singing seventy metres from the first, on the top of another olive tree. His song sounds similar to that of the first male, but it is not identical. Careful listening and tape-recording for a few days would tell us that each male has his own repertoire of several variants of the species-specific song. In our case, the first male has four song-types, the second only three. The song of the male great tit is known to attract a mate to his 1–2 acre territory, and to discourage ever-hungry (for food or mate!) males of the same species from intruding. What use has a male for four slightly different song-types? Why does his neighbour use only three? How does he acquire his repertoire of songs? Is it by genetic inheritance from his parents? By learning from his father? Or maybe from his neighbours? By all these means? Is a certain version of the song more attractive to some female tits? To all female tits? Before trying to answer these questions, let us look at some more tit behaviour.

Our first male is very busy now, foraging among the twigs and leaves of the olive tree for his favourite food – small larval and adult butterflies, moths, beetles and spiders. His collecting techniques are varied and elaborate. He carries out a detailed survey of each twig and leaf that he encounters, including the under-surfaces, which he scans while hanging upside down from a branch. His acrobatic skills are as useful as they are impressive. Many insect larvae are active solely on the under-surface of leaves, and are therefore almost undetectable from above. When our tit finds a small insect or spider he picks it up with his bill and eats it. He has some trouble with a struggling grey moth-caterpillar, nearly three centimetres long. He overcomes it by holding the middle of the caterpillar firmly under one foot, while simultaneously using his bill to tear off and eat first the front, and then the rear ends of the caterpillar. Another moth caterpillar, black and hairy, is ignored. How did the tit know that it is unpalatable? Did he learn it the hard way, from his own bitter experience? Did he learn it from one of his parents, or was he born with an inherent fear of black and hairy caterpillars?

The old olive tree has a twisted trunk covered with heavily cracked grey bark. Many adult insects and moth pupae can be found inside these cracks, but how can they be extracted? For a quarter of an hour, our tit makes a thorough faunal survey of the crevices in the olive trunk. By impaling small, hard-backed longhorn beetles and pulling them out of the crevices, he demonstrates to us, in a most persuasive way, that his bill can be used very effectively, if needed, as a hammer or a chisel. A brown solitary wasp lands on the olive trunk. The tit immediately holds the wasp under his foot and quickly plucks out and discards the sting. Where and when did he learn how to tell bees and wasps from all other insects?

Great tits certainly seem to be resourceful birds. In harsh winters when food is scarce, British great tits have been seen to follow seed-hoarding bird species (for example, coal tits) and steal their food caches. In conifer woods, great tits have been observed to use a conifer needle as a tool to extract larvae from holes in wood. Some individuals in Britain learnt how to remove the lids from milk bottles and take the cream. Since great tits are good at observational learning (especially when food is involved), the habit of opening milk bottles soon spread rapidly among the British great tits. Another talent, discovered long ago by bird-watchers trying to attract great tits to their gardens using food baits, is that the tits are remarkably adept at pulling up lengths of string to which food, such as nuts, is attached. How did they learn such elaborate tricks?

A rather different habit is seen in incubating or brooding females occupying nest-holes. They can produce a snake-like hiss that frightens and drives away unwelcome intruders or predators, including humans. No predator will stick its nose into a black hole and risk a fatal snakebite! Is this hissing behaviour inborn, or did the females learn it from their mothers?

The great tit is a common small bird with many talents. It is one of 8,800 species of birds, and one of 42,000 species of vertebrates.[2] Many vertebrates, and probably most species of birds and mammals, display behaviours as varied and as interesting as those of tits. We want to understand the origins, the effects and the evolution of such behaviours. In the light of the great tit's behaviour, we can re-phrase and amplify the questions we asked at the beginning of this chapter, and now ask: are great tits sophisticated genetic puppets, pulled by the invisible strings of their genes? Are they accurately executing complicated sets of genetic instructions that are designed to take into account some variations in the environment? Can we satisfactorily explain the origin and evolution of their behaviours by assuming that they are solely the products of variations in DNA followed by natural selection? Or are the intelligent and creative behaviours of the great tits largely independent of genetic variations? Did natural selection evolve something that uncouples behaviour from the genes' direct dominion, and, if so, what is it? Do we have any clear criteria that can help us to distinguish and choose between different ways of interpreting the behaviours of birds and mammals? These are difficult questions, but we can start trying to answer them by looking at genetic inheritance, the most fundamental system of information storage and transfer in living organisms.

Genes: fate or challenge?

Genes are the basic units of heredity. They are often described as parcels of information, sets of instructions for building bodies and constructing behaviours, which are passed on, well wrapped within eggs and sperm, from parents to progeny. Specific genes control the development of specific traits such as eye colour, blood group, ear-lobe morphology and, some would also like to add, cognitive excellence, sexual preference and temperament. Modern evolutionary theory is based on our knowledge of genes, their modes of transmission and their variations. For most non-biologists, genes seem like little wizards: they are mysterious, like computer chips, but enormously powerful in their control of what we are and can possibly become. They are made of DNA, a molecule endowed with remarkable properties. The structure of the DNA molecule has become the symbol of modern biology: a long coil made of two strands twisted around each other, forming a double helix. Each DNA strand is made up of a chain of chemical units, or nucleotides. There are four types of nucleotide, each containing a phosphate, the sugar deoxyribose, and one of the nitrogenous bases adenine, thymine, cytosine and guanine. Usually the bases and the nucleotides containing them are abbreviated to A, T, C and G. The two nucleotide chains are complementary, so that an A in one strand always faces T in the other strand, and a C always faces a G. The structure of the molecule makes it easy to envisage how it is replicated: since the two strands of the double helix are structurally complementary to each other, at replication, when the two strands separate, each strand acts as a template for the assembly of a new complementary strand, so two double helices are formed from one. The daughter molecules are (barring mutations) identical to the mother molecule.

The other marvellous property of DNA is also related to the fact that it is essentially a linear molecule built of repeating 'units' or 'modules'. It is common, and in some ways useful, to compare DNA sequences to written language. If we think about the linear array of modules as a written message, and about the four nucleotides as four different letters, we immediately see that many possible messages can be constructed. As in written language, different DNA stretches will be different not because the components differ, but because their sequences differ. Only the length of the molecule limits the number of possible ways in which the four modules can be linearly organised. For a molecule that is 99 modules long, there are 4^{99} possible sequences of 4 different modules, an astronomical number indeed.

The analogy between DNA and written language is helpful because it clarifies other important aspects of DNA structure that are related to its information-carrying function. A written sentence and a stretch of DNA are both meaningless without interpretation. For a written sentence, the interpreter is the human mind, making use of its memory, and knowledge of spoken and written language; for the DNA stretch, the interpreter is a very complex biochemical machinery that is present in every living cell. In the appropriate cellular environment, this machinery translates the sequence of DNA nucleotides into a sequence of very different chemical units called amino acids, which form proteins. It is these proteins that ultimately decide what an animal looks like and does. The analogy between DNA and written language also clarifies how a single change in a nucleotide (a mutation) can sometimes cause a large change in meaning, as in FOG and DOG. But, as with a written message, DNA on its own does not carry semantic information, i.e. 'meaning'. Such semantic information 'resides' in the dynamic interacting system as a whole: in the DNA, in the cellular machinery, in the environment with which the organism interacts; it is this whole dynamic system of interactions that confers functional significance on variations in DNA base sequence.

If DNA participates in the making of proteins within a complex cellular system, what does the common expression that DNA 'controls' or 'determines' traits actually mean? We have already indicated that it could mean that the great tit is nothing but a gene-controlled super-robot, and all its behaviours can be fully understood in terms of the underlying genes. This would not worry most people too much. The idea, first suggested by the French philosopher Descartes, that animals (unlike humans) are but complex machines, is already several centuries old. In one version or another it is accepted by many non-behavioural scientists, and even by some students of animal behaviour.[3] But some people worry because this idea could be extended to human beings too, depriving them of responsibility and a moral sense. Free will would then be denied, and a nightmare is conjured up. It is a futuristic nightmare of computer-generated biological charts that, like astrological charts, serve to predict the future of a newborn baby or even a womb-dwelling embryo. A pregnant woman hands a sample of her embryo's DNA to a large machine which rapidly sequences it, compares it to a huge database and produces a chart specifying the child's characters. Among many other things, the woman is informed:

'You have a girl. She will be blue-eyed and freckled, sensitive to the sun. She has a very good chance of being an extrovert: persuasive, outgoing, decisive, enjoying leadership roles. She will probably be short-tempered and tyrannical. She will also be impulsive and careless, rather shallow and unintelligent. There is a high probability that in very stressful situations she will act obsessively and probably become a drug addict.' And the verdict: 'the chances are that you cannot rely on her in your old age'. The future is no longer drafted in the constellation of stars. The embryo's destiny is written in its DNA.

Is this possible? Geneticists do not believe that genes are destiny, either for great tits or for humans, yet the educated layman often concludes that this is the case for characters ranging from eye colour to assertiveness. There must, therefore, be something misleading in the way genetics is presented. It is such a common, socially dangerous misconception, that it is important to understand how it originated, and to realise why it is false.

Genetics is, indeed, usually presented in a way that implies that genes rigidly and predictably determine traits. This is not done in order to mislead. We believe that there are two major reasons why people have the wrong picture. First, the textbook examples used to illustrate the way a gene affects a character are often biased toward genetic diseases in which a single defective gene has fairly predictable and dramatic effects. Second, elementary Mendelian genetics, which is most people's introduction to genetics, also deals with genes that have large effects and are little influenced by environmental conditions. But both these examples are, in fact, not representative of the way most genes affect the visible characters of individuals.

We are naturally interested in genetic diseases. They are, unfortunately, part of our experience, and seem to excite a mixture of feelings of fatalism and horror, which gives them a tragic fascination. There are about four thousand different identified genetic diseases, over 90 per cent of which are caused by combinations of several genes; the severity of these diseases is greatly influenced by the conditions of life. Only the minority of genetic diseases, such as cystic fibrosis (an ailment that affects the transport of chloride ions into the lungs and gut), is caused by a defect in a single gene. Even in these cases, the severity of the symptoms depends on the other genes the person has. When newborn children have both copies of the gene defective (one copy inherited from father, the other from mother), the children usually develop the disease.

The best understood genetic diseases are almost all like cystic fibrosis: they are disorders that typically have severe effects, and result from a change in the DNA sequence of a single gene. In many textbooks, the genetic basis of sickle cell anaemia is used to explain how a defective gene produces its effects. A change in a single nucleotide in the gene coding for one of the two types of amino-acid chain that make up haemoglobin, the oxygen-carrying protein found in the red blood cells, leads to severe anaemia. A tiny change at the molecular level leads to a tremendous visible physiological effect – to a very sick person. The pathway leading from this gene mutation to the gross physiological consequences of severe anaemia is fairly well understood. Like cystic fibrosis, this is a recessive genetic disease, since two doses of the defective gene are necessary to make the person severely anaemic. The effect of the defective gene is manifest in different environments (though with different degrees of severity), and seems relatively independent of the rest of the genotype.

Unfortunately, there is a strong tendency for people to extrapolate from cases of sickle cell anaemia or cystic fibrosis to other diseases and other characteristics, and to conclude that all genes affect visible traits in a way similar to these disease genes. It is assumed that each gene has a predictable and invariant effect on some feature of the organism. The textbook presentation of elementary Mendelian genetics reinforces this view. We are correctly told that when we mate garden peas, domestic mice or great tits, we are experimenting with sexually reproducing organisms that have two sets of genetic information, one set contributed by the mother and one by the father. Each set of genetic information is carried by chromosomes. The chromosome sets inherited from the two parents are homologous, the members of each pair carrying either identical or different versions (alleles) of the same genes. When genes of a particular pair are identical, the organism is said to be 'homozygous' for this pair of genes, and, when the alleles are different, the organism is then said to be 'heterozygous'. Each offspring is formed from the union of two parental sex cells, the sperm and egg. The formation of the sex cells involves the separation, or segregation, of the two sets of genetic information, so that a sex cell contains only a single set of chromosomes and genes.

The genetics textbooks also correctly tell us that the existence and behaviour of genes was deduced from experiments in which controlled crosses were carried out between parents with heritable differences in a particular trait. The pioneer work was done by Gregor Mendel, who showed how one can infer that a trait-difference is due to a gene-difference. For example, we can infer that a yellow seed colour in the garden pea is caused by a dominant allele, which, even when present in a single dose, determines the yellow colour. The alternative green seed colour appears only when the plant is homozygous for the recessive allele. When Mendel designed his path-breaking experiments, we are told how he was careful to use traits showing sharp differences from each other, which were relatively unaffected by environmental conditions. Gene differences led to predictable character differences in the typical range of conditions in Mendel's garden. Mendel would not have been able to arrive at his theory of heredity and to formulate his laws if the relationship between the genetic constitution of his plants and the visible trait was not stable. If plants with a particular genetic constitution, for example a pea heterozygous for seed colour, were yellow in some of the micro-environments in his garden, but light green or dark green in others, the mapping of traits into genes, and genes into traits, would be impossible. Mendel could not have deduced how the genes that underlay traits behaved.

When geneticists try to explain their trade to non-specialists, they follow Mendel. They show that there is a correlation between the ratios of the characters they observe and the behaviour of genes during the formation of the sex cells and fertilisation. The examples they use to explain Mendel's laws are all based on genes having large and usually independent effects, and a low sensitivity to environmental changes, because only these genes demonstrate in a clear-cut way how one can map traits into genes, and genes into traits. This creates the impression not only that all genes behave in essentially the same way during sex cell production and fertilisation (which they usually do), but also that all genes affect characters in a straightforward and qualitative manner (which they usually do not).

In fact, the frequently discussed genes such as the sickle-cell anaemia gene, the gene affecting the external structure of ear lobes, and the genes used to demonstrate Mendel's laws, are a noisy, and somewhat misleading, minority group. Usually it is impossible to know how a trait will appear from knowledge of a single gene-variant, because the performance of this gene depends on many of the other genes in the organism and on environmental conditions. More than 100 genes contributing to coronary artery disease have been identified, and it is obvious that the interactions between these genes are complex.[4] Many genes influence the development of a single trait, and their effects change in both time and space as the organism develops. Identical mutations sometimes produce different effects in different organisms, both because other genes affect the expression of the trait and because the environment can have long-lasting influences.[5] Even if all the genes of an individual were known in intimate detail, it would usually be impossible to deduce the appearance of a particular trait from the genes, because the expression of genes is often very sensitive to environmental conditions. The many problems encountered when trying to relate differences in traits to differences in genes are the reason why plant and animal breeders, interested in the inheritance of useful traits such as body size, milk yield and height, have to resort to different, indirect methods to study the genetic basis of these traits. They use statistical concepts such as the heritability of a character.

Heritability is essentially a measure of how much of the variation between individuals in a population is the result of genetic variation. If, for example, we are interested in the heritability of the beak length in great tits, we measure the beaks of the parents and their offspring, and determine the correlation between them. If they are highly correlated, we might conclude (not necessarily correctly) that beak length is strongly controlled by genes. If there is no correlation, and the offspring resemble their parents no more than any other birds of their species, we might conclude there is no genetic control at all. To avoid incorrect conclusions, we would also need to measure the correlation between half sibs, or between parents and their adopted offspring, and compare it to the correlation between parents and their biological offspring, all living in the same environment, of course. Such comparisons are necessary because it is essential for geneticists to distinguish the effects of gene differences from the effects of environmental differences. The comparisons between the correlations of differently related individuals could give us an estimate of the heritability of the trait.

The aim of the geneticist is to quantitatively tease out the effects of genes from all other effects, since only the effects of genes are supposed to be heritable. In order to achieve this, statistical techniques like the one described above have been developed. The techniques are all based on the fact that individuals are genetically more similar to their relatives than to non-related individuals. The extent of genetic similarity is used in estimating heritability. If we have individuals who have exactly the same genes, for example, a population of an asexual organism such as the single-celled *Amoeba*, all recently descended from a single individual (so mutations can be taken to be negligible), heritability is, by definition, zero. This does not, of course, mean that genes do not determine traits in the *Amoeba* clone! All it means is that differences between genes are not responsible for the differences between individuals in the clone. Since there is no genetic variation in this population, all the differences between individuals are attributable to environmental factors whose effects are believed to be non-heritable.

The relationship between genes and traits is clearly complex. In order to understand it better, we need to define and discuss some concepts that are an essential part of the vocabulary of genetics, and indeed of modern biology. The important terms are 'genotype' and 'phenotype'. These terms (as well as the term 'gene') were invented by the Danish geneticist Wilhelm Johannsen at the beginning of the twentieth century.[6] 'Genotype' refers to the genetic constitution of an individual. The term is used both for the inherited potential to develop a particular character, such as green eyes or tall stature, and also, more generally, for the sum total of all the genes in the individual – the total developmental potential. The 'phenotype' is the realisation and manifestation of the potential, the actual product. Exactly how the potential is realised depends on the environment. The environment is anything that is relevant to the production process that culminates in the formation of the product, the phenotype. The two examples just given, green eyes and height, can be used to illustrate how the environment may affect the expression of the potentialities of the genotype. In the case of green eyes, the environmental conditions do not seem very important; with certain genotypes (there can be more than one), the eyes will develop a green colour in almost any environment. With height, on the other hand, the way the trait develops is sensitive to variations in environmental factors such as the availability of food. If a child is malnourished when young, she will be of short stature when an adult. With the same genotype she can end up a tall woman if she is lucky enough to grow up in a well-fed family. She has the potential to be tall, but its realisation, the actual phenotype of the grown-up, depends on environmental conditions such as the adequacy of the food supply.

Many useful metaphors have been employed to explain the genotype/phenotype distinction. Richard Dawkins likened the genotype to a recipe for a cake and the phenotype to the actual cake baked, and John

Maynard Smith likened the genotype to a handbook of instructions for building an aeroplane and the phenotype to the actual realised aeroplane.[7] If the recipe specifies how to make an apple-pie, many somewhat different (indeed, sometimes very different) apple-pies will be baked, depending on the ingredients used, the oven, the cook's skill and other 'environmental' circumstances. The same is obviously true for the multitude of slightly different aeroplanes that can be built according to the same plan. Both Dawkins and Maynard Smith stress that only the plan, the genotype, can be inherited: it is the recipe or the handbook of aeroplane construction that is inherited, and not the cake or the aeroplane. Mistakes in the plan will surely be inherited too, but accidental changes in the product that stem from unusual environmental conditions will not be inherited. Translating these metaphors and conclusions into the biochemical language of modern genetics we may say that the DNA, the genotype, is inherited, while the phenotype, its specific context-dependent realisation, is not. This idea is a fundamental assumption of modern evolutionary biology.

One of the important corollaries of the phenotype/genotype distinction is that the visible trait, the phenotype, may not be rigidly specified by the genotype. What an organism becomes depends on the genotype and on the environment in which this genotypic potential is realised. If the environment varies, the phenotype can vary too. But how much can the phenotype resulting from a given genotype vary? How many different phenotypes can a particular genotype support? We have seen that the spectrum for green-eye genes is rather limited. A disease can alter the genotype-specified eye colour, or even destroy the iris altogether, but, for a particular genotype, the eye colour is quite rigidly specified. So, in this case, talking about 'potential' means that the phenotype can only be realised through the processes that take place in normal development, but it does not mean that many variations, many phenotypes, are possible. The range of possible phenotypes is narrow; the trait is rigidly specified by the organism's genotype. Height, on the other hand, is a different kettle of fish altogether: there are many possible phenotypes, many possible heights for the same genotype. In this case, we have a 'wide potential' or, to use the more technical term, 'phenotypic plasticity'. Height in humans is a plastic phenotype, while eye colour is not.

These concepts may become more transparent with the help of an Aristotelian metaphor. The idea that, in order to understand living organisms, one has to accept that something like a 'plan' is involved, occurred more than two thousand years ago to that superb and unsurpassed biologist and philosopher, Aristotle. He often used metaphors to explain how one should think about living organisms. One of his favourite metaphors, used to explain the notion of 'plan' and the relation between the plan and its implementation, was the curing of a disease by a physician. The physician's work, which is done for the purpose of curing people, requires in addition to medicines and instruments, knowledge of the science of medicine. This knowledge is translated in each particular case, for each particular type of disease, into treatment. It always exists as a particular, dynamic, actualisation of the 'plan', as organised activity. Unlike Plato, Aristotle did not believe that the 'plan', or 'idea', was separate from its actualisation.

Aristotle's metaphor highlights some of the complex aspects of the genotype/phenotype relationship. The 'genotype' can be understood as the knowledge of medicine the physician learnt from her teachers and from the books of her library. Her 'medical phenotype' (the way she actually treats patients) is related to the type of medicine she practices, her teachers, her own experience and intelligence, her temperament, her materials and equipment, the patients she happens to encounter and so on. Furthermore, her 'potential' as a physician is not a constant: it broadens as she gains practical experience. The art of medicine is wideranging, its methods are varied and the physician is required to use her judgement, and even her imagination, according to the case at hand. By combining in a new way different methods of medical practice, she may sometimes contribute a novel method of treatment to the already existing body of medical knowledge. In this analogy, the separation between the 'plan' and its execution is clearly much more difficult than in the case of the cake and the aeroplane. Practice and knowledge are intimately interwoven, practice springing from knowledge, and knowledge from practice. In many ways, this more complex analogy is a better description of the relationship between phenotype and genotype than the cake or aeroplane metaphors, because it accords a central role to the agent that is involved in the transformation of the potential into the actual. The intricate relationship between 'genotype' (the medical knowledge) and the treatment given by a particular physician (the 'phenotype') makes it a good analogy for the relationship between the information stored in the organism's DNA and the development of its actual phenotype.

How is the distinction between phenotype and genotype relevant to the great tits in the olive orchard? We have seen that they are phenotypically different from each other: they differ in their songs and in their methods and preferred sites of foraging. They also differ genetically. So, are the differences in genotypes responsible for all the interindividual differences in behaviour? The answer is that theoretically it is possible, but it seems unlikely. It is much more likely that the diverse behaviours of the great tits are not rigidly determined by their genes, but often are responses to the different conditions that individual birds have experienced in the past, or are experiencing at present. In theory, we could show this by comparing the patterns of behaviour of many individual great tits living in two different habitats, and doing cross-fostering experiments involving moving eggs from nests in one habitat to those in another. This would help us to see the relative importance of genetic relatedness and environmental effects. In this case, as in many others, we would almost certainly find that growing up in an environment different from that of the parents affected the phenotype. We would also find that the spectrum of phenotypes that the genotype supports is large, but not unlimited. Yes, the songs of different males differ, but they are all variations on a typical, easily recognised song-pattern. Yes, the methods and sites for foraging differ, but there is a limit to the repertoire of techniques. The basic song-pattern and the range of foraging techniques are, indeed, specified by the genotype.

Going from the tits back to the science-fiction scene of the woman reading the destiny of her embryo daughter from its DNA, we realise that what is wrong with this scene is the assumption that the phenotype is nearly invariant. Clearly, some of the traits of the embryo are more rigidly determined than others: for example her anatomical sex, female, is likely to be accurately predictable, provided no unusual hormonal circumstances occur in the mother's body, and provided the girl does not decide to change sex as an adult. Her character traits, so confidently described by the computer, are a different story: they will depend on the familial, social and cultural environment in which she lives. The psychological development of the girl will depend on what, when and how she experiences and learns. It can be argued that how much she can learn depends on her potential. But how do we define potential in this case? Do we know how to define and delimit the range of possible learning environments? When we talk about phenotypic plasticity we are referring to the way that a particular genotype is realised in different environments. We assume that we can define both the genotype and the range of environments in which it can be realised. For the development of eye colour, the 'plan' is a particular combination of genes, which interact to affect the nature and timing of biochemical reactions leading to the synthesis and deposition of pigments. The 'environments' include the spectrum of normal physiological and biological conditions, which can be defined (with difficulty) in terms of temperature, acidity, interactions between cells and so on. In the case of the development of the girl's character traits, however, we cannot define the 'plan' because we have no idea what genes, how many genes and what interaction-networks are involved. The 'range of environments' means all the social–psychological environments the girl is likely to experience, many of which are partially constructed by her! The number of phenotypes that can be produced as the genotype and environment interact in this case is enormous, and it is impossible to predict the phenotype that will be realised. Even the specification of some general conditions (the girl is going to grow up in a middle-class family) is not very helpful, because psychological and cognitive development depend on a multitude of factors and feedbacks, which are unique and idiosyncratic. The sex of the child, her blue eyes, her freckled skin, can be predicted with a high level of confidence, but her intelligence and character seem impossible to predict.

We have just said that the girl's behaviour is unlikely to be predictable, but can we justify this bold assertion? Some geneticists talk about shyness, religiosity, obsessiveness, homosexuality, as being theoretically predictable from genotypes. (They cannot yet do it, but it is just a matter of time, we are assured.) Does our assertion mean that we can predict absolutely nothing at all about individual human behaviour? Is human nature the nature that has no nature? We know that many patterns of behaviour vary widely in human societies,[8] but we also know that there are traits that are human-specific and universal across human societies. For example, all normal healthy humans living in a society where language is spoken will acquire the common language of their society by the age of four. Within the range of what is humanly possible, how far can we go with our own individual genotype? A famous passage from the writings of the early American behaviourist John B. Watson illustrates the possible irrelevance of the genotype to the development of profession and personality:

> Give me a dozen healthy infants, well-formed, and my own specified world to bring them up in and I'll guarantee to take anyone at random and train him to become any type of specialist I might select – doctor, lawyer, artist, merchant-chief and, yes, even beggar-man and thief, regardless of his talents, penchants, tendencies, abilities, vocations and the race of his ancestors.
>
> (Watson, 1924, p. 104)

Was Watson right? We believe that the answer is a resounding 'Yes', although the 'world' that Watson would need to achieve his goal is probably more complex than he imagined, and the psychological philosophy that underlies his statement is not one we fully share. We believe that the confirmation of Watson's claim is a plain everyday fact, revealed by the sociology and anthropology of poverty and wealth. We also believe that there are invariant human-specific behaviours, and that the plasticity that Watson assumes does not contradict them. It is, of course, important to be more informative: what is the basis of these universal human-specific behaviours? What are the rules of their combination? Do they constrain and channel what is individually and socially possible, and, if so, how? How did they evolve?

The answers to these questions are hotly debated. In the previous chapter, we encountered explanations in terms of exclusive gene selection, exclusive cultural selection, and in terms of co-evolutionary interactions. Does our claim that the girl can develop in very different directions mean that it is possible to think about some behaviour in terms of vast spaces of phenotypic plasticity, unaffected by genetic differences between individuals? A metaphor may clarify the extent to which the genotype sets limits and constrains the phenotype, and how the phenotype can be practically free from these constraints. Imagine flying a man-made red kite. For the kite to be able to leave the ground and fly, a certain minimal length of string is necessary, but the final length of the string that is released can vary. With a relatively short string, the kite is easily controlled. However, the short length of string allows only limited movements and simple manipulations. If more of the string is released, the kite can be manœuvred in finer ways, exploiting the wind. However, once too much string is released, the kite is out of control and seems to act as if it has a will of its own. It is the everchanging pattern of winds, high in the sky, that determines its course. Clearly, the length of the string limits the range of the kite's movements, since it defines the unreachable realms, but it does not specify where it can go within its now vast range. The long string is a constraint on the kite's movements, but it no longer determines them. The kite is free of the tyranny of the string; it is now the kite that determines where the string moves.

The kite, the phenotype, is attached to the DNA string. In the same way that the string is absolutely necessary for the kite to fly, so DNA is absolutely necessary for the existence of the phenotype. The length of the string is, of course, not a physical length, but reflects the number, nature and interactions of genes that are potentially capable of affecting the phenotype; it determines the extent of phenotypic plasticity. The longer the string, the greater the plasticity. Note that this does not mean that the rules allowing great plasticity are not themselves rigidly determined. As plasticity increases, direct genetic control of the phenotype decreases, until it may be effectively lost. DNA becomes a constraint determining the impossible, but having no power to specify the actual. In a given environment, some phenotypes are held by short strings, some by medium-length strings, and some by very long strings. The sex of the child, its blood type, and its eye colour are held by short strings; its height when adult and its physical strength are held by somewhat longer strings, while the different aspects of its character when adult seem to be held by a very long string indeed. These latter phenotypes are uncoupled from genetic variations. They are effectively free.

But what kind of freedom is it? The red kite is free from the tyranny of the string, only to be at the mercy of the winds! Can it manipulate the winds? Can it choose its own course? Here our kite metaphor fails us, and we must change it. If we bring life into the aerodynamic toy by turning the man-made red kite into the biological red kite (a bird of prey), the metaphor springs into life. The freedom of flying now makes much more sense, for now there is a self-regulating entity that interacts with the environment. Of course, there must be some underlying rules that allow and shape functional and flexible flight-behaviour, and the aerobatics of the red kite must obey these 'plasticity rules'. For the bird, the 'string' fairly rigidly specifies its shape, its wing-form and many other traits. But such direct specifications do not determine its flight-behaviour. The red kite can fly straight and level by active flapping, gain height by soaring, glide on extended wings to save energy, and swoop down on its prey. The combination of the laws of aerodynamics, the morphological and physiological characteristics of the kite, its motivations and so on, form a system which allows many types of flight behaviour. Within this range of behaviour, the bird is free.[9]

Plasticity partially uncouples the phenotype from the genotype. One genotype can give rise to many phenotypes – to many types of flight behaviour, many somewhat different songs, many foraging techniques. Moreover, as the tarbutnik example in the previous chapter illustrated, phenotypes can be transmitted to subsequent generations, so the notion of plasticity can be extended into the temporal dimension. Not only can several phenotypes be formed with the same underlying genotype, but different phenotypes can persist for different numbers of generations. How long they persist depends on ecological, developmental and social conditions. This variation in the persistence of behavioural phenotypes over time is an aspect of phenotypic plasticity with enormous evolutionary consequences, because the behavioural phenotype is temporally divorced from the genotype that produced it. The uncoupling of genotype and phenotype is now not just developmental, it is also evolutionary. The inheritance of phenotypes and selection of inherited phenotypic variations can lead to evolution at a new level.

Not only can one genotype support many different phenotypes, but many different genotypes may underlie the formation of essentially the same phenotype. In terms of our kite analogy, the behaviour of the man-made kite would be much the same even if flown with different coloured string. Biologists would say the phenotype is 'canalised', meaning that a typical phenotype is produced even though there are genetic differences.[10] Canalisation means that development is channelled so that, in spite of the many differences between individuals, the developmental path of the great tit produces a typical species-specific morphology, readily recognisable as 'great-tit morphology'. This morphological uniformity is secured by regulatory homeostatic interactions between gene products and the environment, which mask and compensate for deviations, stabilising the developmental path and resulting in the production of a typical form. This means that phenotypic similarity between individuals and constancy between generations do not necessarily stem from genetic similarity and constancy. Both canalisation and plasticity highlight the fact that genes are part of a dynamic system of development, not determinants of development.

Genes do not specify an organism's destiny. Although genes are absolutely necessary in all living organisms, in animals with complex nervous systems, such as birds and mammals, the effect of *gene differ-*

ences on behavioural differences may often be negligible. The development of the organism and its adult behavioural phenotype depend on the interactions with the environment in which it develops, the learnt behaviour of its parents and other group members, the complex, often non-linear relationships among the different factors, the rules underlying these interactions, and its own ingenuity. When the DNA string gets long enough, as it does with birds, mammals and, especially, human beings, it can set the phenotype practically free from genetic determination: variations in genes will yield little information about variations in behaviour. Genes are not fate, they are a challenge.

The problem of information

Darwin's theory of evolution by natural selection is one of the richest and most powerful scientific theories that we have the good fortune to possess. It is the one and only theory that can successfully explain the origin of the array of living forms, for it allows for a progressive accumulation of improvements in many directions. In the previous chapter, we gave a very general formulation of the theory, stating that evolution by natural selection will occur wherever there are entities endowed with the properties of multiplication, inheritance and heritable variations affecting reproductive success (or fitness). The generality of this Darwinian principle led Dawkins to talk about 'universal Darwinism', and to claim that the existence of functionally complex entities in any world would testify to a process of evolution by natural selection.[11] As we know, Darwin was ignorant of the nature of the heritable variations whose existence he assumed, having no idea about genes or DNA, but this lack of knowledge was not harmful for his theory. The fact that the particular mechanism of inheritance that he suggested turned out to be wrong had no effect on his theory of natural selection.[12] The basic structure of his evolution theory is still robust today (in fact, more robust than ever!), after more than 100 years of biological research.

Natural selection depends on multiplication, so whatever the mechanism of inheritance, what is inherited must be a *copy* of what existed before; it cannot be the whole entity itself. As Johannsen pointed out in 1911, your house can be passed on to you by your father, but your nose itself cannot be a legacy from your father, even if the two noses are identical.[13] There are two noses, not one, and an offspring's nose is a *copy* of its father's, it is not the same entity. It is the *information* for developing a particular nose-shape (Jewish, say) that is transmitted by the father and is realised in the offspring. The nose itself is not passed from one generation to the next.

Information has been a leading concept in the biological sciences for over fifty years. As the American psychologist Susan Oyama has pointed out, in a computerised world with a rapidly developing communication technology, it has become common to talk about information 'residing' in genes, in cells, in brains, in communities or in the environment. This use of language has led to 'information' being treated as an autonomous discrete entity, which can be manipulated, rearranged, used and transferred. Information is seen as something that is not changed by the developmental history of the organism.[14] To some extent this view is derived from the digital way in which information is represented in communication technology, where a message is organised as a sequence of digits, which can be altered digit by digit. Our alphabetical written language has the same structure, English having twenty-six letters, the space and punctuation marks. As we saw earlier, DNA can be envisaged in these terms too – as a message with four different modules (A,C,T,G) whose sequence can be transmitted and altered. The focus in this way of looking at information is on the transmissibility of the message, and not on its meaning. Information theory, which was developed by communication engineers following the Second World War, deals with this aspect of information. It is concerned with the fidelity of the transmission of the signal between sender and receiver.

The other way of looking at information is to look at its semantic content, its meaning, which is closely related to the function of the message within the system of which it is a part. It is clear that information in this semantic sense need not be organised in a digital way, it need not be represented as a sequence of modules. Think about 'a cycle of deprivation': although information is passed from parents to offspring, it is not 'modular', but is located in the interaction between the parents and their children. The functional importance of a message within the system of which it is a part is not determined by whether it is modular or non-modular. 'Information' in this functional or semantic sense is diffuse; it is stored in the dynamic interactions of the whole structure, as in a biochemical cycle, a neural network, a genetic network, a set of behaviours, or in the pattern of reliable interactions between the organism and the environment.[15] The 'meaning' of an interacting element is dependent on the system. For a female spider at the start of the mating season, a male signifies a potential mate; if she has already mated or is hungry, the male signifies a source of protein. The meaning of 'male' is determined in the context of the encompassing system. Looked at from the functional and semantic point of view, DNA should be considered as data, not as a program. The program is the organised activity of the genetic network.[16]

The nature of the information, whether it is represented in a modular or non-modular manner, can make a difference to the level at which cumulative evolution occurs. In a modular system, which can be decomposed into independently alterable and transmissible modules, the number of possible messages is very large, even with a small number of modules. A DNA molecule with only ten modules can have over a million (4^{10}) different DNA sequences, all with the same stability in transmission. There is plenty of raw material for natural selection, so the DNA sequence can evolve and accumulate variations. In contrast, a non-modular system, such as a biochemical cycle with ten different interacting molecules, may have only two stable, transmissible functional states (cycling or non-cycling), and therefore only two variants can be subject to selection. Such a system has very little evolutionary potential; it is a limited inheritance system, and no cumulative evolution is possible.[17] However, although the single biochemical cycle may have only two variant alternative states, the cell may have a hundred different biochemical cycles. Therefore 2^{100} different combinations, and hence cell-types, are possible. Inheritance may be limited at the level of the single cycle, but it is practically unlimited at the level of the cell. Cumulative evolution at the cell level can occur. The same argument can be applied to socially transmitted behaviour. Any one pattern of behaviour may be able to vary between only a few alternative forms, but the total behaviour of individuals, composed of combinations of many different patterns of behaviour, allows for many behavioural repertoires. Selection between individuals and between repertoires can occur.

Information is certainly a rather more complicated concept than is commonly supposed. In this book, we are going to use the term 'information' to mean 'any difference that makes a difference', as suggested by Gregory Bateson.[18] We are therefore referring to the semantic meaning of the term, in whatever form (modular or non-modular) it is represented. Since our approach is evolutionary, we need to focus on the way this semantic information is transmitted. We therefore need a general notion of inheritance, one that is not limited to a specific system, or to a particular type of information. We shall use inheritance to mean *the re-generation of phenotypic traits and processes through the direct or indirect transmission of information between entities.*

Natural selection of what?

Information-transmitting systems allow the transfer of functionally significant variations, but how are these variations selected? Often evolutionary explanations of adaptations start by identifying the current function of a trait, and then look at how, beginning from some postulated origin, natural selection of transmitted variations led to that trait. For example, we can explain the highly specific bamboo diet of the giant panda in terms of the environment in which this panda now lives, and reconstruct the evolution of its dependence on bamboo on the basis of our understanding of the diet of the panda's ancestors and the ecological conditions in which they lived. But, even if we have done this, we are far from having a complete evolutionary explanation of the panda's dependence on bamboo. At least three important problems remain. First, have we identified *all* the present and past functions of the trait? Second, have we identified the developmental path that leads to the trait, and the developmental constraints on its past and future evolution? Third, at what level of biological organisation does the transmission and selection of variation occur?

Think about the first problem: why should we think that a trait can have many functions? As every biologist knows, unlike the entities of physics and even chemistry, biological entities, living things, seem to be endowed with purpose. Unlike most physicists and chemists, all biologists use terms and phrases like 'different levels of organisation', 'functions' of the parts of organisms, and 'motivations and goals' of behavioural acts. Because of the intricate nature of the functional organisation of living things, biological explanation is, and must be, different from that customary in physics. Explanations in terms of immediate causes are fundamental to biology but, unlike physics, they are not exclusive. Take an organ like the heart. The structure and activities of the heart can be explained in terms of immediate causes such as muscle contraction and cellular biochemistry. However, a functional explanation of the blood-pumping activity of the heart is also necessary. So, too, are evolutionary explanations of the past origin and the present persistence of the heart's structure and function.

One consequence of the complex nature of biological entities is that structures, processes and behaviours usually have several different functions. Almost every feature, be it an anatomical characteristic such as the black-and-white stripes of zebras, or a behavioural trait such as offspring-licking by a female mammal, has more than one function, and is the result of many selective pressures and serendipitous circumstances. In biology, it is usually a mistake to be satisfied with a single functional explanation, and it is certainly important to take new explanations seriously. A new explanation is often complementary rather than alternative to the already well-established ones, so it is not necessary to abandon the previous explanations. Sometimes, however, a new explanation drastically diminishes the relative significance of previous explanations, and must then, for all practical purposes, be considered as an alternative. In order to understand the role of a particular trait in the life of an organism, it is necessary to look at all its functions, and assess their relative importance. We shall illustrate this with the two examples already mentioned: zebras' stripes and the offspring-licking behaviour of female mammals. Desmond Morris has summarised nine different, but not mutually exclusive, possible answers to the question 'why are zebras striped?'.[19] Here we shall look at just three of these answers in order to illustrate the complementary nature of evolutionary explanations.

Plains zebras are widespread, social, East-African nomadic grazers. They are non-territorial, living in harems with 1 male and 5 or 6 females. The surplus males form bachelor groups. For a few weeks during the rainy season, the small herds sometimes join together to form much larger herds, with hundreds and even thousands of members. Adult zebras remain in their group for life, and are able to recognise all other group members individually. They are day-active, noisy and alert, and rest in exposed localities. They never hide in response to danger: they run away when attacked by lions, but attack without hesitation when harassed by small groups of hyenas or hunting dogs. At night, group members move very little and, while others sleep, one group member acts as a guard. Against the background of this life style, how can one explain the function of the black-and-white stripes?

The once most common explanation was that the stripe-pattern camouflages zebras, thereby protecting them from predators, because stripes help to break up the perceived outline of the body. This explanation is problematic, because zebras' noisy daytime behaviour and clearly visible stripes make them conspicuous. However, under dawn and dusk visibility, when predators such as lions are most active, zebras are much more difficult to spot than similar sized antelopes, so the camouflage explanation seems to hold true. Another explanation is that the stripe-pattern may act as an efficient fly-deterrent! We are told that some parasitic and disease-carrying species of flies refrain from landing on highly contrasting surfaces such as the striped zebras. Consequently, compared to other equine species, zebras are only mildly infested with parasites. A third explanation suggests that the contrasting pattern may act as a cooling device: a series of circular air-currents form between the stripes as a consequence of the differential warming of black and white regions, and these currents blow over the zebra's skin, thereby cooling it.

The three explanations are clearly complementary: during the day the stripes act as fly-deterrents and a cooling device, at dusk and dawn as camouflage. The other six explanations mentioned by Morris are additions rather than alternatives, and may all be parts of the benefits that this seemingly simple character confers. Zebras' stripes evolved and are maintained in zebra populations because of the selective advantages that these multiple functions confer on individual zebras. Evolutionary biologists ask various fascinating questions about these multiple functions: which function of the zebra's stripes was primary, and which functions were accidental and advantageous side effects of the primary function, subsequently elaborated by natural selection? Was the primary function camouflage at dusk? Was it deterring flies during the day? Or was it some other function, not at all related to the present functions? The resolution of such questions depends on an intimate knowledge of the animal's biology, the ancestral ecological conditions and their subsequent changes.

The importance of complementary explanations can be illustrated just as convincingly with a relatively simple behaviour – offspring-licking by female mammals. Any admirer of cats, dogs or rabbits knows that the mothers in these species (as in many other species of mammals) lick their newly born and young offspring frequently, meticulously and thoroughly, paying special attention to the face and the ano-genital region. What are the benefits of such behaviour? In order to understand it, we must remember that the young of these species are altricial: they are born at an early developmental stage, helpless, their eyes and ears sealed, and they cannot walk, regulate their body temperature or void their faeces independently. In most cases, it is the mother who assumes exclusive responsibility for her offspring and their problems. For years it has been well known that mothers constantly lick their offspring to stimulate them to begin breathing, to clean them, to dry them, to stimulate them to urinate and defecate, to mask their characteristic smell from predators, and to help them regulate their body temperature in hot climates by using evaporative cooling from a saliva-covered skin.[20] Offspring-licking behaviour clearly serves a host of complementary functions. However, this list does not exhaust all the benefits of licking! Recent research has shown that the tactile stimulation caused by licking improves and accelerates maturation processes in the nervous system of the young individual. As a consequence, well-licked infants are, in many respects, behaviourally superior to ill-licked ones.[21] In addition, by licking, the mother covers her offspring with her saliva, which contains a high concentration of antibodies, thus providing them with one of several important lines of immunological defence.[22] The 'immunological' and 'tactile stimulation' functions are recent additions to the long list of already well-established functions of licking, and they are probably very important. It would have been a mistake to have assumed that, because the function and evolution of the trait was already 'reasonably explained', the existing explanations were correct and complete.

Even if it was possible to identify all the many functions of a trait, this would not provide enough information to understand its evolution. We would then have to consider the second problem we identified, and look at the developmental path that leads to the manifestation of the trait. It is not only the final product of selection that needs to be understood, but also how it was achieved. At first sight, this may seem unnecessary. For example, if selection leads to a larger body size in flies, and we have identified the various selection pressures correctly, why should it matter to us how this increase in body size was achieved? Selection experiments show that increase in body size in fruit flies can be achieved in two different ways: in some selected lines, size is increased because there is an increase in the number of cells per fly; in others, cell number has not changed, but each cell is larger than those in the ancestors.[23] Similarly, if we see that two groups of mice prefer the same type of grain, it may be for totally different reasons. It could be that the mice in one group have an uncommon digestive enzyme variant that breaks down the components of this grain efficiently, and this leads to a feeling of well-being, which reinforces the tendency to eat this type of grain. The mice in the other group lack this enzyme, but, since in their environment this grain-type is the only source of a vital vitamin, natural selection has led to the development of an association between the taste of this grain and the pleasure centre in the brain, even though eating it to excess causes digestive problems. In both cases, the end-result of selection (for body size or food preference) is identical in the two lines, but the path leading to it is different.

The importance of understanding how the end-result develops is that it can help us to reconstruct the past and predict the future evolution of the trait. It may well be that increasing the cell size of fruit flies beyond a certain point is quite impossible because of the problems associated with the decreasing ratio between surface area and volume, whereas adding more cells to the body is still feasible. Further evolution of increased body size is therefore much more likely in the line of flies with more cells than in the line with large cells, which may stop responding to selection. The second example illustrates another reason why knowing the way in which a trait develops is important. The mice's preference for a particular grain-type is, in the first case, the result of the coupling of the effects of a particular digestive enzyme variant with the result of a general learning mechanism – the enhancement of a preference through positive reinforcement; in the second case, an association between a particular taste and pleasure was formed, without any new enzyme variant. In this second case, we can talk about an evolved domain-specific 'preference module' in the brain, which was selected as such during evolution. In the first case, we cannot. The way that an adaptive trait develops is therefore important for reconstructing the system of which it is part, as well as for reconstructing its past and future evolution.

The third problem – the level at which transmission and selection occur – is readily seen if we extend the mouse example. It may be that the preference for the particular grain-type is neither the product of the evolution of digestive enzymes coupled with a general learning strategy, nor the product of evolutionary changes in taste. It may be that no genetic change has occurred at all, and the preference is traditional: the preference is acquired by the young from their mothers, through the mother's milk, and is perpetuated for many generations through this type of very early learning. If this is the way that the food preference is acquired and transmitted, then understanding the way in which the preference develops is essential for understanding its spread and maintenance through time. As we stressed in the previous chapter, the transmission of a socially learnt behaviour or preference is part of its development! While in the genetic system genes are transmitted irrespective of their function (a nonsensical DNA sequence can be transmitted with the same efficiency and in the same way as a gene coding for an essential enzyme), a pattern of behaviour or a preference is practised and transmitted through learning, all at the same time. Social learning is simultaneously a developmental process and a transmission process. Since the manifestation of a behaviour is necessary for its transmission through social learning, transmission cannot be taken for granted, as it can be with the genetic system; the only way the transmission of behaviour can be understood is as part of a process of behavioural development that involves social learning. The variants that are generated during social learning provide the raw material for the natural selection of behaviours.

We have repeatedly stressed that DNA is not the only system that satisfies the theoretical requirements for an inheritance system. There are other inheritance systems that contribute variations that lead to evolution by natural selection. These inheritance systems are themselves the products of natural selection. They are found not in some green extra-terrestrials from Mars, or in imaginary tarbutniks, but in organisms living now, on our planet. They are part of every one of us. One major group of universal, non-genetic, inheritance systems are the epigenetic inheritance systems (EISs) that underlie what is called 'cellular memory'; they are the systems that ensure that the functional and structural states of cells persist through cell divisions, even though the stimuli that first induced those states are no longer present.[24] The system of inheritance that is the subject of this book is less ubiquitous, but much more familiar: it involves the transmission of behaviour patterns between generations. We isolated this system in our thought experiment with the imaginary tarbutniks. The nervous systems of animals are learning systems that respond (usually adaptively) to the changing conditions in the outside world, and effect changes in patterns of behaviour. They are also storage (or memory) systems: representations of behavioural responses to the environment are stored, to be retrieved and used later in the appropriate circumstances. In addition, a nervous system can be part of an inheritance system: through social interactions with older and experienced individuals, naïve animals may learn the behaviour of the previous generation, and may also acquire developmental or ecological legacies that lead to the repetition of ancestral behaviour patterns. The transmission of learnt information between generations probably originated as an incidental by-product of learning in a social situation. However, once social learning had emerged, it became a direct agent of behavioural evolution by natural selection. Moreover, social learning itself became an important target of biological evolution: the ability to learn in social situations has itself evolved.

Summary

The behaviours of birds and mammals pose many evolutionary questions. How they are answered is closely associated with how the relationship between genes and behaviour is seen. This relationship is clearly very complex. On the one hand, a change in a single gene can sometimes lead to changes in a whole array of phenotypes, and the way a gene affects a particular aspect of the phenotype can be very sensitive to ecological and social conditions. On the other hand, different genotypes often lead to identical phenotypes, which are quite insensitive to changes in ecological circumstances. This uncoupling of variations in genes and variations in phenotypes means that when there is a system of inheritance that operates at the phenotypic level itself, evolution need not involve changes in genes. Since patterns of behaviour can be inherited through social learning, natural selection of behavioural variations that are generated at the phenotypic level have probably played an important role in the evolution of behaviour.

Evolutionary processes that are based on variations transmitted through the behavioural system are different from those based on gene differences. First, unlike genetic variations, if they are to be transmitted, behavioural variations have to be expressed. Acquiring and transmitting variations at this level therefore has to be seen as an aspect of development. Second, information carried by the behavioural inheritance system is non-modular – unlike the genetic system, in which the nucleotides of a DNA sequence can be altered one by one, information transmitted by the behavioural system is contained within the behaviour pattern as a whole. Variation is therefore far more limited than with the genetic system. However, this does not necessarily limit the scope of evolution based on behavioural inheritance, since when the many behaviour patterns that an animal can show are considered, variation is effectively unlimited, and selection can lead to complex adaptations in the overall life style.

Notes

1 The description of the behaviour of the great tit is based on personal observations and Gosler, 1993; Hinde, 1952; Perrins, 1979.

2 The estimated number of bird species is derived from Brooke & Birkhead, 1991, p. 203; the estimate of the number of vertebrates species is taken from Young, 1981, p. 11.

3 What one actually means by 'machine' makes a great difference. As a biological metaphor, the term 'machine' has undergone changes in meaning, sometimes meaning simply that there are no transcendental 'special forces' behind the behaviour and organisation of living organisms. But most people still think about machines as highly determined objects, where the functioning of the whole can be fully understood in terms of the parts.

4 Sing, Haviland & Reilly, 1996.

5 Wolf, 1997.

6 Johannsen, 1911. For a recent discussion of the terms, see Lewontin, 1992.

7 The metaphors of Dawkins and Maynard Smith can be found in Dawkins, 1986; Maynard Smith's metaphor was presented in a televised lecture of the Linnean Society, which was broadcast by the BBC in 1982.

8 Some dramatic examples can be found in Geertz, 1973.

9 Discussing the difficult issue of individual freedom in non-human animals, Hubert Hendrichs, who has devoted much of his research to the study of individual personality in mammals, claims: 'The attention of the mammal is only very rarely restricted to one specific object or to one functional and motivational context. Its attention generally, in various stages of intensity, covers several areas simultaneously. It is thus able to choose several impulses out of many, and to respond to the combination of these selected impulses in a way fitting its program [genetic and cultural program]. It can, in such a case, sometimes choose its action, while possibly not the form in which it is carried out. A mammal even can "load" specific structures and events in the environment with specific significance and meaning, attributing to them specific qualities that make them a source of security or fear, of excitement or tension. In attributing such qualities to specific parts of its environment the mammal can show some kind of what, in humans, would be called imagination, invention, and creativity,' (Hendrichs, 1996, p. 117). In chapter 5, we shall discuss the importance of individual personality in influencing the mate choice and relationships between monogamous mates.

10 The idea of canalisation was developed by Waddington; see, for example, Waddington, 1957.

11 Dawkins, 1986.

12 Darwin's heredity theory (known as the pangenesis theory) was described in his book *The Variation of Animals and Plants under Domestication*, vol. 2. We return to this theory in chapter 9.

13 Johannsen, 1911.

14 Oyama, 1985.

[15] Different types of heritable information are discussed in Jablonka & Szathmáry, 1995.

[16] Atlan & Koppel, 1990.

[17] Maynard Smith & Szathmáry, 1995; Szathmáry, 1995.

[18] Bateson, 1979.

[19] Morris, 1990.

[20] Summarised in Ewer, 1968.

[21] Caldji *et al.*, 1998.

[22] See Janeway & Travers, 1994.

[23] Robertson, 1960.

[24] Most cells in a multicellular organism are believed to have identical genotypes, all being derived from the same single fertilised egg, but, despite this identity in DNA sequence, different tissues have strikingly different cells. Not only are the cells of different tissues different, but these differences also persist following cell division, long after the particular cell-types have been established during development. The systems responsible for this are known as epigenetic inheritance systems, or EISs. EISs ensure, for example, that when skin cells divide the daughter cells will be skin cells, and when kidney cells divide the daughter cells will be kidney cells. This occurs despite the fact that the stimuli that induced the formation of these cell-types during development are no longer present. At some stage during embryonic development, certain stimuli made each cell-type different, and the differentiated state was then 'locked into position' and became permanent and heritable. The skin cells breed true and so do the kidney cells, although both cell-types have the same genotype, the same DNA. In addition to the transmission of epigenetic information in cell lineages within an individual, epigenetic information is also transmitted between generations of unicellular and multicellular organisms. Today, biologists are beginning to appreciate and understand the biochemical nature of these EISs, and their role in development and evolution. Since EISs enable the hereditary transmission of different cellular states, EISs make the cellular phenotype a unit of heritable variation. A review of the role of EISs in development can be found in Holliday, 1990; their role in evolution is discussed by Jablonka & Lamb, 1995.

3   Learning and the behavioural inheritance system

To understand how traditions originate and how they evolve, we must first establish the relationship between learning, memory and social organisation. Not everything that is learnt becomes a habit, not every habit involves social interactions and not every social habit is transmitted across generations. We therefore need to know what learning entails, how patterns of behaviour are memorised and how they lead to the formation of traditions. Our purpose is not to describe the neural mechanisms of learning and memory, but rather to outline the psychological, ecological and social conditions that influence how behaviour patterns are generated, remembered and transmitted. 'Learning' and 'remembering' are not simple and unitary processes, however: different species rely to varying extents on several types of learning and memory. This affects the nature of the habits they develop, and whether or not and in what manner these habits form cultural traditions. To get a better understanding of the different types of learning and their consequences, we will return again to the Judaean hills and observe the behaviour of some of their inhabitants.[1]

It is late spring, and the dry shrubland of the Judaean hills, with its small oak trees dotted among scattered low bushes and wild herbs, is swarming with life. As the daylight fades away, a female orb-web spider, suspended in mid-air on a thin thread stretched between two flowering bushes, is busy constructing her orb-web. Seed-collecting harvester ants move hurriedly along well-trodden earth roads to and from a nearby underground nest. A snorting European hedgehog roots among the leaf litter, cracking snails between his teeth and picking up ants with a flick of his tongue. Not far from the hedgehog, a mother chukar partridge calls her chicks to a safe night shelter in the depths of a bush. A male blackbird, now barely visible, sings his mellow warbling song, the last for the day, from the top of an oak tree.

After twenty-five minutes, the spider's construction is complete. A small but perfect orb-web is stretched between the two bushes, and the landlady sits in ambush, head downward, in the middle of it. The orb-web is basically composed of a frame, radial threads that converge on the centre (much like the spokes of a bicycle wheel), and the sticky spiral. Each of these elements is made of a different type of silk, secreted by one or more spinning glands attached to a spinneret. The frame threads attach the web firmly to the branches of the supporting bushes and also serve as anchors for the radial threads. Both are non-adhesive. The radial threads, converging towards the spider waiting in ambush at the web's centre, serve as spatially accurate lines of communication by transmitting vibration signals to the spider. These signals disclose the exact location of the unfortunate beetle that has just bumped into the web, and cause the spider to rush off in the right direction, along non-adhesive access routes, to inject venom into the struggling insect and finally wrap it in silk to make absolutely sure that it is subdued. Nothing is urgent now, and the spider takes her time (hours!) to suck the beetle dry. The sticky spiral, a thread sprinkled with glue droplets, is highly viscous and very elastic, so that a trapped insect, fighting to free itself, is more likely to end up completely stuck to the web than to tear it and break free.

Twenty-five minutes of spider's work and twenty meters of silk threads have enabled a 400-milligram female spider to sit in ambush in the centre of a 0.4-milligram wonder-web! The web and its owner are certainly spellbinders. From time immemorial, people have wondered at and poets have written about the perfect beauty of the orb-web. Functionally inclined biologists and engineers have admired the simple, economical and efficient solutions (physical, chemical and behavioural) of the orb-web spiders to the complicated problem of 'how to build an optimal orb-web'. J. H. Fabre, the great nineteenth-century French entomologist, succinctly expressed the disparity between the beauty of the orb-web and its lowly function as a fly snare: 'What refinement of art for a mess of flies!' Ignorant of its artistic merit, at dawn the spider eats her precious web, recycling the silk for further use next evening.

The sequence of behaviours used to construct an orb-web and the fine details of its structure are different in different species. Experiments have shown that vision, gravity and experience are irrelevant to the almost automatic construction process. As most orb-web spiders are solitary, short-lived, and never meet their parents, they have no opportunity to learn their masterly art of weaving from others. So, are they nature's ultimate robots? Not quite. If you remove a just-constructed radial thread from the web, the spider will replace it. After this thread has been removed from the web twenty or so times, the spider will suddenly give up her attempts to replace it, and proceed to the next stage in her fixed sequence. Perhaps, after all, spiders can learn something about their environment? Or maybe giving up trying after many trials is just another one of the 'fixed action patterns' that the spider displays, part of the pre-established program? Certainly some spider behaviour does involve learning: a few encounters with a foul-tasting prey such as a stinkbug cause the spider to avoid it ever after. Although a spider does not have to learn how to construct a web, it does have to learn, through individual trial and error, what types of prey are inedible.

Orb-web spiders are found everywhere. Extensive inborn knowledge seems to be the principal key to their success as engineers. The keys to the success of the spider's neighbour, the rooting hedgehog, are somewhat different.

The hedgehog spends the early evening hours carrying out a thorough culinary survey of its environment. It wanders about, following its extremely sensitive snout, covering up to a kilometre in a single night. When it encounters a promising scent or a rustle, it stops and closely examines the site, rooting under leaf litter, turning-over fallen branches with its snout, digging into the soil and pouncing on potential victims. It is a real omnivore, with a voracious appetite, and enjoys almost any species of insect, snail, spider, centipede, lizard, snake or small mammal that it can catch, as well as carrion, leaves and fruit. Its exceptional immunity to many natural poisons and diseases allows it to be less discriminating than most other animals in its approach to potential sources of food. With a minimum of danger to itself, it learns from its own experience, by trial and error, what is good to eat and what is not.

If, for a fortnight, we were able to accompany each of three mature hedgehogs on their nocturnal forays through our area of dry shrubland, we would see that, despite their seemingly indiscriminate diet, each actually prefers a somewhat different supper. One eats mainly earwigs, another beetles and the third snails. Why do they have such different feeding habits? If we test each of the three individuals for their food preferences under field conditions, offering them a choice between earwigs, beetles and snails, we would find that, although they are opportunists and will eat almost anything, they still show different preferences. Are these different tastes innate? Are they transmitted to them by their mother?

The behaviours of four-week-old hedgehogs, who still enjoy supplementary suckling, show us that, although they join their mother in her nocturnal forays and are escorted by her, each youngster forages for itself, unaided and unguided by the mother. One youngster crouches in the middle of a carpet of poppy flowers, surrounded by low spurge shrubs. He sniffs excitedly in all directions, finally showing interest in one particular poppy flower. A dozen chafer beetles lie motionless inside the flower. He first tastes one beetle, then another, and then hurriedly eats the rest. Now he goes sniffing from one poppy flower to the next, until he finds another one occupied by chafers. This time he gobbles them all up without pausing to taste them first. After a few minutes of futile search for more chafers, the youngster's attention shifts to a twig that is spread-out on the ground. It is heavily infested with aphids, which are preyed upon by several ladybirds. The young hedgehog smells and tastes a ladybird, and immediately spits it out. When attacked, ladybirds release drops of a foul-smelling fluid that is distasteful or toxic, even for a hedgehog. One nasty lesson is enough for the young hedgehog to learn to avoid ladybirds hereafter. Now the smells under a fallen log attract him. He turns the log over and finds click beetles, weevils and slugs, which are all tasted and rapidly swallowed. He goes on to explore the peculiar scent of a large predatory ground beetle, which is actively hunting at the base of a spurge shrub. The beetle sharply nips his soft vulnerable snout with her powerful jaws, and he retreats, learning to avoid her and her kind thereafter. His brother, hungrier and therefore more persistent, does not give up so easily. After trying his luck with several predatory ground beetles and earning a sore snout, the brother finally learns that, if he just approaches, grips and starts eating the beetle from the side of its abdomen, it is defenceless. The two brothers have learnt different lessons. The next evening, the first hedgehog follows yesterday's successful route: he goes straight to the poppies and starts searching for chafers, then tries the fallen log, carefully avoiding ladybirds and ground beetles. Only after exhausting the sites of recent success does he let his sensitive snout lead him towards new promising foraging grounds.

When tested for their food preferences at four weeks of age, the young are far less discriminating than either their mother or any other mature hedgehog. However, at four months, when they are fully mature and long since independent, and have gathered a lot of foraging experience, they display definite individual food preferences: one relies mainly on earwigs, another on beetles and a third on snails. A mature hedgehog foraging at night in the Judaean hills has already experienced its own unique series of trials, errors and successes, and has developed its own special food preferences and hunting techniques.

At dawn, as the hedgehog ends feeding, the chukars begin. Early morning in the dry shrubland of the Judaean hills finds a dozen day-old chukar chicks following their mother in single file across a large patch of wiry grasses, watching her closely and listening carefully. The mother stops by a tuft of short grass growing near the base of a mastic tree, lowers her head and explores some blades of grass with her bill, but the chicks do not try to peck at or eat anything yet. Eventually the mother points with a partly opened bill at a crowded colony of aphids occupying the blades of grass, and emits a special call. The day-old youngsters react as if invited to dine, and enthusiastically peck several times at the aphids. More often than not most of them miss their living targets. They will need several weeks of constant practice and self-improvement to perfect the art of accurate pecking. The mother marches on, and, every time she stops, she uses the same audio-visual display to introduce and encourage her youngsters to feed on particular food items, known by her to be both edible and rewarding. She introduces them to weevils on clover, caterpillars and grasshoppers on the grasses and a wide variety of seeds scattered on the ground.

The family food-tutorial march is not without its dangers, and when the mother senses danger, spotting a predator, she frequently emits an alarm call. These calls are different for ground and aerial predators, causing her youngsters to either squat, seek cover or freeze, depending on the type of predator. The alarm calls serve as early warnings of danger, but may also help the youngsters to recognise predators for what they are. Day-old chukar chicks are already physiologically capable of independent feeding, but they are almost completely ignorant both about predators and about how edible and nutritious the potential food items they encounter are. By being constantly tutored by an experienced mother, the chicks gradually learn, over a period of a few weeks, to forage efficiently and securely. They limit their attention to the food items and sites preferred by their mother, adopting her safe list of food and site preferences, without having to resort to the much more dangerous and time-consuming method of individual learning through trial and error.

Although the chukar partridge, the European hedgehog and the orb-web spider share the same habitat, they have very different life styles and very different feeding behaviours. The spider is a sedentary, relatively passive, sit-and-wait predator, dependent to a large extent on the catching competence of its invariant web. It learns almost nothing with respect to web building, little with respect to edible prey and, all in all, relies mainly on innate, fixed patterns of behaviour.[2] The poison-immune hedgehog is a semi-nomadic, actively searching, opportunistic small-game hunter, depending mainly on its ingenuity and great capacity to explore and learn about its varied and varying environment. It relies chiefly on its personal experience, on its own extensive trial-and-error learning, to form its food preferences and ways of hunting for its preferred food. The vulnerable chukar partridge chicks, thriving mainly on an assortment of seeds and insects, can afford to use neither the hedgehog-type of individual learning nor the spider-type innate behaviours. They need a lot of information about the edibility and nutritional value of a host of potential food items, but are not immune to poison, so cannot experiment too much. They rely chiefly on social learning to form their food preferences – on the cumulative, multigenerational, parental experience communicated to them by their mother. Of course, the full behavioural repertoire of every mammal and bird includes all three of the types of behaviour we have described. Innate tendencies, individual learning, and social learning are all mixed together to produce the amazing and rich variety of behaviours displayed by birds and mammals.

Our main concern in this book is going to be with social learning, because it allows learnt information to be transmitted between individuals and across generations. In the previous chapters, we emphasised that, for evolution by natural selection to occur, a system has to show variation, heredity and multiplication. We argued that, when coupled with a social way of life, the nervous system of birds and mammals is part of an inheritance system, because it can store information that can be transferred to other individuals. But does the information that is transmitted also show heritable variation? If evolution by natural selection is to occur, there must be some heritable differences in behaviour. How much individual variation in behaviour is possible, how is behavioural novelty generated, and what are the constraints on the generation of behavioural novelties? We also need to know for how long a new behaviour pattern can be remembered, and in what ecological and social circumstances it is acquired and transmitted. In order to answer these questions, we must first discuss learning, the source of most behavioural innovation.

What is learning?

Learning is so fundamental for us human beings, that we tend to take our great capacity for it for granted. We assume that, because we are so good at it, learning is biologically invaluable. We rarely ask what is so wonderful about being able to learn. Clearly, an outstanding ability to learn is not necessary for evolutionary success. Plants, for example, do not learn in the common-sense meaning of the word, yet their adaptability is legendary, and there is no doubt that their evolutionary success is as great as that of animals. Many invertebrates show complex behaviours that seem completely innate: our female orb-web spider never learnt how to make her beautiful, deadly trap; she was born a master weaver, able to carry out her elaborate and multistaged weaving job to perfection, without prior experience. Spiders are a highly successful group, and any terrestrial habitat contains many more individuals and species of spiders than of birds and mammals. So, why are not all behaviour patterns innate, like the spider's weaving? Why do birds and mammals have to learn so many things and in so doing face the risks of making expensive mistakes? What is learning, anyway?

Learning is usually defined as an adaptive change in behaviour that results from experience.[3] It is associated with changes in the ways nerve cells in the brain interact. These include changes both in the architecture of neural circuits and in the spatial and temporal patterns of brain activity – in where and when activity occurs in the brain. Molecular changes within the brain cells, such as changes in gene expression or in the activity of enzymes and other proteins, may also be involved.[4] And, contrary to common prejudice, which has it that new nerve cells are not produced in adults, learning and post-natal development sometimes involve cell proliferation and cell death in particular parts of the brain.[5]

An interesting picture of some of the things that happen to the brain during learning has emerged from neurophysiological studies of birds, man and other mammals. For example, people who are born deaf and mute, and have learnt to communicate in sign language, with time develop better peripheral (sideways) vision than hearing people. It seems that the more extensive use of vision by deaf–mute signing people, and the lack of competition within the brain from auditory input, lead to the increase in both the size and physiological activity of a particular 'visual' area in the brain.[6] Parts of the brain are modified as a result of experience, and some of the changes are long-lasting. The many studies showing that learning leads to both general and localised changes in the brains of kittens, rats and songbirds, testify to the generality of this conclusion.[7] The different experiences, including sexual experiences, that individual animals have may leave different long-lasting effects on both their behaviour and their brain organisation.[8] However, important as it is to understand the physiological basis of learning, we will not dwell on it here because our present concern is more with the role of learning in the life of animals: about why, what and when animals learn.

The ability of an individual to learn enables it to respond adaptively to the many, often temporary, changes occurring in its environment during its lifetime; it enables the animal to recognise new dangers, and to exploit new resources. Yet learning also incurs costs: mistakes can be made, and precious time and energy can be wasted. In comparison, an innate behaviour, like the spider's weaving behaviour, seems to be both safer and more efficient. However, this very stability and relative invariance limits its usefulness as a solution to the problems of a changing environment. The spider's innate behaviour is a segment of the unchanging part of its way of life, and is underlain by the very long genetic 'memory' of the species. The orb-web spider cannot change the way it weaves its species-specific orb-web, even in circumstances in which it would be better to do so. Learning, on the other hand, allows animals to adapt to changing yet recurring facets of the environment, both by acquiring new behaviour patterns and by forgetting those that are old and no longer useful. If and when necessary, the hedgehog can adapt to new feeding circumstances by forming new foraging habits.

When do animals tend to learn new things? Although in natural conditions learning and updating information are ongoing processes, animals do not drastically change their habits when their environment is familiar and predictable. At the other end of the spectrum, in conditions that are perceived as unusual and unpredictable, animals may be so acutely stressed they are unable to learn. Different degrees of stress frequently lead to very different actions. A bird confronting a snake may mob it and try to peck it, or fly away, or, when cornered and acutely stressed, tremble helplessly. It is the vast middle ground, between the extreme states of placidity and panic, that promotes learning. As long as it is not severe, stress seems to be an important general inducer of learning.[9] For example, mild hunger promotes exploration, often leading to learning how to handle new kinds of food; rainy weather encourages finding and using new kinds of shelter. We have already seen that a mildly stressed, hungry young hedgehog is more persistent in its attempts to feed on problematic prey than is a satiated one, so eventually it learns how to handle a fighting and biting ground beetle. The degree of stress felt in a particular situation differs greatly between individuals. A sand storm in the desert may lead inexperienced human travellers to great anxiety and even hysterical behaviour, while a well-seasoned traveller may sit through it motionless, covered with his garb, enjoying the spectacular sight and alert to new events. What an individual finds threatening and stressful depends on what happened on previous occasions, on what was learnt then. For an inexperienced individual, every minor change in its surroundings is perceived as potentially dangerous, whereas an experienced individual may find the same change only mildly stressful and, confident in its ability to cope with similar changes in its surroundings, will soon start learning. The presence of their parents allows young animals to explore their environment much more readily, since, by providing a protective and comforting atmosphere, the parents act as stress minimisers. A mature domestic mouse or a brown rat is very careful and hesitant when exploring a new territory. A mature chukar partridge testing a new food item is slow and cautious. Youngsters, on the other hand, are much more daring when under the watchful eye of their mother. The mere presence of a mother imparts a sense of security and reduces the level of stress in her offspring, thus facilitating their learning. In addition to the direct transmission of valuable information, parental care seems to provide the young with the emotional pre-conditions necessary to maximise the efficiency of the youngsters' own learning. It also minimises the price of mistakes by preventing them from happening or correcting them before they cause too much damage. There is an obvious, and usually adaptive, association between emotions and learning in both animals and man.[10]

In the wild, learning is a natural and inevitable part of the life of every bird and mammal. In fact, we can talk about an intrinsic, basic urge to learn, explore and control the environment. This urge becomes painfully clear when animals live in zoos or laboratories. The need to make sense of and, if possible, to control their world is so strong that they prefer to expend energy on earning a reward by performing a learnt behaviour, rather than to receive a 'free' gift. If mildly hungry laboratory rats are allowed to choose between earning a food pellet by running down an alley, or getting it free, they prefer to 'earn' the pellet. They will run in a maze past thousands of identical pellets to get to the goal-box and obtain the reward it contains. Many experiments have shown that animals prefer to learn and to predict the outcome of their activities, even when this has adverse effects. For example, deer-mice dislike bright light, and are easily trained to turn the light off every half hour when it comes on automatically; but the same deer-mice, despite their aversion to light, turn the light on when it is turned off automatically. It seems that having control over their environment overrides the dislike the deer-mice have for bright light.[11]

Of course, some learning does not require the animal's conscious attention to its environment at all. It is a kind of 'unaware' learning. For example, rats can learn to associate the particular taste of food that was given to them when they were anaesthetised with the later adverse effects of this food.[12] However, a lot of learning occurs while the individual's attention is drawn or directed to a salient feature of the environment: an alarm call actively directs a bird's attention to an animal that is a potential predator. The role of attention in learning is now being investigated using the new imaging techniques that allow changes in brain activity to be visualised on a computer screen while the individual learns and acts. Active sensing (observing, listening) involves the activity of special attention-networks localised in certain parts of the brain; these networks assign high priority to those computations leading to the analysis of appropriate responses to the stimulus – to planning a move in chess upon perceiving the chessboard, for example. In time, and with a greater degree of expertise, as the learnt operations become automatic, the activity of the attention-networks is greatly reduced, and only the computing areas that are more directly involved with the analysis of the stimulus remain active. The attention-networks are now free to assign priority to another stimulus, so attention can be directed elsewhere. Acute attention to one operation generally reduces the possibility of being highly attentive to another. It is a general finding that, as a task such as riding a bicycle or playing chess becomes habitual, there is a corresponding change in the organisation of the neural networks necessary to perform it. As the task becomes habitual and more automatic, not only do the attention-networks reduce their activity, but the other neural computational pathways change as well.[13] The 'routinisation' of one particular skill allows others to be more readily learnt.

Ways of learning

When experimental psychologists try to understand how animals learn, they explore their cognitive abilities by manipulating the behaviour of the animals under study. Traditionally they have been more interested in investigating learning in animals living in highly controlled laboratory set-ups rather than in those living in natural conditions.[14] A myriad of laboratory experiments in which animals learn to associate a stimulus chosen by the experimenter with a particular response show the endless possibilities for learning that higher animals possess. With one type of learning, known as 'classical conditioning', an association is formed between a naturally occurring behaviour (such as salivation by dogs at the smell of food) and an arbitrary stimulus (such as the sound of a bell). In his now classical experiments, the Russian psychologist Ivan Pavlov showed that if a bell is rung regularly before food is given, dogs learn to associate the sound with food, and eventually they salivate at the sound of the bell alone, even if no food is offered. Another type of learning through conditioning is based on the carrot-and-stick method: an animal is rewarded if it produces the pattern of behaviour required by the experimenter, and punished if it produces unwanted behaviour. This is the traditional method used by circus trainers to make tigers jump through burning hoops, bears dance, and so on. Not surprisingly, most animals quickly learn to obtain the carrot and avoid the stick.

Conditioned behaviour is commonly encountered in nature. As we have seen, the young hedgehog quickly learnt to look for the tasty and easily attainable chafer beetles it encountered, but to avoid the foul-tasting, toxic ladybirds. Insect-eating birds living in savannah or dry shrubland, where fires are frequent, often learn from experience that insects that are usually camouflaged run or fly away from plant cover when fire is approaching, and are therefore much easier to catch. When a fire starts, the birds immediately start searching for fleeing insects. But since these birds have also learnt to associate smoke with fire, they start searching for escaping insects at the sight of smoke, sometimes long before there are any other obvious signs of fire.[15]

The scientific literature of experimental psychology is full of overlapping and repetitive classifications of learning. Happily, we can avoid most of these, since they deal with the precise and detailed relationships between stimuli and responses, which are not crucial to our arguments. It is more important for us to recognise that an animal's learning is organised by processes of categorisation, which makes it disregard some borderline cases and atypical elements of stimuli, and exaggerate the significance of more characteristic features. Categorisation and the application of other rules, such as anticipating that a regularity experienced for a long time will continue, allow animals to make sense of the world around them, rendering it more predictable and more manageable. The construction of the rules that underlie any specific behaviour is bound by such general rules. Ekkehart Schlicht, an economist who has integrated ideas from psychology, history and custom with contemporary economic thinking, has argued that the rules that direct behaviour have to be simple enough to allow inferences about the future to be made on the basis of past experiences. There are therefore fundamental and simple 'rules for making rules' that are embedded in the deep structure of the animal mind and underlie the processes that fashion perception, emotion and cognition. These deep rules, or 'clarity requirements', as Schlicht calls them, underlie the ability to form clearcut categories and to organise the world in ways that make the acquisition, storage, recollection and transmission of information more efficient and more reliable.[16]

Categorisation can be seen in 'stimulus generalisation'. Many experiments have shown that, when animals learn about an important feature of the world, they generalise to some extent, and respond in an essentially similar manner to stimuli that are perceived as sharing features with the original stimulus.[17] For example, when a bird learns that butterflies with a particular red-and-black wing pattern have a disgusting taste and induce vomiting, it will generalise this experience to other somewhat different-looking red-and-black, orange-and-black and yellow-and-black butterflies, and possibly also to other similarly patterned insects. Generally, features of the world that share perceptually striking attributes will be grouped together. Since the perceived environment is never constant, this psychological economy will lead to more efficient responses, because not every feature of the world needs to be learnt anew and stored independently.

An interesting aspect of such stimulus generalisation is stimulus enhancement, in which a stimulus greater than the normal one elicits a greater response. A well-known example is the egg-retrieval behaviour of nesting parent geese, seen when one of their eggs rolls away from the nest. The sight of the egg away from the nest induces an egg-gathering response in which the goose uses its beak to roll the egg towards itself. The larger the egg, the more vigorous the retrieval response of the goose. The same response is seen in other birds, such as gulls and oystercatchers, and is probably quite adaptive since larger eggs often contain larger chicks.[18] This psychological mechanism can be manipulated not only by the experimenter, but also by natural parasites. Chicks of parasitic species such as the European cuckoo induce their foster-parents to feed them by gaping widely and begging loudly, imitating a whole chorus of chicks.[19]

Another way in which animals categorise their world is by generalising their behavioural response rather than generalising the stimulus. In this case, new patterns of behaviour, which are related to the one already learnt, are generated without instruction, and are applied in different situations and in response to different kinds of stimuli.[20] Such 'response generalisation', as this type of generalisation is called, has so far been studied experimentally only in humans, mainly when monitoring the effects of teaching various skills to children or handicapped people, or when trying to correct a behavioural aberration.[21] Response generalisation is very important in animals: behaviours exercised during parent–offspring interactions seem to be a prerequisite for the development of similar behaviours that function differently and in other types of situations, for example during courtship, or during group rituals. A well-known example is the behaviour of a mature, misbehaved dog, as it submissively lies on its back trying to appease its master. This behaviour is derived from the behaviour very young pups show towards their parents, when the latter clean them. Generalising over domains and situations sometimes involves the application of a 'cognitive rule'. For example, understanding the relationship between means and ends on the basis of some particular experience (a phenomenon known as 'insight learning'[22]) often opens the door to effective and rapid learning in many different domains.

The range of behaviour uncovered by experimental psychologists is astounding. All birds and mammals so far tested can learn to respond in new ways to many novel aspects of their environment. During the Second World War, the American behaviourist B. F. Skinner trained pigeons to act as guides and targets for guided missiles.[23] Circus trainers have shown time and again that if wild animals are rewarded when their activities conform to the wishes of the trainer, many species can learn new tricks. The animals can be made to dance, jump through burning hoops, play football and perform many acrobatic feats. Even the European hedgehog, generally considered to be a primitive solitary mammal, can learn to do some rather surprising things. Konrad Herter has shown that these hedgehogs can learn to associate the presence of each of several foods with its specific smell or with the smell of one of the experimenter's shoes. They can be trained to run to food or to their owner when hearing a whistle or their name, to curl up at a word of command, to open simple sliding doors with their snouts or paws, and to rear up on their hind legs and jump up for food. Herter even succeeded in making his hedgehogs use their poor sense of vision: he trained them to distinguish between colours, between shades of colours and even between different intensities of light. Under the right training conditions, even a poor-sighted mammal like the European hedgehog can be trained to show a high degree of visual discrimination, which, surprisingly perhaps, is seldom if ever realised under natural conditions.[24]

But not only can we teach dogs to salivate at the sound of a bell, hedgehogs to jump in the air and pigeons to guide missiles, we can also rear rats that prefer Mozart's music to Schoenberg's! Newborn albino rats, reared for several weeks in acoustic-boxes that allowed the experimenter to control the sounds that they heard, were exposed to the compositions of Mozart.[25] Later, when given a choice, they preferred Mozart's music to Schoenberg's. In fact, rats seem to have a slight inborn preference for Mozart, since even untrained rats, who had never heard the music of the two great composers, displayed a preference for Mozart, albeit not as strong a preference as that of the Mozart-reared rats.

The Mozart-loving rats, like the jumping hedgehogs, are not just an amusing living example of the fantasy world of some experimenters. They also show that animals can form preferences for sounds, smells, tastes, colours and objects that seem to be very different from anything ever encountered in their natural environment. But why are they able to form such preferences? Were these preferences ever of any use to them? We do not know, of course. But the behavioural possibilities uncovered by laboratory studies of learning in birds and mammals certainly give us a glimpse of the unrealised behavioural potential in each species. The behavioural plasticity of birds and mammals is usually very large. There is always some possibility of accomplishing the unusual, the extravagant or the bizarre.

The everyday and the newly learnt behaviours seen in higher animals in their natural habitat are often much more astonishing than those seen in the laboratory or the circus. The building behaviour of a male bower-bird when constructing an elaborate, decorated bower to attract females is a breathtaking example of an intricate set of behaviours that culminate in a great architectural and artistic achievement.[26] The Vogelkop gardener bower-bird, a medium-sized brownish bird, builds the most elaborate structure erected by any vertebrate other than man. Males of one population build huts two meters in diameter, each constructed around a stick tower; the male defoliates the area around the bower, brings a mat of green moss, and decorates his bower and its sur-roundings with fruits, flowers, fungi, butterfly wings and leaves. The decorations are grouped together according to their colours in piles of red, blue, yellow, brown or orange objects. Males from different popu-lations differ in their modes of bower construction and in their prefer-ences for both colours and other features of the display, and it takes several years for a male to learn how to construct a local-style bower. The use of some decorating elements is a unique invention of the indi-vidual, and other elements may be included because they were seen in (and sometimes stolen from!) the bowers of other males. A male learns and often combines elements from many sources in his attempt to build and decorate an attractive bower.

Building behaviour is just one type of complex behaviour that arous-es our admiration; the foraging talents of some animals are almost as amazing as the building behaviours of the bower-birds. During a peri-od of three weeks, when it is feeding its young, a great tit gathers about eight thousand individual insects, which may belong to hundreds of species.[27] It must know how, where and when to find them, be able to distinguish between those that are edible and those that are inedible, and know how to handle the different types of insects. The ability to learn about its prey and to classify them is essential to this small bird. We know that birds do have the cognitive ability to form categories: work with pigeons has shown that they can learn, through stimulus generalisation, to group different shaped flowers into a general concept of 'flower', and different chairs into a general concept of 'chair'. But they can do more than this: they can learn to organise distinct cate-gories into hierarchies and group the flower concept with a chair con-cept to form a more inclusive concept of 'non-moving object'.[28] Judging from what is seen in the field, it is highly likely that, at least with respect to food, the great tit, too, is capable of such cognitive feats.

Conditions and constraints

A close look at bird and mammal learning, especially in nature, but even in the laboratory, shows that, although these animals are capable of many novel behaviours and may form startling new preferences, they are more likely to develop some types of behaviour than others. Some associations are more readily formed, and some stimuli induce a response more easily, or are generalised more readily, or are remembered for a longer time than others. John Garcia showed this in experiments he carried out during the late 1950s and early 1960s.[29] He was interested in the way in which mammalian agricultural pests learn to avoid poisoned baits, and thus avoid extermination. Obviously, this was an economically crucial issue for American farmers. Mice, rats and other mammals that become sick some hours after eating a poisoned bait avoid it thereafter. In his experiments, Garcia gave animals normal, non-poisoned food or drink with an atypical taste or an unusual colour. Several hours after an individual had eaten the food or drunk the liquid, he injected a small dose of lithium chloride into its gut. This treatment had the effect of making the animal ill for a few hours, thus mimicking the effects of poisoned food. When the animals recovered and were offered food or drink with the same unusual taste or colour, they avoided it. In spite of the long time interval between eating the food and the onset of the illness, they had learnt to avoid food and drink with a similar taste or colour to that taken prior to the poisoning. However, when tasty but sickness-inducing food or drink was accompanied by a clicking noise, the animals found it much more difficult to learn to associate the sound with the subsequent sickness. And, of course, there are very good evolutionary reasons for this: the taste and colour of food and drink are, and always have been, associated with the possible dangers of natural food-poisoning, whereas sound cues are not and never have been. The contaminated food or drink of rats or mice rarely clicks! It is hardly surprising that, after many thousands of years of natural selection, animals are readily able to learn to associate with indigestion the normal and familiar features of food, such as its taste and smell, but have difficulty learning that a clicking sound spells gastric trouble.

The ease with which things are learnt depends not only on what has to be learnt, but also on when it is to be learnt. There are certain periods in life, often at an early age, when some things can be learnt with particular ease and are then remembered for a long time. The effect of such brief but commanding early learning is called 'imprinting', implying that the brain has been persistently 'marked' by the early experience. The periods in which this rapid and efficient learning occur are referred to as 'sensitive periods'. A newly hatched gosling will approach, follow and form a special attachment to the first conspicuous moving object it sees (usually its mother). This attachment, known as 'filial imprinting', is the classical illustration of early learning during a sensitive period. It was first studied scientifically in the nineteenth century by Douglas Spalding, an outstanding, self-educated Scottish scientist, who can be considered as one of the founders of modern ethology.[30] In the late 1860s and early 1870s, Spalding started a remarkable series of experiments, trying to understand the role of experience and learning in the development of animal behaviour. In one of his experiments, Spalding put hoods over the heads of chicks and ducklings just as they hatched out of artificially incubated eggs, and observed their behaviour after the hoods were removed. If the hoods were taken off after a day or two, he discovered that the chicks and ducklings would follow him faithfully, as if he was their mother; but after being covered with hoods for four days, the chicks showed marked fear responses upon seeing him, and they did not follow him. Spalding concluded that during early life there is a short critical period in which the young express the tendency to follow the first large moving object they see. The picture of the Austrian zoologist Konrad Lorenz, who studied this phenomenon sixty years later, marching across a green meadow with a single file of newly hatched goslings devotedly following him, is one of the celebrated images in the history of ethology. Lorenz, too, demonstrated filial imprinting by removing the mother goose before the eggs hatched and taking over her maternal role as a moving target for attachment. Like Spalding, he found that once he was identified as 'mother', this identification was difficult to change.[31]

An even more extreme example of imprinting, which is associated with a very long memory, is seen in Pacific and Atlantic salmon. Years after they have left the river in which they hatched, these fish return to it to spawn. Olfactory characteristics unique to the natal river are learnt early, during the first days of the fry's life. An exposure of just four hours is sufficient to engrave the chemical identity card of the natal river in the fish's brain, so that during its return journey, four or five years later, it will easily detect its natal river's unique olfactory features, find the river and use it to spawn in.[32]

It may seem strange to talk about imprinting when our subject-matter is learning. At first sight, imprinting seems to be almost automatic. In the particular examples we have described up to this point, experience seems to be almost unimportant. But it has been found that there is more trial-and-error learning involved in imprinting than meets the eye, and, under some circumstances, an imprinted behaviour can be reversed or 'forgotten'.[33] The dependence of an imprinted behavioural response on experience and learning is more conspicuous when we look at other types of imprinting, such as sexual imprinting, or the imprinting on the paternal song which is found in many young songbirds. Sexual imprinting, the learnt mate preference that young birds and mammals develop on the basis of their early experience with their parents, takes place over a period of several weeks. It is usually very stable: once established the preference is difficult to change. Under some conditions, however, it is possible to override it. For example, when the preferred type of mate is absent, or when a potential mate does not bear the imprinted characteristics but is associated with other desirable breeding conditions and stimuli, such as a suitable nest, imprinting is overridden and a different type of mate is chosen. Sexual preference is sometimes influenced by external rewards – zebra finches who are reared by parents belonging to two different species become imprinted on the parent who feeds them most.[34]

Imprinting is therefore not quite as automatic as it seems. Klaus Immelmann has defined it as early learning that occurs at some well-circumscribed periods in the animal's life and which has a persistent and long-term effect on its behaviour.[35] We will use the term in this way. The behavioural effects cover all aspects of the animal's life: imprinting can be ecological, and involve early learning of food preferences, home ranges and habitats, or it can be social, and involve imprinting on parents or on hosts. It can be thought of as a special type of learning, limited to an early, circumscribed period of life, in which few trials are necessary to learn and elicit a special and sometimes quite complex pattern of behaviour. This learning lays an important and robust foundation for the future of the young animal.

In contrast to imprinting, other behaviours, such as the elaborate nest building of some birds, are learnt over a long period, apparently through a time-consuming process of individual trial and error. The nest of the village weaver-bird is kidney-shaped, made of freshly torn strips of leaf, and has a domed nest-cavity with a downwards-directed entrance. Nests built by first-year males are very untidy compared with those of older males, and the unskilled young weavers often fail to push the pieces of leaf in far enough, or to insert them in the right way.[36] Females never choose the untidy nests of these yearlings, and the males tear them up and try again. Not until they are two years old are male weaver-birds able to weave an adequate nest. Such a lengthy period of learning is typical for the elaboration of many of the complex behaviours seen in birds and mammals. But other behaviours are learnt very quickly, and at any time: one experience with a very obnoxious butterfly causes the blue jay to throw up and to avoid, for at least several months, any species of butterfly displaying similar wing colour-patterns.[37]

Many behaviours are not acquired by individual learning, but are learnt from others – from parents, from neighbours or from peers. In fact, almost all the forms of imprinting belong to this category, with the parents playing a major guiding role. Even in the case of habitat imprinting, in which the young animal becomes strongly attached to the particular habitat in which it was reared and returns to breed in the same or a similar habitat, it is indirectly through the behaviour of its parents and their choice of habitat that the youngster becomes imprinted. We have seen how the chukar chicks become imprinted on particular food items: their mother directs their attention to the food by pointing with her bill and uttering a typical food-call. Maternal guidance is also important for other types of early learning. Cheetah cubs, just three months old, start learning how to catch and handle prey through a lengthy, gradual process of maternal demonstration. The mother releases the live prey she has caught in front of the cubs; usually the prey is a very young gazelle, which is much easier to catch and handle than an adult is. If the cubs do not knock the prey over and kill it themselves, the mother intervenes and kills it in front of them. Over the next months, a growing number of prey is released in front of the cubs, and they kill more and more of it. Thanks to their mother, they have learnt how to handle it efficiently. A similar process of learning through maternal demonstration has been observed and studied in the domestic cat under laboratory conditions, and in other carnivores such as tigers, lions, otters and mongooses in the wild.[38] Although these behaviours are learnt early and are quite stable, they are not referred to as imprinting because the period of learning is not as clearly delimited as in classical imprinting.

When studied in detail, many patterns of behaviour are found to be complex products of several types of learning; they are the result of some inborn pre-dispositions, some individual learning and some social learning. We saw that spiders, hedgehogs and chukars acquire their information about food in very different ways. Spiders rely mainly on innate tendencies, hedgehogs on their own experience, and chukars on their maternal legacy. Yet, even in these deliberately chosen extreme cases, every mature pattern of behaviour results from a combination of several ways of learning. A spider can sometimes learn to avoid new prey that she finds unpalatable after one or more unpleasant experiences; a hedgehog's menu is largely based on its own experience, but there are some innate preferences; a young chukar learns most of its food habits from its mother, but as an adult it also adds here and there to the maternally inherited culinary list, following its own personal experience.

An important question concerns the extent to which patterns of behaviour are genetically constrained. In the first chapter, we referred to the sociobiologists' claim that even in human beings behaviour is constrained by powerful, evolved, genetic factors. It is usually taken for granted that the genetic constraints on the behaviour of animals are even greater. However, it is really not possible to generalise about this. The few examples we have already given show how species-specific and trait-specific the answer must be. The possibilities and limitations of animal learning can best be illustrated by taking a closer look at a single well-researched example. An excellent one, in which innate preferences, social learning and individual learning are all combined, is the intricate singing of a mature male songbird such as the European blackbird.[39] Adult male blackbirds that have been hand-reared and denied the opportunity to listen to the songs of their fathers or other males produce normal but dull songs. In experiments with other species, where males have been deafened as nestlings, the adult songs are similarly normal but dull. Clearly, there is an innate component in birdsong production, but in normal conditions learning is also involved. Although the basic species-specific song can be produced by every young male, in order to construct a full, individual, colourful song, each male must learn from others, and practice singing and listening to its own singing.

A male blackbird learns and modifies its song throughout its life. He constantly listens to the singing of neighbouring birds, mainly of his own species but sometimes also of others, and selects from their songs new motifs which he incorporates, with or without change, into his own song. Even human whistles, flute melodies and mechanical sounds such as the noise of an electric saw find their way into the song of the blackbird. The new motifs are copied with varying degrees of accuracy, since individuals differ in their talents for imitation and improvisation. While a male constantly adds new motifs to his song, he also sometimes discards old ones, so his song repertoire, reflecting both his unique acoustic environment and his individual taste, changes and becomes larger with time and experience. These changes do not obscure the permanent individual style of singing of each male blackbird, which can be identified readily by other blackbirds as well as by experienced human listeners.

Plasticity is the hallmark of the blackbird song, and is also typical of many other species of songbird. There are often individual variations in the songs of birds from the same population, as there were with the two great tits in the olive orchard described in the previous chapter. Frequently, as with corn buntings, there are also variations between populations of the same species, each population having a different local dialect. This dialect is a unique song pattern, which is common to most but not necessarily all members of the population, and is transmitted from generation to generation through vocal imitation. Some songbirds, such as swamp sparrows, learn their species-specific song very early in life and change it hardly at all, while others, for example European starlings, start learning after they are three months old and, like blackbirds, modify their songs throughout life. Marsh warblers can mimic up to seventy-eight other species and incorporate foreign phrases into their own song, with the result that each individual has a unique song. Imprinting, individual learning and learning from fathers and neighbours have all been shown to contribute to the mature song of male songbirds.

Imprinting is a very important component of song learning in most songbirds. Even if some components of the song are later changed, the song heard during a sensitive period in early life, which is usually the song of the father and neighbouring adult males, forms the basis for subsequent song learning by a young male; it also determines the young female's song preference when she matures and mates. The songs of individuals of the same species are often learnt more easily than 'foreign' songs, showing that the song-learning ability is not general, but is channelled towards certain patterns. Even in a species such as the red-winged blackbird, a notorious improviser, the improvisations that the birds introduce are not completely random – some components of the song are altered by improvisation, while others remain constant. Learning the wrong song too readily could lead to a personal disaster, because the male would have difficulty in attracting a mate of his own species, and a female would find herself attracted to males of foreign species. However, in some unusual circumstances, this channelling can be overridden. For example, if males of the same species are not present, and the young bird constantly hears males of a foreign species singing vigorously nearby, he may learn the song of the foreign species. This sometimes leads to hybridisation, as Grant and Grant showed for some species of Darwin's finches living on the Galapagos Islands.[40]

As young birds learn to sing, they rehearse, and in many species the auditory feedback they receive from their own voice is an important part of the learning process. If auditory feedback is denied by experimentally deafening young birds, the song of the deafened bird, like the poor bird itself, is crippled. The song may be incomplete, more like the babbling of a baby, or dull. In normal birds, the song often undergoes a gradual process of selective stabilisation: the initial repertoire is large and not well-structured, but it crystallises and becomes restricted as a result of practising and carrying out singing contests with other males.

During the song-learning period, the bird's brain changes dramatically. Nerve cells located in areas responsible for song-acquisition grow and proliferate, and these brain regions expand and reorganise. Injuries to these regions result in impaired song production. The titres of hormones like oestradiol and testosterone influence song acquisition and song production, and there is feedback between the bird's maturation process, its personal and social experiences, the production of these hormones, and brain changes.

The idea that there is some definite, innate 'song-template' to which singing has to fit is not satisfactory, especially in cases where there is a lot of improvisation and copying of elements from other species. If there is an innate song-template, it must be a very general one, which sets some limits and etches some preferred routes for the production of output but does not specify it. The lengthy bouts of singing by songbirds with large and complex song repertoires often seem to continue without the need for positive feedback from other individuals.[41] Although males with large song repertoires seem to be reproductively more successful than those with a smaller one, the correlation between repertoire size and reproductive success is weak. There seems to be a strong internal motivation for continuously singing different versions of the song, which cannot be accounted for by any obvious functional benefit. The positive feedback from the ongoing activity seems to involve emotional satisfaction – there appears to be pleasure and joy in the continuation of the activity once it is successfully initiated. The feedback also leads to corrective adjustments, fitting the singing with the environment and with the already performed song, enhancing the sense of enjoyment and possibly the sense of beauty.[42]

Birdsong is an excellent example of the complexity of the processes involved in the development of behaviour and the way it is used. Although the basic behaviour does not have to be learnt, learning is essential for its development and sophistication. Through learning, individual variations that are of adaptive significance are often introduced. The range of behavioural variations that members of a species of bird or mammal are able to produce is very large – much larger than is ever seen in a particular population, or even in all the natural environments currently occupied by populations of the species. Animals frequently apply behaviour to new situations and learn new sequences of behaviour, as the pathetic dances of circus bears show. They may even engage in apparently non-functional activities that seem like the luxurious by-products of extensive behavioural plasticity. Yet, despite all this plasticity, it is clear that there are serious constraints on what animals can do. Even the best tutor cannot teach an ass to speak biblical Hebrew without some help from the Almighty. Furthermore, within the range of the possible, some things are more easily learnt than others. In spite of these limitations, the range of variations seen is large enough for us to be satisfied that there is enough 'raw material' for evolutionary change to occur, provided some further conditions are met: the learnt variations must be transmitted to subsequent generations, and have differential effects on the survival and reproductive success of individuals. But, before we come to the transmission of variations, we need to consider what is learnt and how it is remembered.

Remembrance of things past

For most of us, the title of Proust's book, *The Remembrance of Things Past*, captures the essence of memory. Memory is reconstruction of past experiences; it is what gives continuity and coherence to our human feelings of identity and individuality. The devastating effects of memory loss are seen in cases of amnesia resulting from brain damage. In one famous case, H. M., a young man of twenty-seven, lost his memory following a brain operation to relieve epilepsy. He could no longer remember events that happened to him on the same day, and his ability to remember events that happened a decade before the operation was also partially impaired. When describing his world he said: 'Every day is alone by itself, whatever enjoyment I've had, and whatever sorrow I've had.' He also said that he is like a man constantly 'waking from a dream'.[43]

Brain damage leading to memory loss has revealed different types of memory. But although there is some localisation of different types of memory, the 'specialised' brain regions are more like shifting dunes, continuously reorganising, expanding and contracting.[44] One of the first distinctions to be made was between long-term and short-term (or 'working') memory. Some people with brain injuries cannot remember for more than a few minutes, or even seconds, how to do something that they have just learnt, although their memory of the past prior to the brain injury, including their past skills, remains intact. It seems that there are more or less distinct brain regions that deal with short-term memory and others that deal with long-term memory, but the two regions normally communicate. In people whose short-term memory is badly impaired, either the region responsible for short-term memory has been injured, or the transfer of information from short-term memory to long-term memory is damaged.

During the past two decades, forms of memory have been further differentiated on the basis of the types of information they can handle. Studies of both animal and human memory point to a broad distinction between 'procedural' memory and 'declarative' memory.[45] Procedural memory involves the active maintenance of the procedures that underlie skills such as building a bower or riding a bicycle. It is the general procedure that is retrieved when it is recalled, with little reference to any particular content and context. For example, the procedure for riding a bicycle may be remembered without remembering how, when and why it was acquired. The other type of memory, declarative memory, is concerned with the ability to remember places and situations, to orient oneself in space, to construct 'cognitive maps' of the space in which one moves, and to recollect episodes. Procedural memory and declarative memory seem to differ not only in their functions, but also in their localisation in the brain and in the ways that the learning associated with them occurs. Learning procedures like how to ride a bicycle is often gradual, depending on repeated trials and on external positive and negative reinforcement, such as encouragement or social pressure. In contrast, learning the organisation of things in space and time is rapid and depends on what is known as 'internally motivated exploratory behaviour' – on spontaneously generated mental states such as curiosity, which typically lead to the exploration of the environment. The skills acquired by procedural learning and stored in 'procedural memory' usually become stereotyped over time, whereas the cognitive maps of the declarative system are very flexible and can be used in many new ways.[46] Not only are the forms and functions of memory different, but, according to the scientists studying these systems, different memory systems use different types of neural mechanisms, with different rules of operation.

There is an evolutionary reason for the existence of separate memory systems: division of labour.[47] As Adam Smith, the founder of political economics, argued many years ago for social organisation, organisational complexity is always associated with division of labour.[48] While one system (the procedural) deals with generalities, with things that are learnt gradually and incrementally and are forgotten slowly, the other (the declarative) is concerned with specifics, with the unique features of places and episodes, which are rapidly memorised but can also be rapidly forgotten. For optimal performance, these very different functions may need different types of neural activities and neural architectures, and this can best be achieved by separate systems. Indeed, the internal architecture of the hippocampus is claimed to be perfectly fitted for accentuating small differences in inputs of declarative information, and hence for 'storing' unique memories and 'maps'.

The extreme and specialised memories of some birds and mammals demonstrate the distinct features and functions of different forms of memory, and their life styles give us clues to the way that their memory may have evolved. Stories about the remarkable memories of cats and dogs often lack authenticity, but the legendary memory of elephants seems to be more than a legend: an elephant is able to recognise about 600 other elephants as individuals.[49] So, has the elephant's good memory and attention to social detail been selected because of its lengthy childhood and intricate social life? Or did good memory lead to a more complex social organisation? Probably both: elephant evolution is likely to have involved a positive feedback-loop between social structure and the efficiency of social learning and memory: as social organisation became more complex, there was selection for better memory, and as memory became better, a more complex social organisation could evolve.

Remarkable feats of memory are not limited to large-brained, long-lived mammals, however. The excellent memories for song in songbirds and for spatial locations in hoarding birds are examples of two very different types of exceptional memory. A male nightingale has a repertoire of 100–300 song-types, and not surprisingly there is an extremely good memory underlying his musical genius. During the sensitive period, laboratory-reared young birds can learn as many as twenty song-types within five minutes of listening to a recording.[50] As with most songbirds, song is important in sexual display and territorial defence, but for nightingales, who commonly live in thickets where they cannot easily be judged by their colour and size, the song is probably the major criterion by which a female chooses her mate. A long complex song and an excellent memory are probably crucial for a male's reproductive success. It is known that male starlings with large song repertoires attract more females and mate earlier in the breeding season than their rivals do.[51]

The striking and enviable declarative–spatial memory of some food-hoarding birds and mammals has a different role.[52] Many species collect food items such as seeds and nuts, and carry them to places where they are hoarded for future use. Tits and ravens are well-known hoarding birds, and among the mammals the red fox is a famed hoarder. Food hoarding is one of several solutions to the problem of a short or unpredictable food supply, a problem that is not uncommon during autumn and winter, especially in higher latitudes. Some food-storing species, those known as 'larder hoarders', have no need for a particularly good memory since they hoard most of their food in a single place. Others, however, store food items in a large number of scattered sites. Scatter hoarders use the strategy of 'not putting all their eggs in one basket'. Since other animals may pilfer from a 'larder', or even chase away the owner, in a highly competitive environment it is an advantage to scatter-hoard food items over a large area. This maximises the hoarder's chances of recovering at least some of its food. It is these 'scatter hoarders' that need extensive spatial memory, to enable them to find the food they have previously hoarded. There will be strong selection for the ability to remember where their own and their competitors' food items have been stored.

Just how good the memory of some scatter hoarders is has been revealed by studies of some marsh tits living in a natural environment near Oxford, England.[53] During autumn and winter, individual tits were found to hoard in and on the ground several hundred seeds a day, taking just a few seconds to store each item at a separate site. The sites were spaced on average seven metres apart, and were used only once. The hidden seeds were usually recovered and eaten by the tits within two days of hoarding. To recover a food item, a tit went directly and accurately to one of its sites, with a success rate during the first twelve hours after hoarding of over 90 per cent. It is generally believed that the tits form cognitive spatial maps of their environment and commit their hiding places to memory. The memory of the marsh tit must be extremely good, but perhaps it is not as good as that of the Siberian tits. During the late summer and early autumn, these birds harvest and hoard thousands of items of food, mainly seeds.[54] The tits store the food all over the coniferous forest, in places where they will be accessible when snow covers the ground and the treetops are capped with ice. It has been estimated that a Siberian tit may hoard up to half a million items of food per year!

When compared with non-storing species, the brains of food-storing birds such as the Siberian and marsh tits have relatively large hippocampal regions. Although the anatomical and behavioural details vary from one hoarding species to another, it is evident that all scatter hoarders have an accurate, long-term spatial memory that enables them to keep track of and retrieve thousands of items of food from hundreds of separate sites. Their feats of memory are certainly remarkable. We humans, with our large brains, would certainly not remember where we had hidden several hundred items in a forest. But, of course, finding them is unlikely to be a matter of life and death for us. It is for most food-hoarding birds. There has clearly been very intense natural selection for spatial learning and memory in food hoarders. Similarly, there has been strong selection for the ability to create and remember songs in nightingales. Individual birds, who for genetic or other reasons have poor memories, fail to survive or have fewer offspring than do those with better memories. Through natural selection over thousands of generations, the ability to learn and remember has been moulded in different ways in different species.

Thinking in evolutionary terms makes sense of another property of learning and memory: memory often depends on the circumstances in which things are learnt. For example, when human subjects learnt to read a list of words under water, they were able to recall these words better when they were again under water than they could in normal conditions.[55] This dependence on the context in which information has been learnt makes a lot of evolutionary sense, because there is usually some relationship between the information animals learn and the environment in which they learn it. The many experiments showing the importance of context in human and animal memory imply that the essence of memory is not simply the retrieval of 'stored elements'. Neural connections are not merely activated, as in an electronic switchboard; representations have to be actively reconstructed. The emotional state of the individual and a few environmental cues are sufficient to start a process of reconstruction. A total picture of the past can be reconstituted from a very partial piece of current information.

The social transmission of learnt behaviour

In discussing learning and memory, we have already seen how acquired knowledge can be passed on when one individual observes the behaviour of others, or when substances or signs affect behaviour. In the case of salmon, fidelity to the natal river is the result of the amazing olfactory imprinting that the fry undergo during the first days of life. The fidelity to one single stream can be perpetuated for many generations because, when adults return to their natal river to spawn, their offspring smell the same smells as they did. In birds, the idiosyncratic components of the father's song can be transmitted to his male offspring and perpetuated for generations. A novel food preference, avoidance of a new predator, a sexual preference resulting from sexual imprinting on a 'parent figure', a new way of foraging – all these and many other behavioural variations can be passed from one generation to the next as a consequence of the associations of young animals with their parents and other knowledgeable individuals. We now want to look at social learning in more detail, and ask how common it is, with what stability socially learnt behaviours are transmitted, and whether social learning can provide the foundation for evolutionary change at the cultural level.

Can social animals avoid learning from others? All that is needed for social learning is that the presence of one relatively experienced individual increases the chances that a naïve individual will learn a new behaviour. As the British psychologist Cecilia Heyes has stressed, the relationship between the environment and the learnt behaviour is similar for both social and asocial learning; with social learning the 'stimulus' or the 'reinforcer' for the learnt behaviours are simply associated with other individuals: they are the actions or the consequences of the actions of others.[56] Social learning is, in fact, an inevitable consequence of the social organisation of most birds and mammals. Almost all birds and mammals associate, at least to some degree, with members of their own species. Even in the simple social organisation of the so-called 'solitary' mammalian species, maternal care is important, and a crucial part of the young animal's life is spent in the presence of at least one individual who has had prior experience of the world in which the youngster will live. Through the association with its parent, a young animal acquires information that will later affect its behaviour, usually making the behaviour similar to that of its parents. A young male bird has no choice but to hear the song of his father and neighbours, which in many species will form the basis for the development of his own song some months later. A chukar chick seeing its mother peck at particular types of grains will do the same, preferring them over other types of grain, and a young rat will prefer to eat the food whose odour it detected on the breath of its mother or other members of its group. Young monkeys, at first indifferent to the sight of snakes, will learn to fear them after witnessing the panic reaction of adults and becoming infected with their fear. Given the various learning abilities of birds and mammals and their excellent communication abilities, it would be surprising if social learning did not occur. Youngsters usually learn from a parent, but can also learn from another member of their group, from several group members, and sometimes from a member of another population or even of another species.

Patterns of behaviour can spread among the individuals in a population through different types of social learning processes. Experimental psychologists have struggled hard to describe and classify the various types and mechanisms of social learning, with the result that more than thirty different terms are used, many of which overlap.[57] The focus of observation and attention differs in different cases, the cognitive demands can be very dissimilar and so on. Rather than attempting to distinguish between all of the different types of social learning on the basis of the relationship between stimuli and responses, we want to see which types of social learning can support the formation of animal traditions and cultural evolution. We are therefore going to focus on a distinction that, for reasons which will become clear shortly, is commonly regarded as particularly important in this respect – on the distinction between imitation and other types of social learning. We will use the umbrella-term 'socially influenced learning' for social learning that is not imitation.[58] Our argument is going to be that *all* types of social learning, imitative as well as non-imitative, can support animal traditions and cultural evolution if they are coupled with stabilising ecological and social conditions. We shall also maintain that early, imprinting-like learning is of particular significance for the stable perpetuation of many animal traditions, and consequently for their evolution.

In socially influenced learning, the naïve, watching individual (or 'observer') learns about the *environmental circumstances* – the objects, stimuli and events – that elicit the behaviour of the experienced individual (the 'demonstrator', or 'model'). With imitation, on the other hand, the observer learns about the responses, actions or patterns of behaviour of the experienced individual. Some examples may make the nature of socially influenced learning clearer. When young monkeys become fearful of snakes after observing the panic-stricken reaction of the adults, they, too, will avoid snakes. However, what they learn is not the flight behaviour of adults, but rather that snakes have to be avoided. Similarly, when chukar chicks learn to peck objects similar to those pecked by their mothers, they learn that such objects should be preferred over others; they do not learn how to peck them. The fearful behaviour and the pecking behaviour are, in fact, innate behaviours. The cultural spread of the great tits' habit of opening milk-bottle tops can also be explained in terms of socially influenced learning: observer-tits often had their attention directed to the milk bottle as a potential source of food through watching the behaviour of an experienced individual. The method by which the bottle top was removed was not imitated; after trial-and-error learning, each individual tit learnt how to remove the top in its own style.[59] In all such cases of social learning, the observer's attention to the environment or to its own actions is selectively enhanced. The behaviour the observer displays following the model's 'demonstration' is a consequence of the heightened focus on a particular aspect of the environment followed by trial-and-error learning or the triggering of innate or early learnt responses.

In most cases, socially influenced learning leads to similarity between the behaviours of the observer and the model. The model guides or enhances the attention of the observer to the relevant environmental stimulus (such as the milk bottle, or a dangerous predator), and this elicits in the observer an emotional and behavioural response similar to the model's. Since the local circumstances in which the observer performs the behaviour are rarely identical to those of the model, and the observer frequently differs from the model in its personal set of motor patterns, the observer's behaviour may not be an exact copy of the model's actions. Nevertheless, in spite of this individual variation, if the new learnt behaviour is beneficial and the circumstances eliciting it are stable, this behaviour will persist.

With socially influenced learning, an animal learns *what* to do as a result of its association with others; with imitation it learns both what to do and *how* to do it. The observer (or, in this case, the 'mimic') learns about the actual actions of the 'model' and copies them. Some songbirds and parrots are particularly good vocal mimics. When mimicking others, the mimic matches the result of its own vocal performance, which it can hear, to that of the model. Mimicry of movements, of the actions of another individual, is probably more demanding, because the mimic cannot see the whole of itself, and the matching must therefore involve a correspondence between the model's acts and the internally represented image of the mimic's behaviours. It is interesting that the most gifted vocal mimics among the birds, the grey parrot and the European starling, are also reported to mimic quite complex motor behaviours, such as waving goodbye with their wings or feet.[60] Imitative motor behaviours of this kind have been seen in rats, budgerigars, captive and wild chimpanzees, and captive dolphins. The chimpanzee Viki, who was reared from infancy by a human family, learnt to imitate actions such as sticking out her tongue or whirling on one foot when she was shown the action and told 'do this'. She also imitated acts like putting on lipstick and brushing her teeth. Captive bottle-nosed dolphins imitate the behaviours of their human captors, as well as the very different grooming techniques and sleeping postures of the sea-lions with whom they share a pool. But imitation can also take more creative turns. When a man blew cigarette smoke at the transparent glass wall of a pool in the direction of a watching baby dolphin, the youngster immediately swam to her mother, filled her mouth with milk, and blew the milk at the glass towards the man's face, imitating the effect of the smoke. In this case, the dolphin did not, of course, imitate the actual behaviour of the man, but rather reproduced the end-result, a type of social learning known as 'emulation'.[61]

Rats, dolphins, great apes, budgerigars and grey parrots undoubtedly imitate, but in other animals there are very few convincing examples of imitative behaviour that cannot be interpreted as some form of learning through social influence rather than true imitation.[62] This certainly does not prove that imitation is rare in the animal kingdom. What it shows is simply that most experiments have not been designed in a way that would allow us to draw a clear distinction between imitative and non-imitative social learning, which have similar effects and in any real-life situation are usually combined.[63] So, what difference does the way that a pattern of behaviour is learnt make? Some people argue that it makes a big difference, because the mechanism of learning determines the extent to which a particular pattern of behaviour can be faithfully transmitted over several generations, and hence whether adaptive cultural evolution can occur. Adaptive cultural evolution is a cumulative process: the achievements of one pattern of behaviour form the basis for the selection of a modified and better-adapted descendant pattern. Such a process can lead to complex cultural products such as rituals, or to technological products such as a dam. If, as has been claimed, only true imitation has a high enough fidelity to allow cumulative cultural evolution, then it is only in our own species, where imitation is highly developed, that complex cultural adaptations can evolve.[64]

This argument is very problematical, however. There is no good experimental evidence to suggest that the fidelity of transmission depends on whether a behaviour pattern is acquired through imitative or through non-imitative social learning. Heyes has rightly commented that the difference between the fidelity with which information is transmitted by imitation and by other forms of social learning cannot be very large, since it is extremely difficult to distinguish empirically between the two modes of information acquisition and transmission:

> Each time data have been put forward as evidence of imitation, it has subsequently been discovered that they can be explained with reference to social learning; that after observation, the observer animals may have behaved in the same way as the demonstrators because they had learned about the environment, rather than the behaviour, while observing. This indicates that social learning has considerable potential to mediate behaviour and information transmission. If this were not the case then social learning and imitation would have proved much easier to distinguish empirically.
>
> (Heyes, 1993, pp. 1003–4)

If both social influence and imitation have a similar potential for transmitting information, can they support traditions and cultural evolution? We agree with Heyes that, left to themselves, neither social influence nor imitation can do so. We believe, however, that in the same way that maintaining genetic information requires either strong selection or repair, there are analogous processes that maintain the integrity of behavioural information. Selection can be a very potent preserver of a particular type of behaviour. For example, think about great tits opening milk bottles. Each tit has its own distinctive technique, which is learnt by trial and error. However, if competition over milk ever became fierce, those tits that adopted the most efficient techniques would thrive, and those with the less efficient methods would have to learn the better techniques or perish. Strong selection could therefore narrow and channel variation. Thus, a particular method of opening bottles could become established by natural selection, even without imitation. Processes analogous to the editing and repair of DNA can also maintain behavioural stability. Such processes include, first, stabilising feedback-loops formed between a new behaviour pattern and other better-established behaviour patterns; second, stabilising social interactions which lead to behavioural conformity; third, ecological feedback-loops through which the new behaviour changes the environment in ways that enhance the value of this behaviour. These interactions can stabilise a pattern of behaviour by making it part of a durable behavioural package. The crucial point, therefore, is not the precise mechanism of acquiring a new preference or pattern of behaviour, but the circumstances that allow the transmission of this information in a way that ensures its transgenerational re-production. Since animals cannot represent information symbolically, and a behaviour or a preference has to be manifested in order for it to be acquired by other members of the social group, the focus must be the social and ecological conditions that lead to the manifestation and re-generation of essentially similar patterns of behaviour.

Habits and traditions

There is little doubt that social learning, in its many forms, often leads to the establishment of family and group traditions. By a 'tradition' most people mean a pattern of behaviour or a habit, which is observed in a lineage, a group or a population, and is maintained through the transmission of socially learnt information. We have seen that many songbirds have local dialects, and that foraging methods, such as opening milk bottles, can become an established habit in a population. A more recent example, this time of a family tradition, is the observation that some female dolphins off the shores of Australia have developed the habit of sticking pieces of natural sponge on their nose, apparently to avoid damaging this sensitive organ while foraging on the sea floor. The habit seems to be transmitted to some of their offspring.[65] Many different traditions associated with tool use, nest making, sexual habits and social organisation are found among well-studied chimpanzee groups. Here is what Wrangham and Peterson wrote about some of the tool-associated traditions seen among common chimpanzees:

> Chimpanzee traditions ebb and flow, from community to community, across the continent of Africa. On any day of the year, somewhere chimpanzees are fishing for termites with stems gently wiggled into curling holes, or squeezing a wad of chewed leaves to get a quarter-cup of water from a narrow hole high up in a tree. Some will be gathering honey with a simple stick from a bee's nest, while others are collecting ants by luring them onto a peeled wand, then swiping them into their mouths. There are chimpanzees in one place who protect themselves against thorny branches by sitting on leaf-cushions, and by using leafy sticks to act as sandals or gloves. Elsewhere are chimpanzees who traditionally drink by scooping water into a leaf cup, and who use a leaf as a plate for food. There are chimpanzees using bone picks to extract the last remnants of the marrow from a monkey bone, others digging with stout sticks into mounds of ants or termites, and still others using leaf napkins to clean themselves or their babies. These are all local traditions, ways of solving problems that have somehow been learned, caught on, spread, and been passed across generations among apes living in one community or a local group of communities but not beyond.
>
> (Wrangham & Peterson, 1997, pp. 8–9)

A recent survey of cultural variants in chimpanzees at seven different sites in Africa showed that there are no less than thirty-nine different traditions that are found in the groups at some, but not all, sites.[66] But traditions are not limited to our close relatives, the highly intelligent higher apes. The human-like, entertaining and striking traditions of higher apes sometimes obscure the fact that traditions are also extremely common in other mammals and in birds, and cover every mundane aspect of the animals' life from lice-picking, through foraging, to mate choice. As we will show in the next chapters, there are socially transmitted variations in all the basic patterns of behaviour that young animals learn. Often the traditions found in family groups are not given as much recognition as those in large social groups, with the result that the amount of traditional behaviour is underestimated. However, the main reason why animal traditions are commonly thought to be limited in number, scope and importance is probably the assumption that heritable variations in everyday patterns of behaviour are the result of genetic differences. In most cases there is no experimental evidence to substantiate this, and when experiments have been carried out the heritable variation has often proved to be non-genetic. It is always difficult to isolate the causes of variation in natural populations. In most cases differences in transmissible behaviour are caused by a complex combination of ecological, traditional and genetic factors, which are extremely difficult to tease apart. The fact that a variation in behaviour is correlated with ecological or genetic factors does not exclude a role for tradition. Yet, once such a correlation has been found, the tendency has been to exclude tradition from explanations of the spread of the behaviour. The prevalence of traditions and their importance in the development of behaviour are, of course, empirical questions, which can only be resolved by field experiments. However, the information, that we already have about social learning in higher animals suggests that traditions are much more widespread than was previously assumed.

What about the stability and the accumulation of heritable traditional or cultural modifications over time? While the cultural evolution of a human technological product such as a bicycle, or human language, can be analysed bit by bit and can be shown to evolve by the accumulation of adaptive variations, there is no obvious accumulation of heritable variation in, say, bottle-opening by great tits. These, and similar habits, are therefore assumed to be very simple traditions, and rarely (or never) lead to true cumulative evolution. Yet, the long-term study of Japanese macaques on Koshima, which we mentioned in chapter 1, shows that cumulative cultural evolution in animals does occur.

The study began when Japanese ethologists started providing food for the macaques living on Koshima island.[67] To attract the macaques to an open space where they could observe their behaviour, the scientists scattered sweet potatoes along mountain trails and finally on the sandy sea-shore. This innocent trick bore unexpected fruits. A particularly smart eighteen-month-old female, Imo, started washing the potatoes in a nearby stream, thus removing the soil from them before eating them. The new habit spread to other monkeys. Some time later, the pota-to-washing habit began to be carried out in a different place – in the shallow sea by the beach. Imo and other monkeys also bit the potatoes before they dipped them into the salty water, thus seasoning them as well as washing them before they were eaten.

The researchers on Koshima threw wheat on the shore and observed how the macaques dealt with this unfamiliar type of food, expecting them to spend a long time collecting the wheat from among the sand grains. However, the same Imo, now four years old and apparently an Einstein among macaques, found a way round that problem. Instead of laboriously picking up the grains one by one, she threw the mixed sand and wheat into the sea; the heavier sand sank, and the wheat floated on the surface, allowing her to collect it easily. The new habit spread slowly within the group, first from the young to the old, then from mothers to children. Old dominant males, entrenched in their old habits, less attentive to others and having less opportunity to interact with the young, failed or were the last to learn. The habit of bringing food to the sea also had other effects. Infants who had been carried to the sea by their mother when she washed the food inadvertently became accustomed to the salt water, and started playing in it. Swimming, jump-ing and diving, and cooling themselves in the sea in summer, became popular habits, and in time became characteristic of the whole troop, including the adults. More recently, another new habit, eating raw fish, began to spread among the Koshima monkeys. This habit spread from peripheral hungry males to other troop members. Raw fish is not a favourite food, but fish are now collected and eaten when there is noth-ing better to eat.

Since the scientists first started feeding the macaques on Koshima island, a whole new life style has developed. The original potato-wash-ing tradition led to a direct elaboration of this behaviour – to biting the potato before dipping it into the water so that it was seasoned before it was eaten. But the main effects were indirect: it triggered another tra-dition, separating wheat and sand in the water, and the two food-wash-ing traditions, in turn, triggered the tradition of using the sea for swimming and cooling in summer. Bringing food to the salty sea-water and dipping their hands and body in it may have had other effects, such as decreasing the parasite load, which may have further reinforced the value and pleasure of swimming. Each habit reinforces the others, since all are associated with the new habitat, the sandy beach and the sea.

Although there is little modification or variation in any one transmissible habit, the whole life style has evolved by one modification in behaviour producing the conditions for the generation and propagation of other modifications. Through the accumulation of socially transmitted variations over time, the macaques have acquired a new life style.

This example highlights several important points. First, it shows that cultural evolution in animals may involve a whole life style, rather than a particular isolated behaviour. Earlier, we discussed how the accumulation of variation in one pattern of behaviour, the song of a bird, can lead to differing local dialects. Birdsong, however, is somewhat exceptional, because it has a sequential, modular organisation in which the modules can be changed one at a time. The accumulation of modifications in non-modular systems of information is rarely linear like this – it is uncommon to observe cumulative cultural evolution in a single isolated pattern of behaviour. With non-modular information, evolution proceeds through the effects that variations in one socially transmitted behaviour have on another related behaviour, which may, in turn, affect another one, and so on.

Second, the macaque example shows how various related customs can stabilise each other. Food-washing reinforces swimming and swimming reinforces food-washing. This network effect increases the stable maintenance and propagation of each individual pattern of behaviour. Finally, a habit originally acquired by subadults through trial and error or by insight, eventually becomes one that is learnt by infants when their mothers employ the new behaviour. This further stabilises the propagation of the new habit, because early learning is very effective and very persistent. A habit learnt by an infant female is often later repeated and transmitted to the next generation when she becomes a parent. Unless marked environmental changes occur, the new life style will tend to be perpetuated.

Whether or not a particular pattern of behaviour persists obviously depends on its effects on the survival and reproductive success of its bearers, and on its stability and transmissibility, which, in turn, depend on how it is integrated within the total behavioural package. Being able to imitate the way in which another animal behaves can sometimes contribute to the copying fidelity, but this is not necessary for the evolution of culture. True imitation is probably involved neither in the traditions of the Japanese monkeys, nor in the habit of milk-bottle opening by tits.

Summary

One of the ways in which animals are able to adapt to their everchanging environment is through learning. They can track environmental changes by changing their behaviour. Learning, however, is not a simple and unitary process: what, where, when and how things are learnt and for how long they are remembered can all vary. Nevertheless, there seem to be general principles that underlie all types of learning and memory, and enable animals to categorise and generalise about their experiences and hence organise their learning and behavioural responses.

Any complex behaviour includes innate components, which require little or no learning, and learnt components. The various types of ecological regularities and irregularities experienced by different species have led to the evolution of differences in the amount and types of learning and memorising that predominate in each. In some animals, learning is largely asocial – individuals learn on their own, without being influenced by their fellows. In others, particularly birds and mammals, there is a component (and sometimes a very large component) of behaviour that depends on social learning – on learning through interactions with other individuals. Through social learning, behaviour can be inherited. When young individuals learn from older ones (usually, but not exclusively, from their parents), patterns of behaviour, preferences and other types of information are transmitted across generations. This may lead to the formation of local traditions. The scope and role of tradition in the life of animals has been underestimated, first, because stable behaviour is usually assumed to be gene-based rather than tradition-based, and second, because it is believed that the formation of complex traditions must involve special, and allegedly uncommon, cognitive abilities, such as the capacity for imitation. Imitation is not necessary for the formation of animal traditions, however. Traditions can be based on all kinds of social learning, and become richer and more stable as they evolve and become associated with additional patterns of learnt behaviour.

Notes

[1] In the paragraphs that follow, the account of the spider behaviour is based on personal observations by E. A. and the work of Fabre, 1913; Foelix, 1997; and Witt, Reed & Peakall, 1968. The account of the behaviour of the chukar

partridge is based on E. A.'s observations and those summarised in Cramp, 1980, pp. 452–7. The description of European hedgehog behaviour is based on unpublished records of a study (1979–83) by E. A. of a population living in the Judaean hills, and observations and experiments summarised by Burton, 1973; Herter, 1965; and Reeve, 1994.

[2] Some species of jumping spiders that hunt other spiders are known to use a combination of innate, trial-and-error and even planning-ahead behaviours while pursuing their prey. These are probably a small minority group among spiders. See Wilcox & Jackson, 1998.

[3] This definition is taken from Dudai, 1989. For a more general definition, see Shettleworth, 1998, p. 100.

[4] Dudai, 1989.

[5] Gould *et al.*, 1998.

[6] Neville, 1990.

[7] Rose, 1992.

[8] Breedlove, 1997.

[9] The effects of stress on animal behaviour and on learning are reviewed in Sapolsky, 1990, 1992. See also *Stress and Behavior*, volume 27 (1998) of *Advances in the Study of Behavior*, (ed. Møller, Milinski & Slater), particularly the paper by von Holst.

[10] See references in note 9. For recent studies on the intimate association between emotions and learning see Adolphs, Tranel & Damasio, 1998; Hyman, 1998; Morris, Öhman & Dolan, 1998.

[11] Carlstead, 1996; Kavanau, 1964.

[12] Garcia, Hankins & Rusiniak, 1974. The importance to animals of learning during sleep has hardly been explored, yet young animals spend a lot of time sleeping, and it is difficult to imagine that the auditory and olfactory stimuli to which a sleeping animal is exposed do not become associated in meaningful ways with previously learnt responses or stimuli.

[13] The role of attention-networks and the changes in brain activity during learning are discussed in detail by Posner & Raichle, 1994.

[14] For a recent attempt to put learning theory into an evolutionary framework, see Davey, 1989. An excellent and critical discussion of different types of learning and memory and their ecological correlates can be found in Shettleworth, 1998.

[15] Cody, 1974, and personal observations of E. A.

[16] See Schlicht, 1998. Schlicht discusses clarity requirements and their far-reaching consequences for custom formation and economic activities in humans, but his arguments are also valid for the formation of behaviour patterns and habits in animals. Behavioural ecologists have produced similar arguments (see Giraldeau, 1997).

[17] Herrenstein, 1984.

[18] Tinbergen, 1951.

[19] Johnsgard, 1997.

[20] Response generalisation occurs when a particular behavioural response or skill can combine with other behaviours and be used for accomplishing new and often more demanding tasks. This definition is based on Baine & Starr, 1991, p. 64.

[21] Baine & Starr, 1991; McLeskey, Rieth & Polsgrove, 1980.

[22] Köhler, 1925.

[23] Wasserman, 1995.

[24] Herter, 1965.

[25] Cross, Halcomb & Matter, 1967.

[26] Information on bower-bird behaviour is taken from Diamond, 1986, 1987, 1988 and Gould & Gould 1989. It is discussed further in chapter 8.

[27] Gosler, 1993.

[28] Wasserman, 1995.

[29] Garcia, Hankins & Rusiniak, 1974.

[30] Boakes, 1984; Haldane, 1954.

[31] Lorenz, 1970.

[32] Hasler & Scholz, 1983.

[33] Bateson, 1990; Gottlieb and Klopfer, 1962.

[34] ten Cate, 1987; ten Cate, Kruijt & Meeuwissen, 1989; ten Cate & Vos, 1999.

[35] Immelmann, 1975.

[36] Collias & Collias, 1964.

[37] Brower, 1969.

[38] For a discussion of how young animals gradually learn through instruction, see Caro & Hauser, 1992.

[39] The information on the development of male bird song is based mainly on the excellent reviews of Catchpole & Slater, 1995, and Marler, 1990.

[40] Grant & Grant, 1996, 1997.

[41] Catchpole & Slater, 1995.

[42] It is interesting that we call these types of activities – the complex songs of blackbirds and nightingales, the well-orchestrated howling of wolves, the synchronised jumps and dives of dolphin groups, the acrobatic flights of ravens – 'beautiful'. Maybe the pleasure that animals engaged in these activities seem to show corresponds to what we, humans, call a sense of beauty and, as Darwin suggested in *The Descent of Man*, this sense of beauty is shared across species.

[43] The case of H. M. is described in Milner, Corkin & Teuber, 1968.

[44] Rosenfield, 1988.

[45] Memory systems are discussed in Squire, 1987. Up-to-date summaries of different memory systems can be found in Schacter & Tulving, 1994. The argument that different memory forms are subserved by different memory systems and that these memory systems obey different rules of operation, is discussed in both these books. The differentiation of memory systems is more detailed than we have suggested in our account, and includes language-specific memory and other forms of memory that seem to be specific to humans. Here we

have presented only the most general and least controversial distinctions between memory systems.

[46] Nadel, 1994.

[47] Sherry & Schacter, 1987.

[48] Adam Smith, 1776; Maynard Smith & Szathmáry, 1995.

[49] Moss, 1988; Moss & Poole, 1983.

[50] Todt, Hultsch & Heike, 1979.

[51] Catchpole & Slater, 1995; Mountjoy & Lemon, 1996.

[52] The information on food-hoarding birds is based on Källander & Smith, 1990; Sherry, 1985; Shettleworth, 1990, 1993; Smith & Reichman, 1984.

[53] Cowie, Krebs & Sherry, 1981; Stevens & Krebs, 1986.

[54] Pravosudov, 1985.

[55] Godden & Baddeley, 1975.

[56] Heyes, 1994.

[57] Galef (1988) reviewed the various terms and mechanisms of social learning. He discussed 22 terms describing various types and mechanisms of social learning, many of which are overlapping. We have found over 30 different terms that have been suggested for various types of social learning.

[58] For a recent discussion of the mechanisms of social learning see Heyes, 1994; discussions that focus on imitation can be found in Heyes, 1993; Moore, 1992; Whiten & Ham, 1992. We chose the terms 'social influence' or 'socially influenced learning' for all types of non-imitative social learning because of its apparent lack of commitment to a particular mechanism of learning. The term 'social influence' was suggested by Whiten & Ham (1992) and has been used by them in a slightly narrower sense.

[59] Sherry & Galef, 1984.

[60] Moore, 1996.

[61] These examples of imitation and emulation are discussed in Byrne, 1995. Emulation is probably a more sophisticated cognitive process than imitation, since it requires an understanding of the relationship between means and ends. Insight learning is based on such a learning rule.

[62] Tomasello, 1994.

[63] Heyes, 1994.

[64] Tomasello, Kruger & Ratner, 1993.

[65] Smokler et al., 1997

[66] Whiten et al., 1999.

[67] Kawai, 1965; Kawamura, 1959; Watanabe, 1989. Galef (1996) has challenged the generally accepted view that the life style adopted by the Koshima troop is an adaptation that resulted from information transferred among the members through social learning. He argued that asocial learning and reinforcement by human caretakers may be adequate and sufficient explanations for the various new habits that spread in the group. He based his argument on the slow spread of the habits, on some accounts of reinforcement by caretakers, and on the opportunities for asocial learning that were associated with

some of the new patterns of behaviour (e.g. babies who were carried to the water by their potato-washing mothers would spontaneously collect potato scraps from the water). Although we agree with Galef that asocial learning has contributed to the construction of the observed behaviours, we do not think that this is a particularly damning argument against social learning. Asocial and social learning are not alternative ways of adapting to the environment, but are usually complementary and concurrent. The collection of potato scraps from the water is certainly a socially mediated or socially influenced behaviour, since mothers, through their consistent habit of eating in the water, bias the learning of their young so that they are more likely to have behaviour similar to their own. Moreover, the spread of a habit through social learning can be quite slow, especially during the first few generations. Although there is much that none of us will ever know about the Koshima troop during the first years in which it was studied, our own reading of the literature available in English convinces us that this classic example of tra-dition formation provides very strong evidence for the spread of habits among individuals through various processes of social learning. Both the pattern of spread and the direct documentation of social learning by several well-trained researchers suggest that social learning has made an important contribution to the development of the new traditions.

4   Parental care – the highroad to family traditions

Parental care in birds and mammals is so familiar to all of us that it seems unlikely that it can hold any fresh surprises or offer any new insights. However, there are important aspects of parental care that are commonly overlooked when its role in evolution is discussed. Parental care is one of the major routes through which information is transferred across generations. It is largely through the effects of parental care that animal traditions become established. The information transmitted through parental care relates to all the aspects of life; some is used every-day, some only rarely. Information is transmitted through several different but usually interacting channels, and is essential for the survival and reproduction of the offspring. A look at some typical parental behaviour, that of the common domestic mouse, will show the remarkable range and importance of the information that is transmitted from parent to offspring.[1]

Dusk is a good feeding time for village mice. The small, four-month-old, greyish-brown female domestic mouse silently scales the outer wall of the village grocer's warehouse. She enters the warehouse through a small crack in the wall, and quickly slides down to the piles of bags containing pinhead oatmeal and canary seed. This urine-marked route leads safely to the best source of solid food around. It was first introduced to her by her mother, three months ago, and has been used by her ever since, at least twice a day, at dawn and dusk.

The doe is twenty days pregnant, and will give birth any day now. She is hungry, but does not start eating yet. She stands on her hind legs, suspiciously sniffing the air with her sensitive muzzle. Her scent survey discovers no rats, cats or strange mice, so she can now safely dive into one of the bags of oatmeal and eat as much as two grams, almost a quarter of her own weight. The pinhead oatmeal is always her first choice. But why? Mice are omnivorous and will eat almost anything, and canary seed is a well-known mouse delicacy; but, like every other mouse, this doe has some loyalty to the first solid food she ever smelled and tasted. In her case it was the oatmeal of this warehouse.

Having finished her meal the doe hurries back to her partly constructed nest in the far corner of the nearby tool shed, using the same route in reverse. This doe is lucky, since she is the only one of the recent offspring who has been allowed to mature and remain to breed in the deserted, roofless tool shed, within the crowded confines of her father's ten square-metre territory. She has to share these quarters with her everpregnant mother, three other females and their offspring, and her father, the local dominant male, an indefatigable lone fighter against strange male intruders.

Back in the shed she resumes the construction of her nest, vigorously gnawing and shredding grass, straw and paper into small, soft strips, finally shaping them into a fine well-insulated nest for her future infants. At midnight she gives birth, one by one, to six pink, hairless, blind and deaf babies. For each one, in turn, the new mother bites off the umbilical cord, licks it dry, and cleans and eats the embryonic sac and afterbirth. For the next two weeks she will spend long periods of time inside her nest, curled around her young, meticulously licking their bodies and liquid excretions, and suckling them.

Four days have elapsed since the birth. At dawn, the weary, hungry mother mouse is curled around six plump drowsy pups, huddled together inside the nest. Cautiously, the mother leaves nest and pups for a short spell of refuelling with oatmeal at the nearby warehouse. She has to leave the pups several times a day, for even such a devoted mother must occasionally eat. Halfway back to her nest she can already hear her infants' high-pitched ultrasonic distress cries (cries that are inaudible for many predators, or hard for them to locate), calling her back to warm them up and feed them. She dashes for the nest and within a few seconds is again curled around her pups, licking them and pumping warm milk into them.

After three more days, the still blind, suckling pups are almost fully furred. From time to time two of them boldly try to leave the nest and explore the outside world. This is potentially dangerous: the pups are unattended and yet unable to flee or defend themselves, and are therefore liable to fall easy prey to strolling cats or hungry mice. The mother immediately uses her better judgement and her mouth to grab the premature explorers by their abdominal skin, and carry them back to the relative safety of the nest.

At two weeks of age, the pups are open-eyed, fully furred and very curious, sniffing excitedly at everything and everyone in the close vicinity of the nest, and exploring every corner of it. Several times a night the mother leads a tightly packed group of six young explorers on a short-term, short-range excursion about the tool shed. Here the youngsters learn the physical and social topography of their colony, learning in particular to recognise and memorise the multitude of scents of objects and individuals. The youngsters still suckle, but already display real interest in the solid-food items brought in and eaten by their mother. They also start eating their mother's faeces. It is time for them to be introduced to solid food.

At daybreak, a cautious, nervous mother leads a group of stiff-haired, hesitant youngsters up the red brick wall on their way to the warehouse. Suddenly a strong smell reaches their sensitive muzzles, the smell of a brown rat, a notorious mouse-hunter. In a split second the alarmed mother changes direction and leads a scampering group back to tool shed, nest and safety. The youngsters will remember the traumatic smell of the rat for a long time, and know what to do when they smell it again. At dusk, the same team tries again and succeeds, this time without trouble, in entering the warehouse via the well-trodden urine-marked route, and enjoys the pinhead oatmeal. From now on, the warehouse feeding site, and the special routes leading to and from it, will be the youngsters' first choices.

The type of parental care seen in the domestic mouse, in which the father does not participate in the care of the young, is very common among mammals.[2] Maternal care, the least complicated type of parental care, can therefore serve as an example of some of the everyday problems encountered by parents and offspring. Yet even this relatively simple type of parental care raises hosts of difficult questions. What sort of care, and how much should a mother give to her offspring? What exactly do her offspring gain? Breeders of livestock realised long ago that the phenotype of the mother (for example, her body size, her health and her style of care) has major effects on the phenotype and well-being of her offspring, and therefore substantially influences the results of the artificial selection that they apply. But what components of maternal investment and care are important? The proteins sequestered in the egg? The amount of milk? The periods of time spent with the infants? The demonstration of different types of behaviour?

Parental effects

Parents contribute much more to their offspring than their DNA. All mothers provide proteins, food reserves and RNA transcripts in the cytoplasm of their eggs. In addition, both father and mother transmit chromatin marks, such as chromosomally bound protein complexes and patterns of methylation on the DNA, which help to determine gene activity in the next generation. The energy-providing organelles, the mitochondria, are usually passed from mother to offspring, while centrioles are sometimes transmitted only through the father. In mammals, the amount of nutrients in the egg, the quantities of hormones and anti-bodies sequestered in the yolk or in the milk, the type of RNA transcripts in the cytoplasm, the uterine environment, the mother's lactation per-formance and her style of care, all have great and often long-lasting effects on the offspring's development. Paternal effects can also be important, particularly if the father affects the mother's well-being, or provides the offspring with behaviour-related information that goes beyond his genetic contribution. The various non-DNA contributions that parents make affect the degree of similarity between the parents' phenotypes and those of their offspring. Of course, they also affect the offspring's survival and reproductive success – they affect fitness.[3]

The differing effects that pairs of parents have on the development and future success of their young can be the result of either genetic dif-ferences between the pairs, or environmentally induced differences, or both.[4] When parental effects stem from genetic differences, the genes that the parents carry lead to differences in the development of their offspring, but not necessarily because the offspring inherit these genes. The effects of the parental genes are indirect and affect *all* offspring, not just those that inherited particular genes. For example, consider a female who has inherited a growth-stimulating allele from her father. Assume that the gene is expressed only during egg-formation, and leads to the presence in the eggs of factors that promote growth. Because the female inherited the allele from her father, it had no effect on her own growth. But, because its product is present in *all* her eggs, it will affect *all* of her offspring, both those that inherit the allele and those that do not.

Not all differences in the effects parents have on their offspring are the result of genetic differences. Environmentally induced differences between parents are also influential. Two pairs of parents with identi-cal genotypes could have very different effects on their offspring if the environments in which these parents developed induced variations in their phenotypes that subsequently affected the development of their offspring.[5] There are two major types of environmentally induced parental effects.[6] With the first, the parent's induced phenotype affects a trait in the offspring, but this need not result in similarity between parent and offspring. For example, the behavioural decision of a moth-er diamond-back terrapin about where to lay her eggs affects the sex of her offspring, because males develop from eggs incubated in relatively

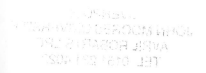

cool places and females from eggs incubated in warmer conditions.[7] The mother thus affects the sex ratio among her offspring. But the sex ratio among the mother's offspring and the sex ratio among her daughters' offspring may be quite different, since their choices of egg-laying site may have been influenced by different environmental conditions. The mother's effect on the sex ratio among her offspring does not lead to the same sex ratio among her daughters' offspring. With the second type of induced parental effects, however, parents transmit a particular environmentally induced phenotype to their offspring, so parent and offspring become more similar. For example, large parents frequently produce large offspring, which may lead to the transmission of large size to the grandchildren and subsequent generations. Similarly, as we have already seen for the domestic mouse and will describe in more detail for other mammals and birds in the following sections, many of the preferences and patterns of behaviour that offspring acquire or learn from their parents make the preferences and behaviours in parent and offspring similar.

Although mathematical models showing how genes with maternal effects could evolve and spread were developed a long time ago, until recently maternal inheritance and other indirect effects have been neglected by most evolutionary biologists. Maternal effects were regarded as factors complicating the distinction between genetic and environmental effects, and hence the estimation of heritabilities. Consequently, efforts were mainly directed towards developing mathematical manipulations that would minimise their influence. The situation changed when, as a result of both theoretical and experimental studies, it became clear that it is impossible to understand the effects of selection and the dynamics of phenotype changes in populations without incorporating parental effects.[8] We believe that parental effects are profoundly important in evolution, so in this chapter we are going to focus on a major aspect of environmentally induced parental effects in birds and mammals – on the way the behaviour of offspring is affected by parentally transmitted information. We therefore need to look at parental care.

The information lacuna

A general, standard textbook definition of parental care is 'A set of behaviours exhibited by parents towards their young and presumed to aid the young in growth, development and survival, both physically and

behaviourally'.[9] This definition is a straightforward ethological definition, but it does not really help us very much when we think about evolution. It does not suggest any common denominator that can be used
to measure or even describe the many things that make up parental
care. Modern evolutionary ecologists have provided such a common
denominator by borrowing time and energy allocation concepts from
the optimality school of human economics, and applying them to animal behaviour. They view each pattern of parental care as the most economical (or optimal) solution to a particular, well-defined problem, such
as for how long a mother should nurse her young. Each style of parental
care is described in terms of the time and energy allocated by parents
to their offspring. This approach also takes into account the effects of
the type of parental care on the offspring's future reproductive success,
and the attempts of offspring to maximise it by manipulating their parents' behaviour. The sometimes differing interests of parents and offspring are included in a balance-sheet of time and energy.[10]

From the point of view of the young bird or mammal, the world is a
complicated and often very dangerous place in which to live. Among
other things, the youngster has to be able to deal successfully with the
basic but complex tasks of finding food, shelter and, later, a suitable
mate. Since at least some of these resources are in short supply, they
may have to compete for them against individuals of their own or other
species. The natural world is also swarming with predictable and unpredictable armies of predators and parasites, ever ready to attack the
unknowledgeable or the unwary. The young will have to contend with
climatic hazards such as frosts, heat waves and droughts, disasters such
as fires and floods, and sometimes even earthquakes and hurricanes. As
we have seen in the previous chapter, a young animal is adapted to deal
with these challenges through a combination of three types of behaviour: it can solve a problem by being genetically programmed to deal
with it, the way the female orb-web spider solves its orb-web construction problems; it can learn how to solve a problem for itself, the way
the European hedgehog learns to forage; finally, it can learn to solve a
problem through social learning, the chukar partridge way, by following the examples set by knowledgeable parents and other experienced
individuals, and adopting their solutions to the problem at hand.

From an evolutionary point of view, the parent is expected to behave
towards the young in ways that will help to maximise its own lifetime
genetic contribution to the next generation. Usually this means rearing

as many offspring during its lifespan as possible.[11] Different species of animals have different ways (or strategies) of doing this. Since parents do not have infinite amounts of time and energy, each strategy is essentially a different trade-off between the number of offspring and the amount of time and energy allocated to each one of them. Evolutionary biologists try to analyse the components involved in this trade-off, and see why a particular strategy is the optimal one. But there is a curious deficiency in the great majority of these optimality analyses. In almost all the accounts of the ecology and evolution of parental care, the transfer of learnt information from parent to offspring, and its evolutionary effects, are ignored. In an otherwise excellent monograph on the evolution of parental care, Clutton-Brock does not once mention the learnt information transmitted by parents to offspring, and does not consider its effects on the evolutionary history of their various behaviours.[12] There is no systematic treatment of the role of social learning in the evolution of behaviour in any of the textbooks on behavioural ecology or evolution. The allocation of energy and time, and their various trade-offs, are treated carefully and in detail. The role of the information that is transmitted through behaviour is absent.

This information lacuna is a puzzle. Many lines of evidence clearly show that there is a constant flood of information pouring from parents to offspring, flowing through a host of hidden, non-verbal channels. The transfer of information that has been acquired by the parent over a relatively long period spent in the same habitat helps the young to prepare for a successful career as an adult. The help covers almost every aspect of life – from immunological defence, through food, mate, habitat and site selection, to anti-predator behaviour.[13] Usually, the help also makes the habits of the offspring similar to those of their parents.

The existence of mechanisms that transmit information about crucial aspects of life has been known for many years to both biologists and experimental psychologists. Yet this knowledge has not been fully incorporated into the main body of ecological and evolutionary thought. While time and energy-based theories flourish, the ways and the importance of transmitting behaviour-affecting information have either been underestimated or disregarded by most theoretical evolutionary ecologists. The main reason for this neglect is probably that, whereas time and energy are relatively easy to measure, the information transmitted is not. If we need to know how much time and energy a mother domestic mouse is allocating to her babies while suckling them, all we have to do is use a good stop-watch to accurately measure suckling times, and collect milk samples to carry out a caloric analysis. On the other hand, proof of the transfer of information from mother to offspring, for example through the mother's milk, is more difficult to obtain. We can directly watch only mice, not bits of information. If we want to find evidence for the transfer of information, there is often no choice but to use indirect and time-consuming experimental manipulations.

As we saw, young mice eating solid food for the first time prefer to eat much the same food as their mother. How can we determine whether this preference is innate or acquired through the mother's behaviours? Since well-fed nursing mouse-mothers readily foster strange pups (reasons for this odd behaviour will be discussed in chapter 7), we can use fostering and cross-fostering techniques, as well as pipette-feeding, to detect any effects of maternal transmission on the future food preferences of the mouse pups. What we find is that, if we exchange newborn pups between two mothers with very different food preferences (cross-fostering), then, when given a choice a few weeks later, every pup clearly displays the food preference of the doe that suckled it, regardless of blood relations. If we let newborn mouse pups spend a normal amount of time with their real mother, but instead of allowing them to suck from her we artificially pipette-feed them with milk from a nursing mother with other food habits, the pups later display the food preferences of the strange female they have never met. The stranger's preferences were transferred to them via the milk. Furthermore, if a novel food component is added to the pipette-milk, this too will later become part of the youngsters' preferred food.[14]

These experiments show that the milk a young mouse obtains from its mother contains more than the food constituents necessary to satisfy its present needs. It also contains information that will help to form future food preferences. Yet the transmission of this type of non-genetic, behaviour-affecting information is almost totally ignored in present research and theories about the ecology and evolution of behaviour. It has no place in the time and energy budgets used to try and understand why animals behave as they do. We believe that this misses a crucial component in evolutionary analysis. We want to concentrate on this neglected aspect of parental care, and look at the way the non-genetic transfer of information can shape the behaviour of animals in the next generation.

Channels of transmission

Milk, as we have seen, is a source of information as well as energy. The results of cross-fostering and other simple experiments with mice show clearly the non-genetic transmission of food preferences through maternal milk from one generation to the next. Such results are also typical for other omnivorous and wide-range feeders among the mammals, such as many other rodents and ruminants.[15] Milk, however, is not the only, nor the earliest channel through which food preferences and other important determinants of future life are transmitted from one generation to the next. An even earlier channel is transfer across the placenta. Mammal foetuses are known to be able to smell semi-volatile liquids that pass to them across the mother's placenta, and later prefer food items carrying these smells. This is not merely a qualitative preference: if the foetuses smell several different kinds of food, when adult they may prefer to eat the foods whose smells they perceived most often.[16] Foetuses are often able to learn food aversions, as well as preferences. A pregnant mother, feeling sick after tasting a repulsive novel item of food, sometimes transfers traces of this food or its smell through the placenta, and this early experience, which makes the embryo uncomfortable, may condition it against the ingestion of this food in the future.[17]

The placenta is also one of the many channels through which immunological defences are transmitted from mothers to their offspring. Several classes of life-saving (or life-extending) antibodies are transferred from mother to foetuses, and later to neonates, via parallel and complementary routes including the placenta, yolk sac, colostrum milk and saliva, but mainly through colostrum milk.[18] The transmitted antibodies endow the vulnerable and immunologically still underdeveloped newborn with passive immunity against the prevalent types of bacteria and viruses. These microorganisms are the causes of disease in the mother's environment, which is obviously also the present, and possibly the future, environment of her offspring. It is not only mammals that use such early lines of defence: birds transfer antibodies to their offspring via the egg. In the blue-and-gold macaw the maternal antibodies persist in the chick until six weeks after hatching, which is when the chick's own immune system matures. In chickens too, maternal antibodies are transmitted to the chicks, and it has been shown that they provide them with resistance to the infections to which the mother has been exposed.[19] In addition to conferring passive immunity, a mother's antibodies can contribute to the future active immunity of the offspring. Maternal antibodies, transferred via milk to mammalian neonates, may sensitise the offspring's immature immunological system against the corresponding antigens, enabling it to form a quick and strong active first reaction in the future.[20] If the mother's infection is a recent one, and some weakened pathogens are transferred along with her antibodies, the sensitisation and subsequent active immunological response may be even quicker and stronger.[21] The mother thus delivers to her offspring a small, highly concentrated army of antibodies, having the primary role of defence, but also acting as an intelligence service, supplying an up-to-date, highly specific assessment of present and future dangers. These antibodies forewarn the young of the common pathogens in their environment, and prepare their immunological systems for efficient future action.

Remarkably, the maternal environment in which the mammalian foetus develops sometimes has effects that can be carried over to later generations. These effects can be surprisingly fundamental, affecting characters such as gender behaviour, the development of ulcers and the sex ratio of offspring.[22] For example, if female Mongolian gerbil embryos develop in a uterine environment in which most of the embryos are male, and they are therefore exposed to high levels of testosterone, they mature late and are more territorial than other females. But this is not all: from our point of view, the most interesting feature of these females is that they, in turn, produce litters with a greater proportion of males. Therefore, their daughters, who usually also develop in a testosterone-rich uterine environment, also mature late. The cycle perpetuates itself! The developmental legacy of the mother is transferred to her daughters, and there is transmission (non-genetic, of course) and repetition of this distinctive reproductive pattern.

The next route of information transfer is no less surprising. Young (and sometimes also mature) individuals of many species of mammals consistently and enthusiastically eat their own and other individuals' faeces. The habit is found in many species of rats, mice, hamsters, guinea pigs, hares, pikas, horses, shrews and the koala. What are the functions of such a bizarre and apparently unhygienic pattern of behaviour? Coprophagy – the habit of eating faeces– apparently fulfils several useful roles. Most of the mammals who eat faeces are herbivores, consumers of cellulose-rich plant material, who have a dense symbiotic bacterial and protozoan gut flora that helps to break down and digest cellulose.

The mature individuals use self-coprophagic behaviour as a means of recycling precious bacteria, protozoa, vitamins such as B and K, and partly digested food, thus promoting the efficiency of digestion. The young of many of the herbivorous species eat the first (partly digested) and often also the second (fully digested) round of their mother's faeces. In this way, they directly inoculate their own guts with a dense, beneficial, maternal flora of microorganisms, a non-genetic inheritance from their mothers.

This flora is an important asset for the young, since it is a pre-selected community of microorganisms adapted to break down the mother's preferred foodstuffs. But, in order to make the best use of the mother's present of microorganisms, the youngster has to eat much the same foods as she does. Do faeces contain the information that enables the young faeces-consumers to identify the kinds and the relative abundance of the foods eaten by mother? The adaptive value of coprophagy is, unfortunately, still an underresearched subject, but the results of some existing studies can be interpreted to mean that one of the functions of coprophagy is to transmit information that establishes later food preferences. More than thirty years ago, in her now classical mammal ethology book,[23] R. F. Ewer reported that in 1937 Minchin described how a mother koala lets her youngster lick from her anus a special kind of soft faeces containing partly digested eucalyptus leaves. This behaviour occurs daily during the month before the youngster becomes an independent eucalyptus-leaf eater. Minchin saw the behaviour as the way in which the youngster acquired digestible 'baby food'; Ewer interpreted it as a way of establishing the youngster's normal gut flora; more recently, Marinier and Alexander reinterpreted this same behaviour as a youngster's way of learning to identify the mother's specific food choices, for its own future use.[24]

Marinier and Alexander's main interest was in the functions of coprophagy in the domestic horse. Mature individuals are selective grazers, each with its own food preferences. A foal starts life as an unselective grazer, grazing for the first six weeks close to its mother, regularly eating some of her faeces. By the end of this period, each foal becomes a selective grazer with a food preference similar to that of its own mother. It is interesting that most or all of the faeces eaten are the mother's, and the highest rate of coprophagic behaviour occurs during the first six weeks of life, a period considered to be the 'sensitive period' for learning food preferences. Since foals do not acquire food preferences either through watching the mother or through smelling and suckling her milk, Marinier and Alexander suggested that each foal learns its food preferences from its mother by eating her faeces. Although as far as we know the experiment has not been done, this idea could be tested by exchanging faeces – by making the foals of two mothers with different food preferences eat the faeces of the non-mother, and observing the foals' future food preferences.

An individual's repertoire of preferred foods is clearly the result of the different types of information about food that it acquired early in life through several different channels. Many studies of wild and laboratory rats and mice have shown the reinforcing effects of learning through complementary channels.[25] For example, Laland and Plotkin found that the food preferences of male Norway rats are socially acquired by smelling or tasting excretions deposited at feeding sites by previous visitors, and by sniffing the breath of an individual who has recently eaten. They showed how a particular food preference was transmitted through a chain of animals, and how the two social learning processes – through excretory marking and gustatory cues – reinforced each other.[26]

Going back to the young mice in the tool shed, we can now reconstruct the way their food preference was established and consolidated. The pregnant doe ate a lot of pinhead oatmeal and some canary seed, and transferred oatmeal-biased liquid traces trans-placentally to her foetuses, thus establishing a preliminary preference for oatmeal. Since the doe continued to eat oatmeal, time and again her fast-growing youngsters received the same reinforcing message to prefer oatmeal. The message was delivered through the successive routes of milk, faeces and, finally, direct demonstration and ingestion. A few weeks of constant positive reinforcement for an unchanging type of food were enough to make the basic food preference of the young mice rock solid and very similar to their mother's. The many routes of transmitting food preferences, which are partly separated in time, supplied the youngsters with cumulative, up-to-date information about the food preferences of their mother and the relative availability of the different foods in their environment.

The parental heritage

Mechanisms of transmission similar to those just described for mice allow inexperienced youngsters of many species of birds and mammals to learn about their environment through the efforts of their parents, without having to resort to learning through their own experience. Individual learning is, without doubt, much more time-consuming, energetically wasteful and dangerous for the youngsters than for their experienced parents. The latter do their best to prevent daring but naïve youngsters from carrying out unattended reconnaissance expeditions at times and to places the parents consider to be dangerous. The way the youngsters are restrained varies from species to species. A well-known method, used by the domestic mouse but also common in other mammalian species with behaviourally helpless (altricial) newborn young, is to transport the young away from the site of danger by carrying them in their mouths.

Greylag goose parents use a rather different technique.[27] They utter action-inhibiting calls when their young prematurely attempt a dangerous solo take-off. Ignoring such a warning may sometimes be fatal. Konrad Lorenz tells how, if the greylag goose youngsters disregard the parental warning call and fly off, the parents quickly act as flying instructors. They take off after the youngsters and assume the lead, finally demonstrating to them the complex art of successful upwind landing in a safe site. The take-off, flight and safe-landing demonstrations by the parents are very important to the young geese. Although the patterns of motor co-ordination they need for flight are innate, they still have to learn through their parents how to estimate heights, distances and wind conditions, and that they must always face the wind as they land. Any goose trying to land downwind is taking the risk of turning somersaults and crashing. Aerial accidents and fatal crashes are much more common among young hand-reared geese, who have never had early parental warnings and aerial instruction, than they are among normal parent-reared geese.

Extreme parental protectiveness is not always in the best interests of both parents and offspring. If they are temporarily prevented from exploring the world independently, bird and mammal youngsters may sometimes miss opportunities for learning new and possibly useful tricks or bits of information that are unknown to their parents. The parents, on the other hand, will do their best to transmit their patterns of behaviour and protect the young, but, if carried to extremes, this would conflict with their investment in future progeny. The emerging conflicts between parents and offspring over the required degree of care and protectiveness, and their resolution, are central issues in behavioural ecology and will be discussed in chapter 6. At present we want to look at how parents hand over to their offspring their store of information about foraging, predators, breeding sites, mates and parental care itself, and hence affect the behaviour of their offspring.

The transmission of foraging behaviour

No bird or mammal is born knowing what to eat, where to find its food and how to harvest it, so every individual has to learn the art of foraging. We have already seen that in seed and grass eating rodent and ruminant species the mother transmits vital information about food and foraging in many different ways. The same is true for predatory species. For example, young polecats learn what to hunt and eat from their mothers: the smell of the prey that a mother brings them, especially during their first 2–3 months of life, determines the prey they prefer when later given a choice.[28] By bringing dead or injured prey to her inexperienced young, or taking them to the prey, a huntress enables her young to learn and memorise the scents and appearances of the common prey species.[29] The ability to recognise suitable prey by scent and sight is an obvious asset to any predator, in particular to the many species that ambush and stalk distantly located prey.

In predatory species, the mother has an additional role – to help her offspring acquire and master a wide range of complex hunting skills, vital to their future success as hunters. A huntress frequently allows her young to practice their budding hunting skills by letting them use the helpless victims she has caught as sparring partners in a simulated, danger-free hunting session, which she carefully oversees. If the injured prey tries to escape, and seems to be out of the youngsters' reach, she will recapture it and put it back into the arena, where eventually it will be killed and eaten by the young. In the previous chapter, we described this behaviour for the cheetah, but it is far from being exotic or exceptional. Any mouse-hunting domestic cat mother may perform this way for the benefit of her kittens on the kitchen floor.[30] In a number of species, such as tigers and cheetahs, the young join the mother's hunting expeditions as junior partners. At first they are merely spectators, staying back and watching intently, but later their mother occasionally allows them to participate in the last stages of the kill.[31]

The complex hunting techniques of predators, in particular the big-game hunters such as tiger, cheetah and lion, take a long time and a

lot of guided and unguided experience to perfect. The youngsters' hunting apprenticeship is usually lengthy, lasting up to twenty months for
the cheetah and tiger, and twenty-four months for the lion.[32] A similar
pattern of extended parental care is found in the many species of birds
that have diverse and complex methods of feeding, or are frequently
exposed to conditions of food scarcity. In species such as terns, owls and
kingfishers, which have small clutches and no second broods, the young
still depend on their parents for food guidance several months after
fledging. Anyone who has watched common garden birds knows that
parental care is often still being given for a week or several weeks after
the birds can fly and feed for themselves.[33] Yet, how this extended
parental care, which often leads to behavioural similarity between
parents and offspring, affects the evolution of foraging patterns in subsequent generations has not been investigated.

In addition to foraging for food, some birds and mammals forage
systematically for medications. These medications are specific health-
improving chemicals that can have preventive or therapeutic effects.[34]
They are usually ingested, but sometimes are applied externally by being
rubbed on the body or placed in the nest. The chemicals serve as stimu-
lants, insecticides or antibiotics, or they can have specific antiparasitic
actions. The best-studied cases are the ingestion of soil and rock by herbivorous and omnivorous mammals, which serves mainly to absorb tox-
ins and detoxify secondary plant metabolites.[35] Another well-known
behaviour is the selective use of antibacterial and acaricidic green
foliage by European starlings, who insert these medicinal plant parts
into their nests.[36] Although the role of social learning in these behaviours has not been studied, it is reasonable to think that the exposure
to the self-medication behaviour of the parents, combined with the positive reinforcement that results from using the medication, leads to the
offspring adopting the behaviour. Cross-fostering experiments could easily test the prediction that foraging for medications, like so many cases
of foraging for food, is socially learnt from parents.[37]

The transmission of anti-predator behaviour
Immunological defence, food habits, foraging techniques, and flight and
hunting methods are passed on from parents to offspring. That is not
all. The risk of predation dominates the lives of most birds and mammals, and one of the most important types of information parents can
transmit is the information needed to prevent the young from being

devoured by predators. The lines of defence include a host of behaviours such as freezing, uttering alarm calls and harassing the predator (mobbing), all of which result in encounters being avoided or made futile for the predator. Uttering alarm calls is an instructive example of the measures taken by parents to minimise for themselves, and for their young, the risks of being taken by a predator.

Alarm calls are multifunctional. A parent bird who spots a perching bird of prey and utters an alarm call is sending concurrently at least three different messages to its partly independent but inexperienced fledglings. First, the alarm call serves as an urgent early warning, causing the young to dive into the nearest cover. Second, since the same alarm call is uttered time and again when the young are already hidden and safe, it may serve to intensify the youngster's attention, and direct it towards both the predator and its parents. Third, as we discussed in the previous chapter, stress, if not excessive, induces learning. An unexpected alarm call certainly qualifies as a stressor, and the degree of stress felt may not be excessive under circumstances of relative safety, so the call may induce the young to learn about both the predator and its behaviour, and about the antipredator behaviour of the parents. The common practice that human parents and teachers use, raising their voice when they want to make sure that an important message gets through, may be a useful way of increasing attention and inducing learning.

Alarm calls are also a way of communicating with the predator. By uttering an alarm call, the animal may communicate to the predator that it has been seen, and that the caller is ready to either fight back or flee. Having lost the element of surprise, the predator should think twice about whether attacking an alert and ready individual is worthwhile. This 'communication' with the predator may also be learnt by offspring and implemented in their maturity.

The alarm calls are sometimes predator-specific. Every place has its own unique community of resident and transient aerial and ground predators. If birds and mammals are to survive and successfully reproduce in a particular location, they have to be able to recognise each predator for what it is and discriminate between it and other similar but harmless species. Following recognition and discrimination, they have to estimate the degree of risk involved and choose a suitable line of action. The need to be discriminating, and utter alarm calls solely towards truly dangerous species, is very important. False alarm calls and misdirected measures are wasteful of time and energy, as well as being potentially dangerous. Loud alarm calls directed by a brooding parent bird towards a harmless individual might attract the attention of a dangerous but previously unsuspecting predator to the nest. It is therefore not very surprising that the mature individuals of many species of birds and a few mammals are able not only to discriminate between a dangerous predator and a harmless species, but also to generalise.[38] They can group several dangerous species, all requiring the same defensive reaction, into the same category.

A good example of this is seen in the vervet monkeys studied by Seyfarth and Cheney.[39] Adults utter and respond in a different way to four acoustically different alarm calls aimed at leopards, martial eagles, pythons, and baboons. Unlike adults, infant vervet monkeys give unnecessary alarm calls when they see other species, and sometimes they fail to respond correctly to a given alarm call. Nevertheless, they seem to possess and use a budding general knowledge of predator classification: the infants' leopard-calls are aimed at terrestrial mammals in general, the eagle-calls at birds, the python-calls at snakes and the baboon-calls at baboons. How do the infants, on their road to adulthood, restrict and optimise their predator classification system? A mass of indirect evidence shows that young vervet monkeys learn to match the right alarm calls to the right species of predator by simultaneously observing both the predator and their parents' reaction to it. The infants learn from their parents both to narrow their functional definition of a dangerous predator, and to choose the right action for a given type of predator and situation. Information acquired through social learning, a non-genetic form of transmission, seems to be superimposed on innate, more general and crude anti-predator reactions.

Whether or not the anti-predator behaviour itself is genetically determined, and whatever the role of very early learning, the young must still learn from their parents and from other experienced individuals the rules about when and how to use each antipredator behaviour. A great tit who, while feeding on the ground, far from cover, spots a distant kestrel that has not yet noticed it, would make a grave mistake if it uttered an alarm call without flying into dense cover first. Uttering a series of alarm calls would probably attract the kestrel's attention, and the tit would pay for its mistake by ending up enclosed in the kestrel's talons. The alarm calls in this case would have been both prematurely uttered, since the kestrel had not yet noticed him, and misplaced, since it was far from cover. Great tits and many other species of birds are known to adjust their anti-predator behaviour to suit different local circumstances.[40] The appropriate behaviour for local circumstances is usually learnt from parents.

The eventual close similarity between the anti-predator behaviours of parents and offspring does not, therefore, necessarily stem from their close genetic relatedness. The resemblance may well reflect the prevalence of accurate and efficient non-genetic mechanisms of transgenerational inheritance of patterns of behaviour. Genetic mechanisms determine the initial range of reactions, but genes are often only distantly relevant to the final outcome, to the realised set of anti-predator adaptive behaviours.

Choosing a home

So far we have described how the mother transmits to her young, by various routes, her own particular arsenal of habits and various types of information about food, predators and pathogens. This arsenal, much of it behaviourally inherited from her own mother and adapted by her to the current environment, is a problem-solving package that fits the unique local conditions of her own dwelling place. Does the mother prepare her young for life on her own site, their natal site? Do the young, when they grow up, prefer to remain and breed in their natal site?

Natal site fidelity or 'philopatry' – the tendency to remain in, or to return for breeding purposes to the close vicinity of the site where they were born – is very common among birds and mammals.[41] Some of the most spectacular examples come from migratory species. We have already mentioned the amazing site fidelity of salmon, who return to breed in their natal stream. Similar loyalty is shown by northern fur seals who, after many months and several thousand kilometres of sea journeys, return to breed at their own specific birth places on North Pacific islands. Many collared flycatchers and colonially nesting common terns return to Europe after an extended journey to Africa, and reoccupy their former natal territories. Massachusetts common terns, returning from their Argentinean wintering range, behave similarly. But philopatry is not limited to migratory species: it is also practised by members of non-migratory species such as great tits and kangaroo rats. About 25 per cent of great tit males and 10 per cent of the females establish their first territory on or adjacent to their natal one. For herring gulls the numbers returning to their natal colony are 77 per cent of males and 54 per cent of females.[42] In these cases, the figures represent just the individuals who have been able to establish themselves near or on their natal site; it underestimates, to an unknown degree, their real overall preference, since the number that failed in their attempt to breed at their natal site is not known.

The saying, 'although you may roam, there is no place like home'[43] seems to carry with it a biological message, relevant for both birds and humans. Every individual is an expert about its own natal site: an expert on the site's best nesting places and foods, the best feeding grounds and feeding methods, its unique aggregate of predators, hide-outs and escape routes, its community of pathogens and a lot more. Much of the detailed knowledge accumulated by youngsters is socially acquired from their experienced parents, and is therefore site-specific rather than merely species-specific. The preference for natal sites is often acquired in early childhood: in collared flycatchers, cross-fostering between parents from different habitats showed that the offspring prefer the site of their adopting parents to that of their biological parents.[44] The advantages of living where you grew up are illustrated by studies showing that in collared flycatchers, sparrowhawks, great tits, lions and kangaroo rats, philopatric individuals are better survivors and reproducers than non-philopatric individuals.[45] Pärt's long-term studies of collared fly-catchers breeding on the Baltic island of Gotland provide a good example. Pärt used several measures to compare the mating success of philopatric and immigrant yearling males. He found that philopatric males mated earlier in the breeding season, occupied higher-quality artificial nest sites, and were less likely (5% versus 20%) to fail to attract a mate than immigrant males. Since nest quality is known to be a good indicator of mating success, Pärt concluded that the enhanced mating success of philopatric males reflects their familiarity with the site, which enabled them to choose the best nesting places.[46]

Yet, although natal sites are apparently superior, not all individuals come back to them to live and breed. There may be several reasons for this. First, remaining or returning to the natal site is often either unprofitable or difficult. It may be unprofitable because of the severe competition with parents, siblings and other individuals, and in migratory species it may be difficult because of the need to find their natal site. Second, a change in the environmental conditions at the natal site can make returning to it disadvantageous. Another site may offer better opportunities of finding mate or food. A third reason, which may explain why dispersal is sometimes favoured, is the conflict between the bene-fits of philopatry and the deleterious genetic effects of inbreeding.[47] If the young always tended to stay in their natal site, the resulting inbreed-ing could be detrimental. There are therefore both benefits and costs to philopatry and dispersal, which vary with the species, population and individual. The balance between costs and benefits determines whether it is better for an individual to disperse or to stay in its natal site.

The costs and benefits of philopatry must also depend on the nature of the habitat in which the young have been reared. In a habitat where most sites are ecologically similar, the cost of dispersal is low, since the disperser's store of knowledge can be useful in a new similar site. The benefits of dispersal will be increased if the natal site is overpopulated. However, dispersal may be far less profitable for species living in a more heterogeneous habitat, where adjacent sites may be so different that the individual's knowledge is of limited use anywhere other than at its own natal site. If and when a disperser faces a choice between sites, it always chooses the one that is most similar to its natal site. This is even true for the sites that inter-continental bird migrants choose as their rest and refuelling stations.[48]

Parentally transmitted information about the natal site is sometimes combined with even more extreme measures of ensuring the offspring's success – the transmission of the real estate itself. In species such as deer-mice, red squirrels, barnacle geese and many primates, the parents transfer to some of their young their territory or grazing ground.[49] In primates the sharing of the natal territory with some of the offspring (often the females) is associated with the transfer of the mother's social status.[50] These non-genetic, intergenerational ecological legacies are enormously important in the organisation of animal societies and in their evolution, and we shall return to them in later chapters.

In the natal site, the animals can usually express the entire behav-ioural repertoire the parents transferred to them: the preference for the local food items, methods of foraging, ways of avoiding the local pred-ators and so on. These diverse habits form a coherent life style. In the same way that the transmission of the same information through many channels enhances its chances of being successfully consolidated and transferred, so transmitted habits that are all related to a particular style of life are mutually reinforcing. The freedom of any particular behaviour pattern to vary is decreased. How stable a particular pattern of behaviour is over many generations depends on the other behaviours the individuals manifest. Examining the stability of a single learnt pattern of behaviour over several generations in laboratory conditions, in isolation from other behaviours, may seriously underestimate the stability of its transmission in natural conditions.

Choosing a mate – the transmission of sexual tastes

The future reproductive success of any young bird or mammal depends on its future mate's genetic and non-genetic qualities, as well as its own. The ability to discriminate and choose a suitable high-quality mate is therefore of great importance. But how does an individual assess the quality of potential mates? The answer seems to be that individuals use a combination of specific phenotypic traits that act as quality indicators. A phenotypic trait is a good predictor of the quality of a potential mate if the way it varies is highly correlated either with the candidate's genetic quality, or with its non-genetic condition. We have seen this principle at work in the previous chapter, when we described the sexual display of male bower-birds, building bowers to attract females. The quality, organisation and diversity of materials used by a male in building and adorning its bower are the clues used by the female to assess his quality. Any male who can build and defend an elaborate, colourful bower, is strong, healthy and cunning.

The same principle can be illustrated by the variations in the brightness of plumage and the pattern of song seen in any male songbird. During the breeding season, male songbirds use both their bright plumage and their song to attract prospective mates, as well as to drive away male opponents. But, as the Israeli ethologist Amotz Zahavi realised, the conspicuous colours and the loud songs are expensive: they use a lot of energy and expose the males to predators.[51] They are extremely costly advertisements – Zahavi calls them 'handicaps', meaning that in the short term the animal would be better off without them. However, because they are truly expensive, such advertisements cannot be successfully used by cheaters, by weaklings who cannot afford them. They are therefore honest indicators of male quality. Indeed, the brightness of male plumage has been found to be positively correlated with the level of testosterone, which is a good indicator of fertility; it is negatively correlated with the male's degree of infestation by parasites.[52] It is therefore in the female's interest to find the brightest male (in every sense!). Similarly, if a male songbird, for example a male great tit perched on a treetop, sings his song at top volume for hours on end, any listening female can be sure that he has surplus reserves of energy, as well as the proven ability to evade predators. He would be a good choice for a mate. The size of his song repertoire is another sign of a male's quality, an indicator of his intelligence and learning ability. The costly advertisements pay off in the long term, because the 'handicapped' male gets the mate.

Bright colours and marathon singing bouts are excellent general indicators of the quality of a male songbird. However, these are not the only clues used, since they are insufficient indicators of all the qualities required in a mate. A male commonly contributes more to the partnership than his genes. In over 90 per cent of songbird species, the male helps to rear the young, so the female has to assess not only his vitality, but also his qualities as a parent and provider. For example, the size and quality of the male's territory are very important. Are there well-protected nesting sites? Is there enough of the right kinds of food?

What a particular female judges to be a good territory for herself and her future offspring may be strongly influenced by her experience at her own natal site. Similarly, her preference for a particular song-style may be influenced by the songs she heard as a fledgling, for example, the songs of her father and other neighbouring mature males.[53] Young brown-headed cowbirds, for example, tend to pair and mate with individuals that belong to their own 'vocal culture', that is, individuals who share the same population-specific calls and preference for these calls.[54] The individual preferences of mature females may therefore differ, with one female's dream mate being another's nightmare. It seems that finding the right mate is not easy for female birds. Why else should each female pied flycatcher visit up to nine singing males before she makes up her mind?[55] The driving force behind a female's choosiness is to find a vigorous mate who also owns a suitable territory and displays what she judges to be attractive behaviour. Because of her preference for a site similar to her own natal site, as well as her preference for a song similar to the one she heard there, her chosen male may often come from an ecological background similar to her own. His similar upbringing makes him, for her, an attractive and possibly also a particularly good partner for a mutual young-rearing enterprise.

In songbirds, even listening from a distance may enable a female to identify the degree of similarity in ecological background between herself and the male singer. The physical structure of a male's song often reflects the nature of the singer's natal environment.[56] The reason for this is that the way sound travels and its perceived quality are affected by plant density, temperature and humidity, and sites can be very dissimilar in these respects. Since the production of the song is energetically costly and the message it carries over distance must be clear, males reared in sites where sound does not carry well learn to enhance the transmissibility of their song. They change some of its physical components, such as frequency, thus optimising their energy expenditure on singing. Consequently, adjacent males reared in ecologically dissimilar sites may have differently structured songs. A listening female may be able to assess the degree of similarity in ecological background between herself and a singer by comparing his song with her memorised version of her father's song.

How do parents help their young acquire the criteria for choosing a good mate? They will influence the future mate choice of their offspring if they ensure that, during the critical period in early life, the offspring become sexually imprinted on them. A female bird or mammal who was sexually imprinted on her father will use his appearance, sounds and smells as a yardstick, and prefer mates who are similar to him. A mature male, previously imprinted on its mother (an Oedipal male), may prefer mates who are similar to his mother. Such sexual imprinting is being found for a steadily growing number of bird species, including ducks, geese, pigeons, doves and finches.[57] Young male mammals also sexually imprint on their mothers. Cross-fostering between sheep and goats immediately after birth has shown that the young males become sexually and socially imprinted on their foster mothers, and that this preference persists even if they grow up with other individuals belonging to their own species and have not seen the species of their foster mother for three years.[58] In goats and sheep, fathers do not care for their young, so it is not surprising that young females, who have no intimate relationship with adult males, are not imprinted on them. The imprinting of young females on their mothers is merely social, and is therefore easier to change.

Cross-fostering experiments in natural conditions show that variations within a species can affect sexual imprinting. In the snow goose, a series of experimental studies made by Fred Cooke and his associates proved that goslings are imprinted on the plumage colour of their parents. These colours can take any shade along the white–blue continuum. When the goslings mature, they strongly prefer to mate with a bird with a plumage colour similar to that of their parents. The differently coloured types are also slightly different in their breeding biology – the white type is better adapted to rigorous environments, whereas the blue type is more successful in warmer conditions. By choosing a mate similar to their parents, the young are choosing a mate adapted to the same environment as that in which they are likely to live.[59] It seems that processes of sexual imprinting and other forms of early learning supply the young with the information they need to identify high-quality and ecologically compatible mates.

Yet, if parental sexual imprinting was very rigid and precise, it could have damaging effects. An imprinted preference directed exclusively towards parents or siblings could lead to excessive inbreeding, which often has deleterious consequences. Patrick Bateson has investigated the effects of sexual imprinting in Japanese quails that were reared with their siblings, and hence sexually imprinted on them. He showed that the quails prefer as mates individuals who are similar but not identical to their sibs. First cousins were preferred over both sibs and non-relatives.[60] This preference for an 'intermediately similar' individual seems to be a good compromise, since it ensures that, on the one hand, there is a good chance of ecological similarity between the mates, and, on the other hand, the problem of inbreeding is avoided. However, as Bateson realised, there was a snag in these experiments: the laboratory quails were highly inbred, so the phenotypic similarity between the sibs and the first cousins was very high. In natural conditions, where inbreeding is less common, the preferred mate may be relatively more closely similar to the imprinting individual (usually the parent) than to anyone else.

Experiments such as Bateson's show that sexual imprinting has more than one function. Certainly, as was suggested many years ago, it helps the youngster with its future identification of members of its own species, and enables it to find a suitable mate. But, in addition, it ensures a higher than average amount of phenotypic similarity between mates and tends to promote incest. Phenotypic similarity may be important in some cases of long-term monogamy, because similar individuals may be well suited both to the site and to each other. High ecological compatibility may ensure that there is good co-ordination between them. But regular incest may lead to all the deleterious effects of inbreeding, and may therefore be countered by selection for some kind of 'negative' imprinting. In their extensive study of Darwin's finches in one of the Galapagos islands, Grant and Grant found that females of both the medium ground finch and the cactus finch need to hear their father's song in order to be able to recognise a mate of the same species, but they prefer males with a song somewhat different from that of their father (and therefore from their brothers, who inherit the father's song).[61] In this way, inbreeding is avoided, which is an important asset in this small island population, where dispersal is limited and ecological compatibility is ensured anyway. It seems that the particular 'sexual tastes' transmitted by the parents have significant effects on the structure and evolution of populations, and may be tailored to the special needs of the individuals within the social and ecological system.

A fascinating case of sexual imprinting has been studied in a group of parasitic African weaver-birds, commonly called whydahs or widow-birds.[62] Female whydahs lay their eggs in the nests of species belonging to the Estrilidine finch sub-family. Each whydah species parasitises one particular species or subspecies of finch. The nestlings of the parasite and the host are very similar in appearance and, most importantly, the parasite nestling has the same species-specific mouth markings as the host nestling, which induces the finch parents to feed the young. A nestling that had deviant markings would not be fed and would die of starvation. There is therefore very strong selection for the parasite to faithfully mimic the host's markings, and very strong selection against any mating that would disrupt the mimic's genotype and lead to imprecise mouth markings. It is the strong imprinting of the parasites on their foster-parents that ensures matings are appropriate. Young female parasite nestlings recognise the host species as parents, form a search image of their appearance, and, when they are ready to lay, look for this host species and lay their eggs in the host's nests. A female whydah becomes reproductively active and ovulates only when she sees the reproductive activity of members of her particular host species. But how does she find a compatible mate, one who was reared by the same parent-species and therefore is certain to have the genotype that will produce the right mouth markings? It turns out that the song of the whydah males is made up of two major parts, one of which contains all the vocalisations of the host. This part of the song attracts a female reared by the same species because, as a nestling, the female became imprinted on her foster-father's song. Sexual imprinting on the rearing father is an integral part of host imprinting in these species.

After a surge of studies on the evolutionary significance of sexual imprinting in the 1970s, interest in the subject declined, but it is now increasing again as studies of social learning in animals become more fashionable. Theoretical work shows that imprinting can have many evolutionary effects, sometimes speeding up the rate of adaptation and speciation.[63] The work on sexual imprinting provides an excellent and non-controversial example of how evolution can be affected by the transmission of non-genetic variation.

The transmission of the transmission mechanism

We have looked at various types of information that parents transmit to their offspring. We can now ask why early learning and parental imprinting are there in the first place, and what maintains these forms of learning in the population. Some of the learning from parents is probably an inevitable consequence of the memory and the learning ability of the young animals. The evolution of more efficient mechanisms of parental transmission can be understood if we recall the many benefits that early, socially transmitted patterns of behaviour confer on vulnerable and naïve youngsters. Any mechanism that promotes the reliable transmission of parental behaviour is likely to be selected. Selection can operate both on accidental mutations in genes, and on any socially learnt and transmitted modifications in parental care that lead to more effective learning in the young.

A gene in the offspring that makes it more receptive to information transmitted by the mother, or a gene in a mother that makes her style of maternal care more influential, will readily be selected if advantageous. But can we think about the evolution of maternal care and maternal influence in terms of the selection of transmissible behaviour patterns rather than the selection of genes? We believe that we can and should. Imagine, for example, a slightly nervous mother monkey who prevents her offspring from eating a mildly toxic food item by forcefully moving them away from it, and creating a lot of fuss. This 'overprotective', assertive, maternal behaviour becomes associated with the adverse effect of the already tasted toxic food. Such a newly invented 'trick', which helps the mother to engrave the avoidance of the potentially toxic foods in the youngster's brain, may make the offspring more likely to repeat the behaviour when they themselves are adults. In time, her daughters will be as overprotective towards their young as she was towards them, and the overprotective maternal style will be reproduced in her lineage. If the behaviour is beneficial, it will also spread in the population at large.

In animals in which the care of young, and therefore their imprinting, is solely maternal, it is easy to see how more efficient early learning and associated patterns of behaviour can spread in the population, even if they have no beneficial effects on the young and their mother. Imagine a population where there is some, but not much, early learning. A variant female, with a 'Jewish mother' behaviour pattern, appears in this population. Slightly altering the previous example, imagine that this female, unlike other females, is fussy over her offspring's eating habits. She reacts violently towards her young if they eat foods which she finds suspicious (not necessarily correctly!), feeds them for a long time with her own repertoire of foods, and shows signs of great satisfaction when they eat them. As a result of the combination of repeated fuss, and both negative and positive reinforcement, the avoidance of the suspicious foods and the preference for her own preferred foods are firmly installed in her youngsters' brains. Whenever they smell the suspicious foods they have a violent aversive reaction, and they continue to avoid these foods as adults and eat the more restricted maternal diet. When the daughters mature, they adopt their mother's behaviour when their own offspring attempt to taste new foods, and behave very much like her. The 'Jewish mother' has transmitted to her daughters both new feeding habits and a new style of maternal care. Studies, mainly on primate mothers, have demonstrated that maternal styles do indeed often differ between individual mothers, and are transmitted to their daughters.[64]

In order to see how the new parenting style will spread, even if it has no selective advantage, we have to focus (in this mammalian case) on daughters, who are not only imprinted with their mother's various preferences and skills, but also transmit her maternal-care style; sons are irrelevant here, for, although they are imprinted, they cannot perpetuate her maternal-care style. Imagine that, whereas the female offspring of mothers who imprint less well have a certain chance (say 50%) of having the maternal-care style of their mother, the female offspring of the 'Jewish mother' have a higher probability (say 80%) of showing both her maternal-care style and the behaviour patterns socially transmitted through it. If the more transmissible 'Jewish mother' style does not have serious deleterious effects, it will spread in the female population very rapidly, in a manner similar to the spread of a gene causing parthenogenesis in a sexual population. A gene causing parthenogenesis makes a female clone herself in her offspring, and ensures 100 per cent transmission of itself. If the gene does not have serious detrimental effects on fitness, it will spread because it will be present in all, instead of only half, of her offspring. In a sense, a mother capable of transmitting her style of maternal care clones in her female offspring both her maternal-care style and the many patterns of behaviour transmitted by this style of behaviour.[65] The same type of argument is also valid when both parents care for the young. The spread of a parental-care style is always affected by its own transmissibility, even if the patterns of behaviour that are transmitted through it are somewhat detrimental.

Sexual imprinting presents a more complex situation. Through sexual imprinting, a young animal chooses a mate phenotypically similar to its parent. The chance that genes similar to the parent's, including the imprinting gene, will be chosen is also somewhat increased, because similar phenotypes may have genes in common. A father causing his daughters to mate with a male phenotypically similar to himself is increasing phenotypic homogeneity in his lineage. If the daughters mate with a male who also shares the father's imprinting tendency, the accumulation in the lineage of similar phenotypes associated with the imprinting gene will be further increased. This would lead to increased differentiation and divergence between different paternal lineages.

The evolution of the transmission of mechanisms of transmission is of central importance for the evolution of learning and behaviour. It is akin in importance to the evolution of parthenogenesis, obligate inbreeding or obligate outbreeding for the genetic system in living organisms. We shall therefore come back to this aspect of behavioural transmission in different contexts throughout this book.

Evolving family traditions

An environment may sometimes change, either gradually or suddenly, and the organisms living in it must either adapt to the new conditions, disperse or perish. What would happen, for example, if one day the suckling mouse doe discovered that there was no food in the warehouse? She would be mildly stressed, and would have no choice but to give up the family food tradition inherited from her mother and revert to finding a new source of food through the time-consuming practice of individual learning. She would have to search the surroundings until she found a good alternative – perhaps a store of cultivated wheat or barley seeds, or chicken-feed that she could steal from the dishes of unsuspecting hens. From then on, new information about sites, tastes and relative abundances of wheat, barley and chicken-feed, and the safe routes leading to and from them, would flow to her youngsters through her milk, saliva, breath and faeces, and, eventually, be communicated by direct demonstration. The new information would quantitatively override the old, now obsolete, information about an oatmeal-based diet. The youngsters would quickly adopt the up-to-date version of their mother's food habits, a version already adapted to the new, different set of ecological conditions.

The ability of parents to transmit to their offspring information about the use of a new source of food may become the key to the ecological success of a population, leading to the occupation of a new habitat and to the establishment of new habits. The fascinating research of Aisner and Terkel on pine-cone opening behaviour by Israeli black rats is a beautiful illustration of this point.[66] Israeli black rats have recently extended their range of habitats to include Jerusalem-pine forests. The sole possible source of rat food in this habitat is the large number of pine seeds, enclosed within pine cones, situated on the upper branches of the trees. The seeds can be obtained by using an elaborate pine-cone stripping technique. The results of a lengthy set of experiments have shown that black rat pups, but not adults, are able to learn the stripping behaviour by practising on partly stripped cones provided by their experienced mothers (or any other experienced individual). The ability to strip pine cones is the result of maternal transmission of behaviours, and not a novel genetic adaptation. Experimental cross-fostering of pups between experienced stripping mothers and 'naïve' (non-stripping) mothers showed that pups learn from a skilful mother, irrespective of genetic relations.

It is not clear how this behaviour became established, but it is not too difficult to imagine. It is reasonable to suppose that the new food source and the accompanying technique were found by a hunger-stressed female rat, living on the edge of the forest. In her search for food, she finally overcame her usual food and site neophobia, and ventured into the nearby pine forest. Switching to cautious individual trial-and-error learning about any object that smelt promising, she eventually found, probably not without some preliminary nauseating disappointments, a gastronomic gold mine of tasty pine seeds, hidden under a pine-cone cover. Trying to strip the pine cones as fast as possible, she gradually learnt a basic technique for achieving this goal. Possibly her first technique was simple, and not very effective. In due course, her pups learnt from her the new pine-cone stripping technique. The technique was perfected in the subsequent generations, and the efficiency and speed of the stripping increased. Natural selection of more efficient patterns of stripping behaviour ensured the spread of better techniques in the population, and social transmission ensured their stable inheritance within the maternal lineages.

This is, of course, an entirely imaginary scenario, for we have no idea how the pine forests became occupied by black rats. However, what is clear is that the black rats studied by Terkel and Aisner were very lucky. Jerusalem-pine forests are free of squirrels, so the invading rats found an ecologically vacant food niche, and did not have to face competition from squirrels over the limited supply of pine cones. Such competition would have limited the prospects of ecological success. In both genetical and behavioural evolution, luck is sometimes an important component of local success.

The cultural transmission of a new food preference and of a new feeding technique illustrates some of the distinctive aspects of behavioural and cultural evolution discussed in the previous chapter. First, it is clear that the transmission of new patterns of behaviour can lead to a new habit being established in a population. It is also not very difficult to see how the new habit can evolve by selection for increasingly more elaborate and efficient techniques derived from a relatively crude ancestral one. Furthermore, the habit is not isolated from the encompassing web of relations with the environment in which the animal lives. Other new habits that are beneficial in the new surroundings must become established, and old ones must be changed or abandoned; the new habit indirectly causes the selection of other behavioural variants, and may alter the general behavioural profile of the animals. The new technique of cone stripping has led to new food and site preferences, to a slightly altered physiology of digestion, and possibly also to new techniques of nest building and new strategies for avoiding a different spectrum of predators. All these interconnected new traits and patterns of behaviour are likely to spread in the population of pine forest rats by social learning. Useful, easily remembered and easily transmitted behaviours are likely to become the new norms of behaviour in the population. However, because of the common biases in behavioural transmission, it is quite likely that the newly established techniques are not optimal ones in the engineering sense. If a certain technique is more easily remembered than an alternative, it may well spread in the population even if it is somewhat inferior in efficiency or speed to the alternative technique. Whether and how often a particular technique of foraging and the associated food preference are established will depend on the habit's selective advantage, and on the ease with which it is acquired, remembered and transmitted.

There are both similarities and differences between the fate of genetic and of socially transmitted adaptations when the environment changes. When conditions alter, the previously genetically adapted individuals have to rely on other genetic adaptations or on phenotypic plasticity in order to deal with the new environment. Similarly, socially transmitted adaptations have to be modified, or other adaptations used, when the environment changes. However, it takes many generations for genetic variants to disappear from a population, and if the previous environmental conditions reoccur, individuals who have retained the old genetic make-up will easily readapt. In contrast, the persistence of the socially acquired adaptations depends on the continued performance of the actual behaviour, and not on stable, underlying, genetic resources. This seems to mean that, after one or two generations, a change in the environment that results in a change in socially transmitted behaviour will lead to the final disappearance of the old behaviour. All the advantages accrued through the natural selection of the previous socially transmitted behavioural adaptations are thus lost.

However, for such a loss to occur, the environmental change would have to be very drastic. It would need to be not only very rapid, so that no adjustments of the established behavioural adaptations are possible, but also quite wide-ranging. Think about the black rats in the Jerusalem-pine forest: for the stripping technique and the life style associated with it to disappear, it would be necessary for not only the pine forests that the rats presently inhabit to be destroyed, but all adjacent pine forests as well, so that rats could not find a similar alternative habitat. Such drastic and wide-ranging catastrophes would almost certainly lead to the extinction of the local population, with both its non-genetic and genetic adaptations. If the change was not drastic, social transmission of behavioural modifications that compensate for the change might preserve some of the previous achievements. For example, a deterioration in the population of pine trees might lead to modifications in behaviour that enable the black rats to find the hidden seeds of other locally abundant but previously unused plants.

There is no doubt that the requirement that behaviour has to be displayed in order to be transmitted diminishes the potential persistence of socially transmitted adaptations. Can the cumulative effects of selection of a socially transmitted adaptation nevertheless be preserved? It is reasonable to assume that they can. Think what would happen if selection acted in the same direction for a long time. Assume, for example, that the pine forest environment persists and the population of black rats survives and prospers. Cumulative evolution of socially transmitted behavioural adaptations will occur. The population may well colonise new similar habitats, and, if it does, the adaptations are unlikely to be lost unless there is a wide-ranging ecological catastrophe. The cumulative effects of natural selection can also be preserved if the adaptation is advantageous under a wide range of conditions. Imagine, for example, that an animal learns to build a crude nest for its offspring. In most conditions of life, building a nest will be advantageous over no nest at all, since it offers more protection and many other benefits to the young and to their parents. Up to a point, which is determined by the learning ability of the young of the species, cumulative evolution of nest building through the natural selection of socially transmitted behaviours will occur. Other complex behaviours, like excavating burrows, or using a sequence of signs for communication with family members or for sexual display, can also initially evolve through the natural selection of behaviourally transmitted patterns of behaviour.

If there is long-term selection in a particular direction, as in the cases just described, some of the control of the behaviour will probably be transferred to the genetic inheritance system. This would happen through straightforward Darwinian selection: there would be gradual selection of the combinations of genes which lead to increasingly more rapid learning, depending on fewer and fewer trials. The result would be an increase in the relative importance of the innate component of the behaviour.[67]

The parental transmission of behaviour patterns has been described here as a fairly straightforward process. However, we have hinted at some possible complications: the interests of mates may clash, parents and offspring may have conflicting interests, and siblings may compete with each other. In the next two chapters, we shall look more closely at these complications and at the effects of social learning on their resolutions.

Summary

There is almost no aspect of a young animal's development that is not profoundly affected by the phenotype of its parents. At the earliest stages, the development of the young is affected by the quality of the mother's contribution to the egg. If the mother invests further in her offspring, they will be influenced by her physiological state and by the quality and nature of her care. A father may also have important effects, especially if he provides the offspring with care, but also if he affects the environment in which they develop. All these parental effects are influenced both by the genotypes of the parents and by the environmental effects on the parents' phenotypes. Frequently they lead to similarity between parents and offspring.

The effects of parentally transmitted, behaviour-affecting information on the survival and reproduction of offspring has been a neglected aspect of evolutionary biology. Yet, in birds and mammals, learning from parents is ubiquitous and essential. Information can flow from parents to offspring through multiple and complementary channels – through the yolk and placenta, through milk and faeces, through incidental and deliberate demonstrations of behaviour. The information acquired from parents affects all spheres of life, from foraging and avoiding predators to selecting a home, from resistance to parasites to caring for the young. Variations in the patterns of behaviour that the young acquire from their parents can often be transmitted to subsequent generations. Family and local traditions can be formed, as pine-cone stripping Israeli black rats demonstrate. For many behavioural traits in birds and mammals, failing to take into account non-genetic variations that are transmitted by the parents misses the most influential factor affecting the frequency of these traits in the population.

Notes

[1] This account is based on observations of village-dwelling domestic mice in the Judaean hills near Jerusalem made by E. A. during 1976–78, and information in Berry, 1981; Crowcroft, 1966; Crowcroft & Rowe, 1963; Meehan, 1984.

[2] The female mouse in this case cared for her young by herself. Domestic mice occasionally display co-operative parental care, with two or even three females, usually sisters from the same litter, giving birth together and caring for their offspring together. See König, 1993; 1994.

[3] Kirkpatrick & Lande, 1989.

4 We discuss the interaction of genetic parental effects and environmentally induced parental effects in chapter 9.

5 In chapter 1 we painted a scenario in which two lineages of imaginary, genetically identical tarbutniks diverged following environmentally induced differences in their behaviour. Parents transmitted their behaviour to offspring and others through social learning. In the real world, it is unlikely that individuals will be genetically identical, except as a result of twinning, but close genetic similarity is possible if there has been a lengthy history of inbreeding, i.e. mating between close relatives.

6 Lacey (1998) distinguished between three categories of induced parental effects, with the first leading to effects that do not necessarily lead to similarity between parents and offspring, and the second and third categories leading to transgenerational transmission of parental phenotypes. The difference between the second and third category is that in the second category parental transmission leads directly to offspring similarity, whereas in the third category parental transmission leads to the offspring's reconstruction of the environment in a way that leads to the emergence of offspring phenotypes that are similar to those of the parents. We consider the third category as a special case of the second.

7 Roosenburg & Niewiarowski, 1998.

8 These issues are reviewed in Mousseau & Fox, 1998.

9 This definition is based on Rosenblatt, 1987, p. 362.

10 Clutton-Brock, 1991.

11 Alcock, 1993; Pianka, 1994.

12 Clutton-Brock, 1991.

13 Some notable examples of social learning, including learning migration routes and mobbing, have been omitted here. Since parents as well as non-parents contribute to the learning of these behaviours, we shall discuss them in chapter 8, along with other socially transmitted group behaviours.

14 Many examples can be found in Hughes, 1993, and references therein.

15 See Galef & Sherry, 1973; Hepper, 1988; Provenza & Balph, 1987. Unfortunately, little research of this type has been done on mammals other than rodents and ruminants.

16 Provenza & Cincotta, 1993.

17 Smotheran, 1982.

18 Beer & Billingham, 1976; Gill, 1988.

19 Lung et al., 1996; Smith et al., 1994.

20 Janeway & Travers, 1994. See also references in note 18.

21 Janeway & Travers, 1994.

22 For information on ulcers, see Skolnick et al., 1980; Huck, Labov & Lisk (1987) looked at parental effects on growth and sex ratio in hamsters, and Clark, Karpiuk & Galef (1993) discuss parental effects on sex ratio and gender behaviour in Mongolian gerbils. For a brief review of prenatal influences on sex ratio and reproductive behaviour, see Clark & Galef, 1995.

[23] Ewer, 1968, p. 274.

[24] Marinier & Alexander, 1995.

[25] See, for example, Galef & Beck, 1990, and references therein.

[26] Laland & Plotkin, 1993.

[27] Lorenz, 1978.

[28] Apfelbach, 1973.

[29] Ewer, 1968, pp. 277–8.

[30] Kitchener, 1991; Leyhausen, 1979.

[31] Caro, 1994; Schaller, 1967.

[32] Estes 1991; Macdonald, 1984.

[33] O'Connor, 1991.

[34] Clayton & Wolfe, 1993; Lozano, 1998.

[35] Johns & Duquette, 1991; Kreulen, 1985.

[36] Clark & Mason, 1985, 1988.

[37] Learning to forage for medications could affect an animal's ability to find new sources of food. If an animal is able to cure any ill-effects of a newly tasted food by applying self-medication, it is more likely to try out new foods and may even be able to add foods with potentially adverse effects to its diet. Animals such as the European hedgehog can explore a very large variety of potential foods because of their genetically based immunity to poisons; others may do the same because they have innate or socially learnt self-medication behaviour.

[38] Many birds have two major classes of alarm calls, one for airborne predators and another for ground predators; see Marler & Evans, 1994. See also Leger, Owings & Gelfand, 1980, for alarm calls of California ground squirrels; Slobodchikoff *et al.*, 1991, for the alarm calls of Gunnison's prairie dogs.

[39] Cheney & Seyfarth, 1990.

[40] Great tits have varied foraging methods, adjusted to their habitats. While feeding on the ground is relatively uncommon in most places in England, where sparrowhawks are common predators, in Israel, where this predator is rare, feeding on the ground is quite common. For details of locality-specific adjustments in the alarm calls of other species of birds see Gyger, Marler & Pickert, 1987.

[41] Greenwood, 1980.

[42] *Ibid.*

[43] This saying is based on a song from an opera of 1823, *Clari, the Maid of Milan*, by John Howard Payne. The original is: 'Mid pleasures and palaces though we may roam, Be it ever so humble, there's no place like home.' It is not too surprising that this original 'meme' did not last.

[44] The work on collared flycatchers by Löhrl, 1959, is summarised in Welty & Baptista, 1988, pp. 255–7.

[45] Jones, 1986; Newton & Marquiss, 1983; Pärt, 1991; Pusey & Packer, 1987.

[46] Pärt, 1994.

[47] It is a common (but not universal) finding that the offspring produced after

several generations of mating between close relatives are less viable and/or fertile. The reasons for such 'inbreeding depression' are not fully understood, but, since close relatives are likely to carry identical alleles for many genes, a major contributory factor is probably that inbred animals are homozygous for detrimental recessive alleles.

[48] Baker, 1978.

[49] Harris & Murie, 1984; Larsson & Forslund, 1992; Price & Boutin, 1993; Woolfenden & Fitzpatrick, 1978.

[50] Chapais, 1992.

[51] Zahavi, 1975; Zahavi & Zahavi, 1997.

[52] Hamilton & Zuk, 1982; Zuk, 1992.

[53] Catchpole & Slater, 1995.

[54] Freeberg, 1998.

[55] Dale *et al.*, 1990.

[56] Catchpole & Slater, 1995.

[57] McFarland, 1987; ten Cate & Vos, 1999.

[58] Kendrick *et al.*, 1998. It would be interesting to follow the maternal style of the fostered female offspring when they become mothers.

[59] Cooke, Mirsky & Seiger, 1972.

[60] Bateson, 1982.

[61] Grant & Grant, 1996.

[62] Nicolai, 1964, 1974. We shall come back to this case in chapter 8 when we discuss the speciation of hosts and parasites.

[63] Laland, 1994*a,b*.

[64] Berman, 1990, 1997. See also the review by Fairbanks, 1996.

[65] Avital & Jablonka, 1994.

[66] Aisner & Terkel, 1992.

[67] We return to this topic in chapter 9, where we discuss the co-evolution of genes and socially learnt habits.

5   Achieving harmony between mates –
the learning route

The mother-mouse portrayed in the previous chapter worked hard to rear her offspring, providing them with all the essentials: with food and warmth, with information and with security. As a typical mammalian single mother, she was not assisted in her labours, and hence did not enjoy the increased reproductive success that the help of another individual, such as her mate, might bring. But in some species of mammals and most birds, the mother is not the only caregiver; frequently the father participates in parental care and contributes to the offspring's 'education'. Paternal involvement is not without complications, however, and sometimes there are conflicts between the parents over who should care for the youngsters, how much care should be given and for how long. Mates may also disagree over copulation frequency, fidelity and the level of commitment to the relationship. Indeed, our everyday experience of the relationships between human mates, as well as observations of monogamous birds and mammals, testify to frequent disagreements. The great Scandinavian playwright August Strindberg, one of the most bitter and eloquent writers on the struggle between the sexes, described the conflict between human males and females as being as old as sex itself and fundamentally insoluble. But what does this ancient conflict mean for biologists? Can we interpret family disputes as a reflection of conflicting evolutionary interests? How is the regular and often spectacular co-operation between mates achieved?

In this chapter we are going to discuss co-operation, especially long-term co-operation, between mates. We will argue that there are aspects of this co-operation that cannot be fully explained in terms of 'conflict theory' – as the results of an evolutionary compromise between the conflicting interests of males and females. We believe that learning about and with the mate is an essential part of long-term partnerships, and that such learning introduces additional factors that need to be incorporated into evolutionary explanations of the stability of co-operative relationships.

Male–female conflict

The standard approach to the evolution of social interactions among family members is to interpret them in terms of underlying evolutionary conflicts.[1] In the same way that we talk about the evolution of predator–prey or host–parasite interactions in terms of the conflicting interests of the organisms involved, we can talk about the evolution of interactions between family members in terms of the conflicts stemming from their different fitness interests. At the genetic level, conflicts lead to arms-races: the spread of genes that increase the fitness of one type leads to the selection in the other type of genes that counteract the effects of those in the first. This evolutionary arms-race results in conflicting 'behavioural strategies', as the sets of behaviours that are shaped by natural selection are called.

Here we are concerned with how the behaviours of interacting males and females have been explained as products of evolutionary conflict. The conflict stems from the selection in each sex of behaviours that lead to the production of as many offspring that survive and breed as possible. In other words, conflict stems from the interest of the members of each sex in maximising their own fitnesses.[2] The implementation of these fitness interests is shaped by the different reproductive biology of males and females, and this is what leads to conflicts. A male bird or mammal can (theoretically) produce many more offspring than a female, since he is limited only by the number of accessible females willing to copulate with him, his stamina and his sperm reserves. The number of offspring a female can produce is much more limited, but a female can be much more certain than the male that the offspring she tends are her own.[3] When a male and a female form a partnership that goes beyond mere copulation and fertilisation, their reproductive differences do not go away, and the way these differences translate into behaviours may lead to conflicts. For example, the male may try to copulate with as many additional females as he can, and leave most of the parental chores to the female. The female may evolve behaviours that minimise the possibility of such 'selfish' male behaviour, and make it difficult for the male to spread his parental efforts at her expense. On their part, females may evolve behaviours that are to their benefit, such as being extremely choosy about their mate and, like the male, being as lazy as possible, thus enhancing their own survival and chances of breeding in the next season. Obviously, male behaviours evolve that minimise the reduction in fitness that such 'selfish' female behaviours could cause them. When translated into genetic and evolutionary terms, this means that genotypes promoting choosiness, laziness, early desertion or infidelity in one mate, lead to the evolution of counter-measures in the other mate that oppose these tendencies and ensure that the mate works hard and is faithful.

The consequences of male–female evolutionary conflict are seen most clearly in those species in which the male merely inseminates the female, donating to her no more than his sperm. Since in these cases the male and the female hardly interact beyond a brief copulation, there is little chance for overt quarrels and disputes. Nevertheless, in such polygynous species, many aspects of male and female morphology and behaviour can be seen as the outcome of evolutionary conflict, with each sex trying to 'outwit' the other using evolved strategies that are to its own benefit, even if they are to the detriment of the other sex. There is an arms-race, with the evolution of ploys and counter ploys. The strategies produced by these arms-races can be seen in many mammals, where usually the male does not care for the young. The male strategy is to try to father as many offspring as possible by inseminating as many females as he can with his abundant, readily available, sperm. The fathering possibilities of such a successful male are huge, as the harems of elephant seals testify. A successful male elephant seal can readily father fifty offspring per year.[4] Of course, the extraordinary success of such a harem-holder is very much at the expense of other males, who must remain celibate. Not surprisingly, there is fierce competition among the males for sexual access to fertile females.

The female partner in this system has rather different interests: she is much more limited in the number of offspring that she can parent, not least because she has to produce energetically expensive embryos and care for them single-handed.[5] Hence, the female is concerned with the quality rather than the mere quantity of her offspring. She wants the best genetic father for them. She is therefore very choosy, scrutinising males for markers of genetic quality. A mate is not just any male with some sperm – it must be the best carrier of good genes that she can get, and she is expected to do her best to ensure that she gets such a mate. In the previous chapter, we described how in many polygynous species a female looks for a mate with the most extravagant traits – the most colourful male, the biggest, the strongest, the one presenting the most elaborate and demanding ritual. In some polygynous birds, including the peacock and a variety of game birds and waders, males gather together in advertising congregations called 'leks', where they show off, and females come to look them over, compare them and choose the best male. When the female preference is not innate, and discernible markers of male quality change with circumstances, inexperienced young females may face a problem: they cannot easily tell who is the superior male. In such cases, young females seem to observe the behaviour of older and more experienced females, and copy their choices.[6] By copying, young hens probably make the most successful choice of partner. The consequence for males is that an already successful male becomes increasingly more popular!

The male-quality criteria that are important for females vary, but in all cases they should indicate the male's genetic quality. The peahen chooses the peacock with the largest and most colourful tail, and middle-aged males, who have proved their ability to survive, are often preferred to young, inexperienced ones. The brightness of the male's plumage reflects his health, since parasite-infested males tend to have dull plumage. Plumage therefore serves as a good indicator of a male's genetically based immunological ability. A dull (and therefore unhealthy) male would be a poor choice of mate because, as well as contributing a poor set of genes, he might harm either the female or her offspring by transferring parasites during copulation. The brightness of the plumage is also an index of fertility, because it is often correlated with high concentrations of the male hormone, testosterone.[7] In addition, as Zahavi has suggested, when brighter plumage makes the male more conspicuous to predators, it serves as an honest advertisement of his quality, because it shows he has the ability to survive in spite of his provocative appearance.[8]

Female choice means more than assessing the quality of the male from his appearance and behaviour. Attractiveness *per se* is also important. As Fisher pointed out many years ago, a female who chooses a male considered to be attractive will probably have attractive sons. These sons will attract and mate with many females, and consequently produce many grandchildren. If attractiveness does not reduce the survival of the males too much, then in each generation the most attractive males, and the females most susceptible to their charms, will be selected. This can sometimes lead to evolutionary escalation – to exaggerated female preferences and inflated male traits that have more to do with fashion than with quality.[9] The magnificent tail of the male peacock and the complex and lengthy song of the nightingale are famous examples of traits that probably evolved both as extreme advertisements of genetic worth and as escalated fashions. It seems that both good sense and fashionable strictures drive female preferences.

Female choice is clearly multifaceted and includes many exacting and sometimes unexpected ways of assessing male quality. We are only beginning to be aware of the complex criteria that females employ for choosing the 'best' mates. For example, sometimes the quality of the males' sperm seems to be tested by the female. In many species, females copulate with several males in rapid succession. The males' sperm are therefore subject to ruthless competition within the females' reproductive tracts. The winners are the ones that fertilise the females' eggs.[10] If the fitness of the offspring is positively correlated with the fertilising success of a package of sperm (an assumption that is only partly substantiated), testing male sperm by forcing it to compete is a sensible female strategy.

The male and female strategies just described are appropriate for species in which the male contributes little more than his genes, and the female alone cares for the young. However, when the male also contributes a breeding territory, the quality of his territory becomes another highly important mate selection criterion for the female. Following Dawkins, we see the male's territory as part of his 'extended phenotype', informing the female of his genetic quality.[11] But, in addition, the female must assess the compatibility of the male's territory with her own preferences and aptitudes, which are based on her experience in her natal environment. If the male also invests in parental care, as he does in most monogamous species, the interests and the efforts of the mates become more obviously coupled. Co-operation becomes important. This is most apparent when pairs show long-term monogamy, as in albatrosses and the many goose species in which the male and female are faithful to each other throughout their lives, and fully share the work of raising their common offspring.[12] The reproductive differences between males and females remain, of course, and the mates' behaviour towards each other reflects the fitness conflicts arising from these differences. Their behavioural strategies vary, depending on the stage of development at which the young are born (whether they are helpless or not), and on social and ecological circumstances.[13] Exactly how the relationship develops, and how much each partner invests, often depends on local conditions.

The variety of reproduction-related strategies, and their dependence on ecological conditions, is well illustrated by the dunnocks. These small songbirds (also known as hedge-sparrows) have an unusually flexible and opportunistic family organisation.[14] The female builds her nest, defends her feeding territory against other females and incubates her eggs, all without male assistance. However, males do help to feed the young. Males have their own territories, which they defend against other males, and these territories overlap those of the females. If a female's territory fully overlaps a male's territory and is of a size that the male can effectively defend, the result is a monogamous arrangement in which one male mates with and helps one female. Alternatively, two males can co-operate to defend a territory that overlaps the territories of two adjacent females. In this case, the result is multi-mate polygamy, with both males mating with the two females and feeding their offspring. If the female's territory is small but rich in food, a single male can defend more than just this one territory, and splits his paternal efforts between two or more adjacent females. This situation leads to the polygamous breeding system known as polygyny. When the feeding territory of the female is larger than that which a male can guard (the most common situation in most populations of dunnocks), the result is another kind of polygamous system, polyandry, in which two or more males guard the territory and together care for the female's young. In this case, the amount of care that a male offers to the young that he and other males help to raise is proportional to his expected genetic shares in the young: the male who copulated most often with the female is the one who cares most for the offspring, while a male who had fewer copulations during the egg-laying period helps less and sometimes even tries to destroy the eggs. The female allows all males to have some copulations, however, so all will have a vested interest in her nestlings. The breeding system in dunnocks is clearly very plastic: all types of family organisation can be found within a single population.

Although the dunnock male's investment in parental care varies with ecological and social conditions, both parents care for the young. Nevertheless, a female may still try to manipulate as many males as she can to care for her young, and a male may try to copulate with and distribute his parenting efforts among several females, thus potentially increasing the number of his offspring. The selfish tendencies of each mate have to be curbed. After all, they have a common interest: to rear their joint offspring successfully by feeding and protecting them, and transferring to them all the knowledge and skills necessary for their future success. Too many conflicts over the nature and the amount of care would be harmful to all family members. Co-operation has to be achieved.

So how are co-operation and harmony between mates achieved? Since conflict seems inevitable, it is not surprising that biologists studying the evolution of the behaviours seen in families have focused mainly on this aspect of the relationships. They have interpreted the striking co-operation that is so often observed between monogamous mates as the outcome of the resolution of evolutionary conflicts – as a fragile evolutionary compromise reached as a result of the arms-race between conflicting behaviour-affecting genes in males and females.[15] The co-operative social relationships between mates are remarkably stable, however, and in this chapter we want to focus on the evolutionary reasons for this stability. While accepting the compelling logic of evolutionary conflict theory, we will argue that the stability of long-term relationships is related to the information that the mates acquire about each other and with each other, and to the relationship-specific skills and behaviours that they build together. The more detailed the information they acquire and the habits they form, and the more personal and idiosyncratic they are, the more likely it is that mate choice will be successful, and that a robust, co-operative relationship will be built up. Information exchange increases the advantage of long-term co-operation between mates, and thus makes the evolution of co-operation more likely. This, in turn, reinforces the selective advantage of information exchange, which is therefore expected to evolve wherever there are long-term relationships.

Mate choice – the use of learning

The relationship between monogamous mates starts with mate choice and is followed by a stage of consolidation and active partnership. At the initial choice-stage, before the mates are committed to the relationship, each seems to attempt to obtain all the information it can about the other in order to be able to make the right choice. Once this stage is over, the mates start the second stage – consolidating their partnership by learning to co-ordinate their activities, forming adaptive pair-specific habits and preferences, and acquiring environment-specific and pair-specific skills. During this ongoing process of mutual learning and adjustment, the mates seem to be 'training' each other through their joint activities.

We will begin at the beginning, at the mate-choice stage. Since our focus is on the relationships between monogamous mates, we will concentrate on birds, where 90 per cent of species have this type of breeding system; among mammals, monogamy and paternal care are relatively rare. In monogamous birds, the female's task of choosing a mate seems a formidable one. The male is expected to share the burden of parental duties, so the female must be able to choose a healthy carrier of good genes who is also an able, dependable future provider for a hungry family. The chosen male must also be a good match for the particular female, fine-tuned to her individual preferences and abilities, so that the pair can work well together and form a good young-raising team. There is therefore a personal element in the process of choice. There is no single 'ideal male' that all females fancy, because each female has somewhat different preferences, and will bring into the partnership a different package of knowledge and past experiences. It is this package that shapes her mate selection.

Although usually anecdotal, bird and mammal watchers and breeders have provided many accounts of individuality in mate-selection criteria. Bird breeders in particular are well aware of the need to match what they call 'emotionally compatible mates' if breeding is to be successful. But, in spite of the outstanding example set by pioneer ethologists like Konrad Lorenz, who studied emotional individuality in birds and mammals, systematic studies are still scarce, and we are not sure what emotional compatibility entails in non-human animals.[16] In the next sections we will suggest that the colour pattern of the male, his song, his natal origin, and his age and size not only are general indicators of quality, but also frequently contain important ecological information.[17] This information may help the female to evaluate not just how suitable he is as a mate, but how suitable he is *for her*. The female may even force the male to go through some kind of aptitude test. A common avian version of such a test is courtship feeding.

Courtship feeding
When watching birds during the breeding season, we often see something rather strange – one adult bird feeding another. A closer look shows it to be an adult male feeding an adult female. The female behaves in a seemingly regressive manner, adopting chick-like behaviours such as crouching, wing shivering and giving squeaking begging calls, demanding to be fed. The male responds by flying back and forth, each

time bringing food and feeding it into the eagerly awaiting beak of the begging female. This set of male and female behaviours resembles a chick-feeding session so closely that some inexperienced observers may mistake it for one. It is very common in many groups of birds, from gulls and terns to birds of prey, from pigeons to jays, from European robins to flycatchers.[18] Usually it starts before mating and is a pre-requisite for it, but the intensity of courtship feeding goes up as the egg-laying period approaches and declines after its completion. The male's offerings are not a mere ritual, since they form a substantial part of the female's total food intake – from 33 to 40 per cent in tits and up to 100 per cent in birds of prey such as the osprey, where for several months females are fed almost exclusively by their mates.[19]

What are the functions of this extraordinary behaviour? What is its ecological and evolutionary importance? In the half a century since David Lack first drew attention to these questions, many functions have been suggested. For example, it has been suggested that courtship feed-ing increases the female's reserves of fat and improves her physical well-being, thus allowing her to lay more and larger eggs and making her more fit to perform the time and energy-consuming role of a mother; that it promotes the female's fidelity; that it decreases the chances that the male will desert; that it advances the time of reproduction; and that it allows the female to assess the capabilities of the potential mate as a future provider for the family. As we shall see, there is evidence sup-porting most of these suggestions.

The food contribution of the male is certainly translated into breed-ing success for both sexes. First, the food is an inducer for copulation: the more food the male brings, the greater the tendency of the solicit-ing female to copulate with him. Second, fed females tend to be faith-ful to their mates. Females that have been fed more than usual because, in addition to natural courtship feeding, they have been given supple-mentary food by an experimenter, are even more faithful to their mates.[20] Third, there is evidence from some species that the more food the male brings to the female, the larger the eggs and the clutch size.[21] Both sexes therefore benefit from courtship feeding.

Some of the effects of courtship feeding may be long lasting. The food the female demands and receives may affect not only the success of the brood she is about to rear, but also her reproductive success in the more distant future. This is because females who are well provisioned through good courtship feeding frequently lay their eggs earlier in the breeding season, and consequently stand a better chance of rearing a second brood.[22] Therefore, through courtship feeding, the male has indirectly been forced to invest in the female's future broods. It has been suggested that this is either a reflection of the female's dominance, or is her insurance policy against male desertion.[23] It is argued that, in a short breeding season, a male will probably have more chance of rearing further offspring if he stays with the female in whom he has already invested, rather than deserting and starting a new relationship.[24]

If, as has been suggested, courtship feeding is a reliable indicator of future paternal care, we expect to find that good courtship-feeders are good fathers. This does seem to be the case. In the common tern, there is a good correlation between the amount of food the male supplies to his mate during courtship feeding and the amount he later supplies to the chicks.[25] In domestic budgerigars, where the male feeds the female and, in some nests, also feeds the young, males that fed their mates most often were found to be the most eager to feed their nestlings.[26] Courtship feeding in terns and budgerigars therefore reflects the male's future degree of parental commitment.[27] But do we know that females use the quality of courtship feeding as one of their criteria when choosing a mate? Since males are known to differ in how much food they bring to females, it is reasonable to infer that females assess a male's feeding performance, and that, if it does not meet her standards, she initiates a break-up of the relationship.[28] We would therefore expect to find that females let many males feed them before they make a choice. In common terns, some prospective couples do indeed break up after a short period of courtship feeding, but the evidence that during courtship feeding females gain information about the likely quality of the suitors' chick-feeding is only circumstantial. Certainly they gain food when 'sampling' different males, so we can ask why the males should be so generous. The answer seems to be that a male has to courtship feed the female in order to copulate with her, and his only option is to try and get the female to make her decision about him as early as possible during the period of courtship.

Courtship feeding may give the females more than food and information about the male's future paternal behaviour. We believe that other types of information are exchanged. For example, courtship feeding may provide information about the ecological preferences of the male, and allow the female to evaluate the similarity of his ecological preferences to her own. It may also give the female information about the availability and abundance of various types of food. The range (and not merely the quantity) of food types that the male brings may be important for her future offspring, because the father often influences their food preferences. For one or two weeks after fledging, young frequently forage in the company of their father, from whom they probably learn foraging locations and techniques.[29] By judging the quality of his courtship feeding, the female may be able to assess the male not only as a future provider, but also as a future 'educator' for her young. It is also possible that, as the male is courtship feeding, he is learning to respond to the specific food preferences of the female, thus becoming increasingly more compatible with her. If so, courtship feeding is not only a choice-stage, it is also a pair-consolidation stage.

Unfortunately, the role of learning has usually been neglected by biologists thinking about the development and evolution of mate choice and parental care. There are no studies that examine whether information, as well as food, is transferred during courtship feeding, although it is not difficult to design experiments that would test the idea. If ecological and social information is indeed transferred through courtship feeding, then this behaviour should be seen as being to the mutual benefit of the mates, and not merely as the fragile resolution of a conflict between an extorting female strategy and a male strategy aimed at giving as little as possible. The choice of an incompatible mate in a monogamous partnership is to the detriment of both sides.

The colours of the natal environment
Female birds certainly do not rely solely on the knowledge gained through courtship feeding when choosing their mate. They have other sources of information. One is the colourful plumage of many males, which contains more information than meets the naïve human eye. Plumage colour is frequently based mainly on high concentrations of carotenoid pigments. These pigments, often yellow, orange and red, are found in varying concentrations in plants, insects and shellfish, but they cannot be synthesised by birds.[30] Birds obtain them from their diets. Since most birds cannot digest cellulose and therefore do not eat leaves (a major source of carotenoids), they are left with an indirect, limited and sometimes rare supply of some of these pigments in the form of herbivorous insects, which vary in availability between seasons and habitats.[31] However, eating a carotenoid-rich diet is not the only thing needed for bright and colourful plumage. Males infested with parasites

cannot extract and deliver carotenoids to the target tissues, so a young male chicken that is sick with a parasitic disease has a duller comb than a healthy male, even if both have eaten the same amount of carotenoids.[32] The carotenoids that give birds their colour are therefore also sensitive indicators of health. In addition, they have a role in defence against cancer, and influence colour vision: the quality of colour vision depends on the presence and quantity of several carotenoids deposited in the coloured oil droplets of the retinal cones of birds' eyes.[33]

The American ornithologist Geoffrey Hill claims that local and seasonal differences in the access males have to certain carotenoids can explain ecological and geographical variation in male plumage. He found that female house finches strongly prefer the reddest male available. Since in the diets of these birds red pigments are less abundant than yellow, the reddest male advertises his superior ability to forage for the rarest commodity. When the carotenoid content of the diet of captive house finches was varied, individual male plumage corresponded extremely well with the content of the diet offered, leading Hill to claim that 'house finches are what they eat'.[34]

The plumage of the great tits we met in chapter 2 also reflects their diet. They are mainly insect eaters, and moth and butterfly larvae are their main source of lutein, a common yellow plumage pigment. Great tits live in both deciduous and coniferous woodlands, habitats that differ in the availability of lutein. Since the lutein-containing larvae are more common in deciduous woodland, it is not surprising to find that males from deciduous woodland are frequently decorated with a richer yellow than their coniferous counterparts.[35] Experimental exchange of eggs between nests from different habitats has shown that the plumage colours of individuals are related mainly to their food, and not to genetic differences. Furthermore, nestlings that were experimentally fed with a higher proportion of the lutein-rich butterfly and moth larvae were yellower than their sibs.[36]

Bright colours may therefore be more than good indicators of male quality: a particular hue may also disclose a male's ecological origin. Females who, when young, have been imprinted on the local male plumage may prefer similar plumage in their mates. Males with such coloration are likely to be ecologically compatible, so the females would benefit by choosing the brightest plumage within their own particular 'natal zone' of coloration. Any male who displays bright, rich plumage colours is popular among females, because he manifests his fertility, health and vigour, and also his skill as a forager of useful carotenoids. He provides the female with carotenoid-rich food during courtship feeding, and later he feeds carotenoids to his offspring. Since fledglings commonly forage with their father, he may also demonstrate to his offspring how and in what type of location to collect carotenoids. In these ways, he biases their food preferences, and helps to turn his sons into healthy, bright, richly coloured 'sexy sons', who are attractive to females; his daughters benefit too, since carotenoids improve their health and colour vision.

The sounds of home

In the previous chapter we discussed the song of male birds, and pointed out that the structure of the song varies with the habitat of the singer. Eurasian great tits, Argentinean chingolo sparrows and North American summer tanagers all show a good match between the structure of the song and various vegetational and climatic features of the habitats they occupy.[37] A female listening to a male unconsciously learns about his geographical and ecological origin, and this information may direct her choice of mate. Female cowbirds and corn buntings have been found to prefer males whose songs have features similar to those of males originating from the areas in which the females grew up. An adaptive explanation in terms of genes has been suggested for this preference: it has been argued that birds may benefit from mating with individuals from the same population and habitat, because in doing so they preserve co-adapted gene complexes. However, since the song indicates ecological origin, choosing a mate with the same song as was heard in the natal area makes evolutionary sense if ecological compatibility between mates is advantageous. Unfortunately, the evidence that females prefer the natal song is conflicting at times. For example, in the thoroughly researched case of the white-crowned sparrows, some researchers claim that females prefer natal songs, whereas others show either that there are no clear preferences, or that the females prefer unfamiliar songs.[38]

Why do different researchers working in different areas get such conflicting results? The answer may lie in the nature of the environment in which the birds they studied were living.[39] In a homogeneous environment, a male's song is a good indicator of his natal origin, and can therefore be used as a valid criterion for ecological compatibility. However, when the dialect area is heterogeneous, containing several different habitats, the song cannot reflect the properties of the natal territory, and is therefore a poor indicator of ecological compatibility.

From the female's point of view, a male may sing the 'right' song (the one she heard when young) even though he lives in the 'wrong' part of the dialect area. We therefore suggest that the confusing results obtained by different investigators may have something to do with differences in ecological diversity within dialect areas. When the dialect area is ecologically diverse, the song-type is not a good criterion for choosing a compatible mate, and females do not use it, whereas, in more uniform areas, they do. Since females use more than just song characteristics when selecting a mate, and since there is no reason why the criteria used should be the same in all parts of the species' range, females may use different criteria in different areas.

Familiarity as a basis for co-operation between mates
Learning about a male's ecological origin from the colour pattern of his plumage or the dialect of his song may seem somewhat indirect ways of assessing the compatibility of a mate. Barnacle geese have a more direct approach.[40] The barnacle goose is a long-term monogamous species that ranges from Greenland to Russia. It feeds on grasses and sedges, winters in grasslands and salt marshes around the coasts of Britain and the North Sea coast of mainland Europe, and breeds in the arctic tundra during the summer. The small breeding colonies, each of about 100 pairs, are located on small islands and cliffs overlooking the sea. The young follow their parents in their long-distance journeys to and from the arctic breeding grounds and the more southern wintering grounds. They first breed in their third or fourth summer, after having formed pairs the previous winter. They choose a mate from among the densely packed individuals of their wintering colony, which in Scotland, for example, can be twelve thousand birds strong.

Barnacle geese choose their mates carefully. During the winter season, they sample one to six potential mates before settling with a permanent partner. With each potential mate they form a temporary partnership, which lasts from a few days to several weeks. It enables the would-be partners to assess how well they succeed as a pair in the highly competitive conditions of the winter colony, which is packed with thousands of other geese with whom they have to compete for food and space. This 'winter engagement' gives the partners a chance to get used to each other, and is typical of many other species.[41] But what criteria do these geese use when choosing a mate? Who are the potential partners they test?

Among the thousands of unpaired geese available in the winter colony, a barnacle goose very clearly prefers a mate who is of similar age and size, comes from the same natal territory and is familiar. Since both feeding and nesting conditions vary between breeding sites, mates who are familiar with the same local area and with each other may be better adapted to local conditions and better co-ordinated in their future breeding and feeding activities. Choosing a familiar mate reduces the time spent looking for a suitable partner, and increases tolerance and co-operation between the mates.[42] One of the most obvious benefits of familiarity is that it may save time and energy during courtship. The first phase of pair formation often includes courtship displays, parts of which are aimed at overcoming the initial antagonism between potential mates that are complete strangers.[43] Old, well-established couples do not need these parts of the display; their courtship is therefore shorter, and they breed earlier.[44] We believe that potential mates who are familiar with each other from early life may treat each other more like old mates than total strangers, and therefore breed earlier than other young pairs. Barnacle geese that pair with familiar individuals do tend to bond earlier than others, and consequently gain a higher social status in the breeding colony and better access to the best sources of food.[45] This is important in species in which high-quality food and nesting sites are in short supply, as is commonly the case at high latitudes or altitudes.[46] Familiarity not only allows the young pair to breed earlier and more successfully, it is also a good starting-point for establishing an enduring and conflict-free long-term relationship between mates.[47]

Early life familiarity clearly has much merit as a criterion for mate choice for long-term monogamous species living in harsh and competitive conditions. How common it is is not known. However, for many species, early familiarity cannot be a criterion of mate choice, because members of one of the sexes (in birds it is usually the female) leave the natal territory and settle elsewhere. Nevertheless, even in such cases, individuals seem to attempt to choose partners who in some respects are as similar to themselves as possible, and hence are also compatible. We have already mentioned Patrick Bateson's finding that the preferred mate of a female Japanese quail is one who is phenotypically similar, but not too similar.[48] According to Bateson's optimal discrepancy theory, the similarity ensures compatibility, while the discrepancy indicates that the mates are genetically different enough to avoid the deleterious effects of inbreeding. Resemblance that is based on similar developmental or ecological experiences may be a good way of ensuring optimal compatibility.

Choosing a mate, the first stage in establishing a family, is obviously as important for monogamous birds as it is for humans, so it is not surprising to find that individuals often use several different but functionally overlapping criteria for mate choice. A female bird may use the song of the male and his coloration, and test the male's future parenting and foraging ability while he courtship feeds her. Both sexes may use the most direct clue of all for ecological compatibility – early life familiarity – or use clues indicating similarity in age and strength. Using several criteria increases the reliability of their choice. Many of these criteria are based on information that the animals acquired early in life, in their natal surroundings, and this learnt information will be the basis of the compatibility and co-operation between the mates when they become parents.

Once a mate has been chosen, and a new family is started, co-operation is essential. Nevertheless, each partner may still attempt to 'exploit' the other, for example by leaving it with more than its fair share of parental duties. However, we feel that these rather 'negative' aspects of family life have been overemphasised. If evolutionary conflict theory is accepted, the wide scope and the apparent stability of mates' co-operation are puzzling. The male and female in any monogamous couple are, of course, locked into an intimate partnership, where close co-operation helps both of them to maximise their reproductive success. But, if a little selfishness can be beneficial in the short term, why is co-operative behaviour not gradually eroded during evolution? We think that the stability of the co-operation between mates is related to the 'personal' aspects of mate choice. When mate choice is not based solely on the 'universal' criteria that indicate 'good genes', but includes more individual criteria indicating ecological origins or idiosyncratic features, it is much more difficult to replace a compatible partner than it would be if only 'universal' criteria were used. Moreover, the qualities that make a mate relatively 'unique' for its partner also lower its attractiveness to many other potential mates. The cost of desertion may therefore be very much higher than it would otherwise be, and is expected to become even greater with the duration of the relationship.

The monogamous family system, especially when coupled with individual-specific compatibility, is probably one in which a mate who co-operates with its partner has, on average, a higher fitness than one who cheats and exploits its mate. In the jargon of evolutionary biology, the 'pay-off' – the sum total of the costs and benefits to an individual's fitness – is greater for the co-operator than for the exploiter. This type of co-operation, in which an exploiter has a reduced fitness relative to a co-operator, has been called 'by-product mutualism,' and 'no-cost co-operation'.[49] Since we shall use the latter term, we need to emphasise that 'no-cost' means that the *co-operative function* of the behaviour is not costly; it does not mean that the behaviour itself is cheap. We believe that in monogamous families no-cost co-operation is very common. Although it probably evolved as a strategy that maximises the fitness of each of the partners, and not as an anti-cheating measure, when mate choice is tailored to the idiosyncratic needs of the individual, it increases the costs of cheating and desertion, and thus promotes co-operation. Subsequent learning about and with each other, as the pair establishes, develops and maintains the relationship, further increases the cost of desertion or cheating for both sides and, as we argue in the next section, is an even more important route promoting long-term, stable co-operation.

The parental team – co-ordination and division of labour between mates

The choice of an ecologically and behaviourally matching mate is a good starting-point for a successful family life, benefiting both partners. But it is really only the start of a climb. After the choice is made, the mates must get over any aggressive tendencies towards each other, get used to each other's close proximity, and learn to co-ordinate their rearing efforts. They must learn to work smoothly as a team to protect the nest and the young from parasites and predators, forage efficiently, and feed their offspring. Division of labour, efficient chore-sharing and combined actions are the strategies used by parental teams in birds, mammals and humans. These strategies enhance the reproductive success of the offspring as well as that of both partners. As they establish their relationship, mates seem to learn with and about each other and form their own, pair-specific habits and routines.

There are, of course, considerable species-specific differences in the pattern of chore-sharing between mates. If we take incubation in birds as an example, we see that, in the majority of species, both sexes incubate the eggs according to some species-specific, regular (though not necessarily equal) schedule of shifts. In most other species, it is usually the female alone who incubates, although in some it is the male.[50] Shift-work requires co-ordination between the two parents. This co-ordination has to be closely tied to the local ecological conditions, about which the mates learn as they go along, and it requires an efficient communication system. Pair-specific interactions, such as the pair-specific vocal and visual displays heard and seen in many species, serve to co-ordinate the activities of the pair members.

Age, experience and breeding success

We have already described the 'winter engagement' in barnacle geese, which is an early test of compatibility between already familiar individuals. Choosing a mate with whom the individual is well acquainted promotes future co-operation. Another way of achieving success in family life is by choosing a mate who is more experienced than oneself, and learning from him or her.

Females of some bird species are known to prefer middle-aged, experienced males as partners.[51] Such males have many advantages for the female: they are frequently at the peak of their reproductive performance,[52] and have proved their vigour, health and good sense just by getting to this ripe age. A middle-aged male has probably learnt a lot during his life and can put it to good use. But a young monogamous female may be doing more than choose a high-quality mate when she chooses a middle-aged male. We believe that middle-aged males may also play a 'parental' guiding-role for young females. By choosing a knowledgeable mate from whom she can learn, a female may accelerate her rate of improvement as a forager, mate and breeder. Unfortunately, although there is a good positive correlation between foraging ability and age in several bird species,[53] as far as we know no one has shown that young females do indeed learn from an older mate and thus enhance their long-term fitness. We expect that the knowledge of middle-aged and even old males may be particularly significant in species with serial monogamy, where the older male may be one of a series of partners. Any disadvantage caused by his age-related loss of vigour is likely to be small relative to the informational benefits he confers on the inexperienced female. If social learning from an older mate is important, we expect that in species where serial monogamy is common and the male contributes to care and defence, young females will have a greater preference for older males than they will in species with long-term monogamy, where reciprocal learning may be more important. An older mate may also be preferable in species that live in complex environments where a large array of foraging techniques must be learnt, and where proficiency is limited mainly by experience.[54] In such conditions, learning from the more experienced partner (usually the male) may speed up the processes of acquiring knowledge by the younger partner, and consequently enhance its long-term fitness.

Co-ordination between mates – duetting

In long-term monogamy, the length of the time the partners have been together has a clear positive effect on breeding success.[55] Partners become more tuned to each other's messages and actions, and can communicate and respond to each other more effectively. One type of behaviour that seems to contribute to this improvement in some birds and mammals is duetting.

Duetting, a 'practice of mutual displaying between members of an established pair',[56] is well known in many tropical and some temperate species of birds, and is probably more widespread among both birds and mammals than is currently appreciated.[57] It involves well-coordinated singing between and with mates, or mutual visual displays, or both, but most research has been concentrated on the relatively spectacular vocal duetting in species of tropical shrikes and gibbons. Duets that do not involve learning seem to act as co-ordinated threats to other individuals or couples, and as inter-mate signals that reduce the chances of a misdirected attack against the partner.[58] Such non-learnt duetting is thought to have evolved into the more complex duetting that involves mutual learning and is associated with tightening the pair bond. It is these learnt duets and their co-operation-promoting effects that are of most interest to us.

The studies of African bou-bou shrikes carried out by the English ethologist W. H. Thorpe and his collaborators have become classical illustrations of learnt vocal duetting. In each species, after a few weeks of learning, a pair develops a highly pair-specific version of the species' song that contains elements from the song of each mate. Each of the mates not only contributes to the mutual song, it can also recognise its mate's contribution and, if the partner is missing, can sing both contributions as a unified solo performance. The co-ordination between the songs of the two mates is achieved in different ways. In some shrike species, the duet is composed of an alternating pattern, with different notes emitted by each mate. In other species, one bird repeats the notes made by its mate, and in yet others various differing sequences are exchanged.

According to the German ethologist Wolfgang Wickler, duetting may be a pair-bond cement.[59] When it involves a complex process of learning, each mate invests considerable time and energy in learning the individual version of a potential mate's song at the start of the breeding season. It then uses it, in concert with its mate, to enhance the efficiency of territory and nest guarding, to synchronise its physiological state with that of its mate, and possibly also to co-ordinate shift activities such as incubating and feeding.[60] By repeating the duet after it is thoroughly learnt, the mates seem to be reaffirming as well as advertising their bond, and are thus actively maintaining it. Individuals who stay together for the next breeding season do not have to create a new duet, and hence they gain some reproductive advantage over mate-changers, who must invest in forming a new duet with a new partner. Wickler sees the heavy investment in forming a duet as an insurance policy against mate desertion. Unlike insurance policies such as courtship feeding, the investment is mutual and non-transferable. A female who has been courtship-fed can desert the feeding male and choose another mate without losing the energy gains that she has acquired. But if a female or a male who shares a learnt duet deserts its mate, it would have to learn to duet with a new partner. The cost of desertion is equal for both sexes. This feature of duetting is thought to contribute to the stability of co-operation between mates, because each would have to pay a high price if it defected: looking for a new mate during a short breeding season and investing in a lengthy learning process is costly.[61]

Information about the effects of the parents' duetting on the behaviour of their offspring is surprisingly sparse. It is possible that duetting serves to increase the attention of the young to their parents, and they become conditioned to respond selectively when they hear certain familiar notes. They may learn the parental pattern of song co-ordination. In American crows, a partially migratory species that has extended parental care, the complex duet is learnt and practised by the young, and it seems to maintain the cohesion of the pair and the family during its seasonal migration in large winter flocks.[62]

Co-ordination between mates – time and chore-sharing

In many monogamous birds, parents become more successful at rearing their young as the length of their partnership increases. For several species, this improvement is known to be far and above the age-related improvement of each partner.[63] Many factors contribute to this. Some are related to the interaction of the mates while rearing their young, but others are associated with the establishment and maintenance of the pair bond itself. Tenured mates are frequently less aggressive towards each other than new mates, and spend less time on aggressive and appeasement displays.[64] Nelson's classical studies of gannets and boobies, and Pickering's work on the wandering albatross show that the displays of partners towards each other are shorter and better synchronised in tenured pairs than in new ones, and result in earlier and more successful nesting.[65] Tenured partners devote more time to other activities: kittiwakes spend more time feeding and grooming, and well-established pairs of barnacle geese spend more time on the feeding grounds.[66]

Mates need to co-ordinate their activities in order to achieve reproductive success. As well as duetting and performing other social displays, birds groom each other to get rid of parasites, and have to share parental chores such as incubation, protecting the nest against predators, and feeding their young. Co-ordinating the changeover of incubation duties is essential for successful hatching,[67] and co-ordinating chick feeding is just as important. Great tits provide a particularly spectacular example of co-ordinated chick-feeding behaviour. When both parents are feeding the brood, they frequently supply the same type and even the same species of prey; if one mate switches to a new type of food, the other mate often switches too. This happens even though mates usually do not forage together – while one is out looking for food, the other is guarding the nest. The co-ordination cannot be attributed to similarity in the prey items available to each parent, because the food brought to the offspring is frequently not selected according to its availability.[68] It seems that this provisioning behaviour must involve some transfer of information between the mates. It has been suggested that established pairs may be better at this co-ordinated choice and use of food than newly formed pairs, and that co-ordination develops with the growing mutual experience of the pair members.[69] But why should both parents feed the young with the same kind of food? One reason may be that the young learn what is good to eat more readily when different types of food are not mixed up. As we discussed in chapter 3, repeated positive reinforcement promotes learning.

Unfortunately, it is not clear how co-ordination between the partners is brought about in great tits. In most cases of obvious co-operation between mates, detailed comparisons between established and new pairs have not been made, so the extent to which co-ordination is due to mutual and reciprocal learning is unknown.[70] In fact, there are only a few studies that have looked in detail at who is doing what and when during the nesting cycle. One study that did explicitly address these questions looked at the way European starlings shared parental duties. It was found that the female, who is larger, supplied the young with more of the food than the male, but, if either was handicapped, the other partially compensated for it by working harder.[71]

Information sharing and the evolution of co-operation between mates

We have suggested that mate choice and the subsequent relationship between monogamous mates involve the transfer of information between them and mutual learning. Acquiring, transferring, and sharing information are integral parts of their co-operative offspring-rearing enterprise. In recent evolutionary thinking, there has been a tendency to regard most co-operation and collaboration between individuals as an evolved resolution of underlying conflicts, with co-operative behaviour being perpetually in danger of crumbling because of the selfish interests of the individual partners. We agree that looking at social behaviour in these terms is important, because it alerts us to the Achilles' heels of social interactions, those weak points that can be exploited by 'selfish' individuals and lead to the disintegration of the social system. However, it tends to ignore those properties of the interaction systems that make them robust. Monogamous mating systems that involve co-operation between mates are often remarkably stable social systems,[72] and we believe that one of the reasons for this stability is that the transmission of information through individual and social learning has properties that reduce the benefits of selfishness, save energy and make cheating very difficult or pointless. For example, some acquired information, such as those aspects of plumage colour and song that reflect the natal origin of a bird, have no special energetic costs, are difficult to alter, and therefore are difficult to fake. Cheating over natal origins is almost impossible.

There are other ways in which the intrinsic properties of information use and transfer reduce the likelihood and benefits of cheating or desertion. Acquiring, sharing and using information have features that make them very different from acquiring, sharing and using calories. Unlike calories, which once used are gone for good, information may be non-depletable – once acquired, some information, such as that which leads to a preference for a particular type of mate or food, is remembered throughout life and can be used time and again. Learnt information about the environment, such as the knowledge that certain types of food are poisonous, can sometimes be generalised and transferred to other environments, but other types of acquired information are useful only in specific circumstances, and, although they can be used repeatedly, they are non-transferable. For example, learnt duetting in birds involves non-transferable information – the duet is specific to the particular pair, and a new duet would have to be learnt, at considerable cost, if one pair-member changes mate. Similarly, the knowledge about the unique features of the territory, which have been learnt by the territory holder, is non-transferable, as is the detailed knowledge an individual acquires about members of its social group. When information is non-transferable, the cost to the animal of changing its physical or social conditions may be very high.

Unlike the sharing of energy, sharing some types of information may be cheap, and information can be given away to a large number of individuals without being dissipated. Demonstrating what to eat to eight rather than four chukar partridge chicks does not require a proportional increase in their mother's expenditure of energy. Furthermore, the more behavioural information an individual gives away (i.e. the more it 'demonstrates'), the more stable its store of information becomes, because repeating a behavioural act is a process that enhances and stabilises information storage, and secures memory.

We want to stress again that we do not believe that learning about and with a mate originally evolved to restrain selfishness, although this may well be an important current function. Courtship feeding did not originally evolve to decrease the chances of desertion. But if a decrease in desertion was a by-product of this pattern of behaviour, it may have contributed to its subsequent evolutionary elaboration. The same may be true for the evolution of learnt duetting. Non-learnt duetting enables mates to recognise each other and keep track of each other's movements at a distance and in conditions that preclude good visual communication, such as dense plant cover or mist.[73] Non-learnt duetting also seems to serve as a territorial advertisement, which deters neighbouring pairs from trespassing and facilitates the co-ordination of joint territorial defence.[74] In addition, it may enhance the efficiency of breeding activities, since in some cases it seems to synchronise the mates' physiological states through vocal–hormonal feedback.[75] There are therefore many reasons for non-learnt duetting to be selected – reasons that have nothing to do with selfishness-curbing benefits. However, once non-learnt duetting had been established, it had additional effects: by disclosing the whereabouts of the mate, it reduced the chances of extra-pair copulation and enhanced the attentiveness of the mates to each other. The need for individual recognition in conditions of limited visibility or in large flocks may have led to selection for an increase in the complexity of the duet, and to an enhanced role for learning during this process. Duets could then become pair-specific, idiosyncratic and non-transferable – properties that increase the cost of desertion. The stability of the co-operation between mates is therefore a by-product of selection for securing the selfish interests of each mate!

There may be more to the evolution of the relationships between mates than this, however. So far we have focused on the selected or incidental advantages to each individual in the partnership, but an established pair can be seen as a well-integrated functional unit. In fact, where the compatibility between individual partners is important and idiosyncratic, the properties of an individual may not predict its reproductive success. For example, pinyon jays prefer mates who match their size.[76] Although as an individual a large male pinyon jay has the benefits of being dominant over other males, as a partner he may be inadequate for most females (who are smaller) because his large size leads to more aggressive encounters with them. When the fitnesses of mates are non-additive, as is likely in such a case, selection might operate mainly at the pair level.[77] The more compatible and co-operative pairs will produce more offspring than other pairs. They will therefore contribute to the population relatively more offspring with genotypes that promote learning-based, pair-specific co-operation (for example, genotypes conferring better learning ability and social attentiveness). Moreover, through social learning, offspring may acquire the co-operation-promoting behaviour and caring style of their parents. Since co-operators produce more offspring, the chances of two 'co-operative' individuals meeting will be enhanced in the next generation, even if mates meet

and bond accidentally. But, as we have seen, pairing is far from accidental, and non-accidental mate choice may increase the chances that
a 'co-operator' will choose another 'co-operator' as a mate. A solid partnership between co-operators is likely to lead to more effective rearing
of young, and to a more effective transfer of information to them than
would occur with less co-operative pairs. The spread of co-operation will
therefore be enhanced, because co-operators are better transmitters.

Summary

In trying to understand the way mates behave towards each other, evolutionary biologists have generally focused on how the selfish genetic
interests of males and females differ, and how these conflicting interests are resolved in different types of mating systems. In this chapter,
we have argued that the behaviour of monogamous mates is easier to
understand if the information that is transferred between them or
gained together is also taken into account. Much of this information is
individual or pair-specific, and is of little or no value in relationships
with most other potential mates. It therefore increases the benefits of
co-operation with the existing mate, and decreases the genetic gain that
could come from desertion or infidelity. The more 'personal' and idiosyncratic the knowledge and related behaviour are, the more stable the
relationship is likely to be.

The use of individual-specific behaviour often begins with a 'personal' choice of mate. This may involve recognising morphological or behavioural features, such as the colour or song of a bird, which reflect the
ecological or social origins of the potential mate. As the partnership is
consolidated, pair-specific habits, skills and preferences are formed.
Sharing information through learning from and with each other not
only saves energy, it is also often self-reinforcing, because practising bolsters memory. Through learning, mutual recognition improves and the
division of labour becomes more effective. The various forms of co-operation between monogamous mates are commonly functionally related,
and induce and enhance each other. For example, the co-ordination of
feeding shifts leads to co-ordination in nest-guarding shifts; vocal duetting is often accompanied by co-ordinated visual displays.

Through the personal nature of the choice of partner, and the network of interactions that form between mates, the partnership becomes
more robust and reliable. The components of this system of stable

co-operation may evolve even further, both through individual selection and through pair selection.

Notes

1 See Trivers, 1985.
2 The 'interests' animals have are not, of course, necessarily present in their minds. Rather, their biology is shaped by evolution in a manner that maximises fitness 'as if' they psychologically pursued such interests.
3 Although the range of variation in the number of young a lone female can parent is much narrower than that of a male, when the sex ratio is 1:1 the average success is, of course, the same for both sexes, because each offspring must have both a mother and a father. Males and females simply have different strategies for ensuring their own reproductive success.
4 Andersson, 1994, p. 117–18.
5 There is a common misunderstanding about the evolution of parental care. The greater care female mammals show for their offspring is often, and incorrectly, explained as the consequence of the greater investment of the female in her gametes. The female's eggs are large and immobile, and she produces relatively few of them, whereas the gametes of the male are small and mobile, and are produced at a high rate. In fact, maleness and femaleness are defined in terms of the type of gametes individuals produce: individuals producing large, immobile gametes are females, whereas individuals producing many small and mobile gametes are males. Therefore, by definition, the female's investment in terms of material and energy per gamete is much higher than that of the male. The fact that the female initially invests much more in her gametes, providing for them not only genes but also energetically costly cytoplasm, is supposed to predispose her to invest more in the resulting offspring. The reason for this is said to be that continuing to care for the offspring may be cheaper for her than deserting and starting all over again, investing once more in large, cytoplasm-rich eggs. It is argued that, for the male, with his smaller initial investment, the cost of desertion and producing the next round of cheap gametes is smaller, and therefore the benefits from deserting his present offspring and investing in future offspring are often higher. In a sense, the female is caught in a vicious circle: the more she invests the more likely she is to continue to invest. The greater investment of females in their young is explained as the evolutionary outcome of this vicious circle, which starts with expensive gametes.

  There is a clear fallacy in this argument, however. The investment in gametes should be assessed *per fertilisation,* not *per offspring.* For every egg the female produces, the male produces thousands or millions of sperm, and his overall investment per fertilisation may be as large, if not larger, than hers. Furthermore, in mammals the female has all her eggs almost ready for fertilisation at birth, whereas the male has to keep the machinery for producing and replacing sperm running throughout life. It is likely that the cost of running the sperm factory is rather high, and this cost should also be included in the balance-sheet of male and female reproductive investments. The fallacy in the 'investment in gametes' argument can be seen in the fact that, in the majority of the fish species that show parental care, it is the male rather than the female who is the caring parent, guarding the eggs and sometimes going to great lengths to care for them by fanning them and building nests for them. This care occurs even though the male produces many more gametes, and at a much higher rate, than the female.

6 This seems to be the case in the black grouse, sage grouse and other lekking species (see Dugatkin, 1996); in the sage grouse the traits preferred by females vary at different times and in different places (Gibson, Bradbury & Vehrencamp, 1991).

7 Zuk, 1994.

8 Zahavi & Zahavi, 1997. See also chapter 4, p. 125.

9 Fisher, 1958.

10 See Birkhead & Møller, 1992.

11 Dawkins, 1982.

12 Black, 1996.

13 Clutton-Brock, 1991.

14 Davies, 1992.

15 For a female-focused view of conflict theory, see Gowaty, 1996.

16 Köhler (1925) emphasised individual personality in higher apes. One of the few ethologists studying individuality in mammals is Hubert Hendrichs (see, for example, Hendrichs, 1996).

17 Birds can see, hear and memorise extremely well, so are able to recognise the individual differences that reflect ecological and social origins (Beecher, 1988; Beer, 1972).

18 Lack, 1940.

19 Green & Krebs, 1995.

20 Brooke & Birkhead, 1991, p.259.

21 This correlation has been found in the common tern and in other species. The most successful male providers among the common terns were found to be those capable of switching from one kind of prey species to another. In this species, during courtship feeding, each male selects the largest and the heaviest fish it can carry and brings it back to the female. See Hume, 1993; Nisbett, 1973.

22 According to Daan et al., 1988, earlier breeding in marsh tits and other species is also seen when supplementary food is offered to the females.

23 Smith, 1980.

24 The idea that because one has invested a lot in a relationship one is 'committed' to it and should continue to invest in it, despite an advantage in deserting and starting anew, has been called the 'Concorde fallacy', because the reasoning is the same as that used by British politicians to justify spend-

ing further funds on developing a supersonic aeroplane – spend more so that the past expenditure is not wasted (see the discussion in Clutton-Brock, 1991, pp.188–90). Although such reasoning may indeed be fallacious, past investment reduces the ability for future investment and can provide an index of future expenditure, so past expenditure is not irrelevant to future investment. Once a relationship or a project is long-term, there are likely to be transient fluctuations in the costs of investment, which will not be revealed by an optimality analysis that does not take into account the *whole* duration of the relationship. We therefore expect that, when long-term relationships are beneficial, natural selection will favour a psychological commitment effect that inhibits desertion during transient 'downs'.

[25] Nisbett, 1977.

[26] Stamps *et al.*, 1985.

[27] Of course, we also expect the male to be able to assess the likely quality of the maternal behaviour of his potential mates, especially if he is going to invest a lot of energy and time in their offspring. However, in most cases, much less is known about the criteria males employ when choosing their mates (see Andersson, 1994).

[28] Hume, 1993; Nisbett, 1973.

[29] See Skutch, 1976, chapter 25, and Gosler, 1993, chapter 9.

[30] Goodwin, 1950.

[31] Brush & Power, 1976; Goodwin, 1950; Hill, 1992.

[32] Zuk, 1994.

[33] Varela, Palacios & Goldsmith, 1993.

[34] Hill, 1993, 1994.

[35] Slagsvold & Lifjeld, 1985.

[36] Brush & Power, 1976; Hill, 1992; Slagsvold & Lifjeld, 1985.

[37] Morton, 1975; Wiley, 1991.

[38] Baker & Cunningham, 1985; Nottebohm, 1972; Searcy & Yasukawa, 1996.

[39] Catchpole & Slater, 1995, in particular, chapter 9.

[40] This account is based on Black & Owen, 1995; Choudhury & Black, 1993, 1994; Ehrlich *et al.*, 1994.

[41] Coombs, 1960; Gorman, 1974; Rowley, 1983; Weller, 1965.

[42] See Janetos, 1980; Prop & de Vries, 1993. For information on the importance of familiarity in ringed turtledoves see Erickson, 1973; for kittiwakes see Chardine, 1987; for gannets see Nelson, 1965, 1972.

[43] Black, 1996.

[44] Rowley, 1983.

[45] Black *et al.*, 1992.

[46] Gill, 1995.

[47] For the effects of familiarity on barnacle geese, see Black & Owen, 1988; Owen, Black & Liber, 1988.

[48] Bateson, 1982.

[49] Brown, 1983; Dugatkin, 1997.

[50] Gill, 1995; Skutch, 1976.

[51] We prefer to talk about 'middle-aged' rather than 'old' males, since, in most species of birds and mammals, individuals in the wild rarely reach 'old age' in the human sense. They die from disease or accidents long before they approach the maximum species lifespan. Evidence of the females' taste for middle-aged males is given by Burley, 1981; Enstrom 1993; Kokko & Lindström, 1996; van Rhijn & Groothuis, 1987.

[52] See Forslund & Pärt, 1995, for a short review; more extensive data can be found in Clutton-Brock, 1988; Newton, 1989.

[53] Martin, 1995; Wunderle, 1991.

[54] Wunderle, 1991.

[55] For evidence of the positive effect of the duration of the relationship on breeding success see Bradley, Wooller & Skira, 1995; Cézilly & Nager, 1996 (and references therein); Rowley, 1983. For a typical example see Perrins & McCleery, 1985.

[56] The definition is taken from Smith, 1977a.

[57] For evidence that duetting is widespread, see Farabaugh, 1982; Thorpe, 1972; Smith, 1994. The role of visual duetting is unclear, and is probably underestimated; see Armstrong, 1947; Malacarne, Cucco & Camanni, 1991.

[58] Arrowood, 1988.

[59] Wickler, 1980.

[60] Farabaugh, 1982; Wickler, 1980.

[61] Although the idea that learnt duetting enhances the benefits of co-operation is at present speculative, it can be tested. Comparing the mate fidelity, territorial defence and offspring-rearing ability of control pairs with those of pairs whose learnt duet has been experimentally disrupted would indicate whether learnt duetting is important as a pair-bond cement. A crude estimate of the costs of learning duets might be obtained by comparing the reproductive success after the breakdown of partnerships in species with learnt duets and in related species in which the duet is not learnt.

[62] Brown, 1985.

[63] A summary of many detailed case studies can be found in Black, 1996.

[64] Kavanau, 1987.

[65] Nelson, 1965; 1972; Pickering, 1989.

[66] See Chardine, 1987, for information on the influence of pair status on the breeding behaviour of the kittiwake, and Black, 1996, for information on barnacle geese. Comparing pairs in which one of the mates has been recently widowed and taken a new mate with long-standing couples of the same age might help to distinguish between improvements in parenting due to age and those due to the length of time the partners have been together.

[67] For co-ordination of incubation routines in Adélie penguins, see Davis, 1988. Bost & Clobert, 1992, have shown that, when the change-over of incubation duties in gentoo penguins fails, the pair usually 'divorces' and a replacement clutch is not laid.

[68] Royama, 1970.

[69] Gosler, 1993; Perrins & McCleery, 1985.

[70] Black, 1996.

[71] Wright & Cuthill, 1990, 1992.

[72] It is the *social relationships* between monogamous mates that are stable. Many studies have shown that sexual monogamy is not strictly observed, because extra-pair copulations are quite common.

[73] Thorpe, 1972.

[74] Harcus, 1977; Hooker & Hooker, 1969; Seibt & Wickler, 1977; Smith, 1977*a,b*; Todt, 1975. Attentiveness is discussed by Smith, 1994.

[75] Kunkel, 1974.

[76] Phenotypic resemblance between mates may sometimes be the result of the *different* choice criteria of males and females. In pinyon jays, birds frequently pair for life and divorce is very rare. Male jays prefer females who are both larger than the average and matched to them in age and reproductive history. Females prefer bright-plumaged males that are smaller than average, and matched in age. These mutual preferences lead to maximal similarity in size and maximal compatibility in experience between the mates. Pairs matched in size, age and experience are more successful than non-matched pairs; see Marzluff & Balda, 1992.

[77] This would be a case of group selection, with a group consisting of two individuals – the two mates. We discuss group selection in some detail in chapter 8.

6 Parents and offspring – too much conflict?

As with the relationships between mates, the focus of most evolutionary studies of the relationships between parents and their offspring and between siblings is conflict. This is not really surprising. Human beings have always been fascinated with family conflicts, as our myths, literature and gossip show. The Old Testament is a rich testimony to the centrality of conflicts in our lives: think about the bloody dispute between Cain and Abel, which culminated in the murder of Abel and the stigmatisation of the human race; think about Rebecca's maternal manipulation of the rivalry between Jacob and Esau over status; think about the story of Joseph and his brothers. But family conflicts are not limited to humans. Animal life is also full of sibling rivalry and parental attempts to control their unruly children. The interests of siblings often clash, and frequently those of parents and offspring seem not to coincide. As we know all too well, the joys of family life are marred by many problems.

Although learning is an essential part of the ambivalent and intricate interactions between parents and their offspring, evolutionary interpretations of these interactions have failed to take into account the limitations and possibilities that learning introduces into the relationship. In this chapter, we will try to show how incorporating learning into the evolutionary scheme provides additional and alternative explanations of many aspects of parent–offspring relationships. But first let us return to the Judaean hills at springtime, and look at the chick-rearing activities of a pair of great tits, whose family life illustrates interactions that are typical of many bird and mammal species.[1]

By mid-April, the old olive orchard has virtually turned into a tit nursery. Adult males, each holding a caterpillar in its beak, can be seen sneaking silently through the trees, each individual eventually entering the small hole in an olive-tree trunk that leads to its nest-cavity. Soon the male reappears for a moment at the nest entrance, before fading silently into the nearby caterpillar-rich foliage. A short while later, cautious as ever not to disclose the location of the nest entrance, he brings back another morsel, one of about four hundred that he will provide during the course of the day.

In one nest, seven chicks have just hatched, all within a short time of each other. The chicks are blind and naked, and are really little more than guts and gaping orange mouths that are fully occupied in turning the insect protein brought by the diligent male provider into a great tit. At first they cannot regulate their own body temperature, so are brooded by the female, but within the next five days the mother will gradually become an equally important food provider.

Every time the male lands on the nest entrance he is greeted by a noisy chorus of seven widely gaping chicks, begging for food. The chicks are all roughly the same size, but some of them stretch and gape higher, beg louder or are more persistent than others. The male either offers them the caterpillar, or passes it to the female to feed to the chick who gapes highest or begs most persistently. If this chick does not swallow the caterpillar fast enough, the parent offers it to another chick. On the next parental visit some other chick will act the hungriest and win the juicy meal. Since food items are brought in in rapid succession and replete chicks tend to settle down to digest, every chick gets a more-or-less equal share of the food. In addition, nest sanitation behaviour helps to ensure equality: each chick has to void its droppings almost immediately after being fed, and do so in a way that does not soil the nest. When it is very young, a nestling changes its position to enable its parent to take away its dropping. When a bit older, it is able to move to the periphery of the nest and void its droppings without parental assistance. Because of this nest-sanitation behaviour, a nestling often loses its favourable feeding position to another chick, and the chances for a bully to monopolise the parental food distribution are minimised. Only in lean seasons or situations, when the most persistent chicks have voided and are hungry again before the least persistent ones have been satiated, will the weakest chicks starve.

A week later we find the parents working in concert to feed their fast growing, ever-hungry chicks. Both frequently bring in and feed the chicks with the same type of prey, both switching after a while, as if by mutual agreement, to another type: moth, caterpillar, beetle, moth, caterpillar, spider. The frequency of parental visits is influenced by the level of the chicks' begging – the louder the chick chorus, the more frequent the feeding visits. Are the parents influenced by the noisiest individuals, or is it the combined volume of noise that affects their visiting frequency? Do the parents discriminate between the young according to sex, size or other characters? Observations combined with tape-recording have shown that parents match their efforts to the loudness of begging by the brood as a whole, and not to the loudness or condition of individuals. In another tit species, the blue tit, in which the chicks hatch asynchronously and therefore differ in size, the mother feeds the smallest and weakest offspring preferentially, whereas the father, like the great tit, feeds them according to their begging intensity and accessibility.

After a fortnight of hard work by the parents and fast development by the chicks, the young are almost fully fledged. Some of them are bigger than others, more active and reactive, and push the others aside when moving around the nest. The parents do not always bring the food into the nest, but put their head in and let the first chick that jumps up to the hole win the prize. They also start to entice the young to leave the hole and fly, by showing them food at the nest entrance and then flying away with it.

The chicks jump around in the nest and try out their wings, preparing for their 'solo' flights. In the following days, the parents bring less food, and it is the 'bullies' amongst the chicks who get more than others. During this period, towards the end of the nestling stage, we see that the female has altered her behaviour and now feeds the smaller offspring more, while the male feeds the slightly larger ones. Is the female, who spent more time brooding in the nest, more familiar with the nest and her young, and therefore more sensitive to their nutritional requirements and less susceptible to loud begging that may not be indicative of need?

As time passes, the food brought by the parents becomes insufficient, even for bullies, and eighteen days after hatching, hunger drives the youngsters, one by one, out of the nest. But they are not yet independent. The fledged brood stays with the parents, particularly with the male, and together they explore the parental olive-orchard territory for another week or two. While foraging together, parents and young frequently utter metallic-sounding contact calls. During this period in early May, the young still beg vigorously for food, and the olive orchard is filled with the loud begging calls of the youngsters and their wing-shivering displays. The parents still feed them, but, with each day, the youngsters beg more, receive less and learn to manipulate more prey on their own. By the time they are a month old, they are independent.

Who really controls the allocation of resources in the tit family? Obviously it is the parent who distributes the food, but are parents blackmailed, through their offspring's begging, to feed vigorously begging individuals at the expense of others and themselves, and against their own better (evolutionary) judgement? Do the loudness of a chick's begging and the intensity of its display honestly reflect its hunger, or do the chicks 'cheat'? How much competition is there between the siblings in the nest, and what do the parents do about it?

Gene-centred evolutionary conflict theory offers ready answers to these questions. As with the interpretation of male–female interactions discussed in the previous chapter, the basic assumption is that behavioural clashes reflect underlying evolutionary conflicts caused by different and non-matching fitness interests. In what follows, we want to look at this commonly held view, and then see how it is affected by our assumption that learning is central to the relationship between parents and offspring. We will argue that the evolutionary conflict interpretation of the parent–offspring relationship is problematical, first because the terminology is misleading, second because the data are inconclusive, and third because, in many cases, hostile patterns of behaviour may not reflect evolutionary conflict. We will show that taking into account the social learning that occurs within families leads to somewhat different interpretations of behaviours that are traditionally thought to result from the different fitness interests of members of families.

Conflicts, quarrels, and the co-evolution of parent–offspring relationships

As we saw in the previous chapter, evolutionary conflict is said to occur when the spread of a gene that increases the fitness of its carriers, while lowering the fitness of other individuals in the same population, leads to the evolution of genetic counter-measures. For example, a mutant gene that causes a young individual to act in a selfish, bullying way towards its sibs, extracting more than its fair share of resources from its parents, is expected to increase the reproductive success of its carrier. The selfishness-inducing gene therefore spreads, and sibs compete with each other for parental care. But the fitnesses of parents depend on the survival and reproduction of all their offspring – future offspring as well as present offspring. The interests of parents thus clash with those of offspring who act selfishly. As parents and offspring have such conflicting fitness interests, selection leads to changes in offspring that make them attempt to maximise their fitness at the expense of their parents, and to parental counter-measures and manipulations that prevent this. An evolutionary 'war of attrition' between the interests of competing offspring and their parents ensues. According to conflict theory, the story of the co-evolution of parental care and offspring responses should be told in terms of such an evolutionary arms-race.

The familiar instances of sibling rivalry certainly provide good examples of competition that could drive evolution. Sib competition is truly competition for survival and future reproduction. Sibs share a resource, unique to the family set-up, whose distribution will have repercussions for the rest of their lives. They have to share parental care. Bird and mammal youngsters who are partially deprived of this resource are likely to die or to become physically and emotionally enfeebled and crippled. When mammalian young compete for milk and maternal attention, it is not only the immediately obvious aspects of care – food, warmth and security – that they are after. Milk, as we have seen, contains both immunological defences and information about preferred foods, while maternal licking and handling during lactation accelerate and regulate the proper rate of maturation of the nervous system. No wonder that sibs compete, often ruthlessly, for parental care and attention. In some rodents, competition between the pups for access to the mother's nipples has led to the evolution of a tenacious grip on the nipples that is associated with specially adapted incisors that enable the pup to lock on.[2]

An extreme and exceptional example of competition between offspring is seen in siblicide. There are several species of birds in which active siblicide is an integral part of the life style.[3] One of the best-known cases is that of Verreaux's eagle. This spectacular raptor lives in the rocky hills, mountains and gorges of northeastern and southern Africa. Between April and June, the female lays two eggs in a nest built on a cliff ledge. These two eggs are laid and hatch about three days apart, so the eaglets differ in size and strength, the older chick being much larger than its younger sib. Soon after the younger sib hatches, the larger eaglet launches brutal attacks on it. In the great majority of cases, these attacks lead to the death of the younger sib. The parents passively watch this drama, never interfering. The only chance of survival for the younger sib is if the older one dies of disease or accident. Although the older chick certainly seems to benefit from the death of its younger sib, getting exclusive care from its parents, it is not at all self-evident that its selfish behaviour conflicts with the evolutionary interests of the parents. Whatever the parents' investment in the young chick, it is usually a short-term one, and it provides an insurance policy against total loss if the first egg fails to hatch or something happens to the older chick.

Murders in the family, although dramatic and striking, are the exception rather than the rule among birds and mammals. Does this mean that parents are usually successful in curbing the selfish, siblicidal tendencies of their offspring? Why are bloody clashes so rare? Since the 1960s, when William Hamilton presented his theory of kin selection, it has been recognised that ideas about selfishness and altruism have to be moderated by considerations of relatedness – by how closely related genetically the interacting individuals are. Hamilton argued that a mutant gene that leads to the altruistic behaviour of an individual might spread if it benefits the survival and reproduction of kin. For the gene to spread, the cost to the carrier of the 'altruistic gene' must be less than the overall benefit to all the kin enjoying the carrier's generosity, weighted by their relatedness to it.[4] For example, suppose the average number of offspring an individual can potentially parent is two, but some individuals have a gene that causes them to give up reproduction and instead help their parents raise, on the average, three additional full sibs. Individuals are, of course, related to their own offspring to the same extent as to their full sibs, sharing 50 per cent of their genes with them. The altruistic gene will therefore become more common if its carrier invests in rearing three full sibs rather than only two offspring. The effect of the altruistic behaviour on the frequency of the gene influencing this behaviour is therefore no longer just the result of its direct effect on the fitness of the carrier. The frequency of the gene depends on the impact of this gene, through the carrier's behaviour, on the fitness of the carrier *and* those family members (who are likely to carry the same gene) enjoying the benefits of its altruistic behaviour towards them. Hamilton thus defined a new and broader concept of fitness – inclusive fitness – which is more appropriate for describing the effects of genes affecting social interactions.

Hamilton's insight lead to an elegant solution of a long-standing enigma – the reproductive self-sacrifice observed in many bird and mammal families and, most dramatically, in social insects such as many bees, wasps and ants, where most workers are at least partially sterile (unable to produce daughters), and seem to work 'for' the good of the large family unit, the colony. For a long time, it had not been clear how such altruistic behaviour could be stable through evolutionary time. Why is it not obliterated by genetically selfish individuals, who reproduce themselves and exploit the more altruistic members of the colony? Surely genes for selfish behaviour, not altruism, should spread? But, as Hamilton pointed out, individuals in a colony are related, and their relatedness reflects the probability that a particular individual and its kin share the same altruism-affecting gene. When such a gene leads to an increase in the overall reproductive success of individuals within the kin group who carry copies of it, it will spread. This may happen even if the cost to *some* individual carriers is high, and even if most carriers forgo reproduction altogether, as do most worker ants, social bees and social wasps. Hamilton's theory demands that we move backwards and forwards among three levels of analysis: the gene, the individual and the kin group. The frequency of the behaviour-affecting gene rises or falls, particular individuals reproduce more or less successfully within the kin group, and the kin group is more or less thriving compared to other kin groups. By looking at evolution from the point of view of a gene affecting social behaviour, Hamilton provided a viable explanation for altruism and co-operation in the family.

Of course, altruism need not be expressed only in dramatic demonstrations of lifelong celibacy or suicidal self-sacrifice. Consider a family system with long-term monogamy. In such a family, an offspring is related to its future full sibs and its own future offspring to the same extent – it shares 50 per cent of its genes with its offspring and with its sibs. So if an individual helps its parents to produce more offspring than it itself is likely to produce during the same period, then (at least for a while), in terms of its inclusive fitness, it gains more by being a helper than it would from being a reproducer. We have to bear in mind that the prospects of very young individuals becoming successful reproducers are often smaller than those of their parents, who have already proved they can successfully acquire and hold a territory, find a mate and raise young. If the parents are in their prime, healthy and successful, the inclusive fitness of a youngster may be increased by spending time helping its parents before it becomes a reproducer. How much and for how long the youngster should help depends on the ecological opportunities open to it, the likely reproductive success of parents and young, the chances of the death of one of the parents, and the parents' loyalty to each other.

In the 1970s, Trivers looked at the empty half of the cup – at the dissonance and selfishness in families rather than the harmony and altruism. He noted that Hamilton's theory does not suggest that there will be no conflicts of interest between parents and offspring and between sibs; in fact, the theory predicts their existence, and sets definite limits to the extent of altruism. These limits are dictated by how closely related individuals are, and lead to theoretical predictions of the extent of conflict between different family members. For example, since an individual shares 50 per cent of its genes with its own offspring and 25 per cent with the offspring of its full sibs (its nieces and nephews), it is expected to have a strong, but not a boundless, interest in producing its own offspring rather than helping its sibs to produce theirs; it might even behave selfishly, at its sibs' expense. When, by failing to act selfishly, an individual suffers a fitness cost that is greater than the overall benefits to its kin weighted by their relatedness, selfish behaviour will be favoured by natural selection, and overt conflicts among kin are expected. Hence, since an individual is more closely related to its own offspring than to its nieces and nephews, it will tend to promote its own reproduction at the expense of that of its sibs. But, since parents care equally about the survival and reproduction of all their progeny, they will try to curb the selfish behaviour of any offspring that harms their other offspring too much. Conflicts between parents and offspring, as well as selfish conflicts between sibs, are therefore expected.[5]

Trivers was impressed by the seemingly aggressive encounters between parents and offspring that take place around the time at which bird and mammal offspring are gradually becoming independent of parental feeding and other forms of care. He described the behaviour of free-living domestic pigeon parents towards their offspring: at first both parents encourage the chicks to feed, but later, as the chicks grow, the parents become more ambivalent, and eventually they try to avoid the begging chicks. This sequence of behaviours is typical of many species of birds, and is functionally quite similar to the gradual decrease in food provisioning in the great tits that we looked at earlier. Trivers also noted how two-year-old human babies, young baboons and langur monkeys display temper tantrums when the mother refuses to suckle them or let them ride on her back. The young utter piercing cries, roll on the ground and kick vigorously. Pink-backed pelican chicks seem to behave in a similar fashion: a begging, frustrated chick may occasionally attack other chicks and bite its own wing, growling and shaking its head all the time, and may even throw itself on the ground, beating its wings wildly.[6] Since in mammals these patterns of behaviour appear around the time of weaning, Trivers interpreted them as behaviours reflecting the conflicting genetic interests of parents and offspring: parents try to increase their own fitness by weaning the young as early as they can, so that they can save energy for the next reproductive season, while the young try to continue to exploit parental resources by psychologically manipulating the parent to extend the care period.

From just-so stories to just-so strategies?

If genetic relatedness as well as social and ecological conditions are taken into account, it is possible to explain almost any behaviour, from the amazing acts of self-sacrifice of functionally sterile honey-bee workers that are ready to die for the defence of their colony, to the murder of sibs by Verreaux's eagles. But interpreting the clashes between parents and offspring in terms of evolutionary conflict is not straightforward. One reason for this is that the fundamental asymmetries between parents and offspring have to be taken into account. Theoretical models of parent–offspring co-evolution have shown that, when genetic relatedness is the sole consideration, the fitness interests of offspring and parents do not coincide; the resolution of the conflict is a compromise in which offspring extract more than the parents want to give, but not as much as the offspring would like to get.[7] But these models are too simple. Parents and offspring are not equal contestants. Fitness calculations have to incorporate in some way the different ecological opportunities open to parents and young, and to take into account which party controls the allocation of resources. Since parents very often control resources, they are in a position of power within families, and thus are more able to forcefully impose their 'point of view' in spite of their offspring's attempts to manipulate them. They can regulate the numbers of offspring (by killing some, or allowing sib competition to accomplish brood reduction), discriminate in favour of some rather than others,[8] evade a whingeing offspring, and force a youngster to behave more altruistically towards its sibs and towards the parent. Taking into account this asymmetry in the relationships between parents and offspring greatly alters the evolutionary outcome of the simple symmetrical model.

Another problem with conflict interpretations of parent–offspring relationships is that it is not at all clear that 'cheating' by the offspring is a viable evolutionary possibility. Why should parents succumb to their offspring's selfish demands? An offspring's begging is a message indicating a need for more care (usually more food), which the parents must evaluate. In previous chapters, we have discussed Amotz Zahavi's idea that communication between individuals must be costly for it to be reliable, and that extravagant traits, such as the peacock's tail and the nightingale's song, evolved as costly, reliable signals between mates. Zahavi has applied this idea to the evolution of the communication between parents and offspring. He argued that the message the offspring sends to its parent must be honest, or else genes that lead parents to be able to expose fraud and ignore extra begging will soon spread. Since an honest message must be costly, for otherwise cheats may make use of it, offspring in real need are expected to beg in an exaggerated, costly manner. Thus, according to Zahavi, the excessive begging of truly hungry offspring costs them extra energy or puts them at increased risk of predation. Nevertheless, it is still worthwhile for them to beg in this way, because they have more to gain by their costly begging than offspring who are well fed. Hence, Zahavi argues, begging must be costly and exaggerated, and its intensity is an inevitable result of the evolution of honest signalling rather than a sign of psychological blackmail.[9]

The difficulty with Zahavi's hypothesis is that it is far from clear how costly begging actually is. The begging of nestling songbirds can, in fact, be energetically quite cheap, yet could still be honest.[10] It is also possible that an offspring demands as much as it can because it does not know how bountiful its parent actually is, and the only way to find out is to ask for the maximum.[11] Moreover, infrequent begging or begging that is not noisy enough may be taken by parents as signs that these offspring are weak, and imply that their condition is beyond help.

The problems of who has the power in the family and the honesty of signals mean that deciding how the behaviours seen in parent–offspring relations have evolved is difficult. In the jargon of modellers, 'there are many confounding variables' in the real world. But we want to try to go beyond the tiresome statement 'life is complicated'. There is, we believe, much good sense in the assumption that strategies of parental care and offspring behaviour have been co-evolving – it is quite clear that the behaviour of the parent constitutes a major aspect of the selective environment of its offspring, and vice versa. It is equally clear that taking into account genetic relatedness is important in evolutionary explanations of social behaviour. What is far from being clear, however, is that parent–offspring co-evolution always entails conflicting fitness interests, and that the clashes observed between parents and offspring are overt expressions of their different fitness stakes. Proponents of the conflict theory rarely specify in advance the predicted set of alternative behavioural strategies, and with the addition of enough 'confounding variables' such as ecological circumstances that promote or suppress co-operation, it can remain the explanatory scheme for the evolution of parent–offspring relationships whatever the observed behaviour. As we know too well, by varying our assumptions or adding some more, even an erroneous theory can be made to explain everything. The Ptolemaic system of astronomy, which was based on the assumption that the sun revolves around the earth, could describe the known behaviour of the planets in the solar system very well indeed when enough 'confounding' assumptions about the movements of planets where added to it! Have evolutionary biologists moved from the just-so stories used in the past to explain every adaptation, to just-so strategies?

Evolutionary conflict is certainly not the only explanation for the agonistic behaviour observed in families. Mock and Parker have warned that quarrels in the family do not necessarily represent *genetic* conflicts. Using the word 'conflict' to describe both quarrels and opposing fitness interests does not make them causally related. In the previous chapter, we saw that the evolutionary conflict between males and females is usually not expressed as overt behavioural disagreements or hostility between them. The opposite is also true – a behavioural conflict need not imply there is an evolutionary conflict. In fact, in spite of the huge amount of theoretical work on parent–offspring conflict, for birds and mammals there is hardly any empirical data that unambiguously lead to the interpretation of quarrels in terms of genetic conflicts between parents and offspring. After an extensive review of the studies that tested the parent–offspring conflict hypothesis, Mock and Parker concluded that although 'many squabbles have been interpreted as evolved manifestations of POC [parent–offspring conflict], few convincing tests support such claims'.[12] Patrick Bateson, who discussed the dynamic interactions between mammalian mothers and their maturing offspring, came to the same conclusion. He showed that mother and offspring are very sensitive to environmental conditions and to each other's abilities and needs, and respond in a way that is optimal for both. He concluded that the fitness interests of mothers and offspring are usually similar.[13]

In addition to the glaring lack of data supporting the interpretation of family quarrels in terms of evolutionary conflict, there is another problem. It is the problem of the language that is used and the confusion that some of this language tends to induce. Natural selection requires the existence of heritable variations that affect fitness, so any and every outcome of natural selection is a consequence of fitness differences. To talk about all selection in terms of 'competition', or to call the genes 'selfish', provides us with no additional insights, but may lead to a lot of misunderstanding. For example, to interpret the evolution of co-operation in terms of 'competing', 'selfish' genes, just because the genotypes affecting co-operative behaviour spread more successfully than the alternative genotypes, does not help us to understand the evolutionary process leading to co-operation. This approach entices us to explain co-operation in terms of the resolution of evolutionary conflict, and to ignore other possible explanations. But other explanations do exist and should be considered. For example, as we discussed in the previous chapter, in monogamous relationships some forms of co-operation may be inevitable and adaptive side-effects of learning with and from mates. When we transfer terms from the level of the observed phenotypes to that of genotypes, we tend to interpret the genetic evolution of the behaviour in the same terms that we use for describing the phenotypes.[14] Translating agonistic behaviour among family members into 'evolutionary conflict' may mislead us. The assumption that opposing fitness interests invariably underlie agonistic behaviours is neither logically necessary, nor is it founded on much convincing data. It is just one of several possible explanations for observed behaviours.

So what is the alternative to the standard, conflict-based interpretation of parent–offspring squabbles? We are going to argue that many family clashes form an integral part of social learning and behavioural maturation processes, and that the intimate interactions between parents and offspring are usually to their mutual evolutionary benefit.

Weaning conflicts or parentally guided maturation?

One of the main inspirations for the interpretation of family relationships in terms of parent–offspring genetic conflict was the very obvious, and sometimes dramatic, clashes between parents and offspring that are often seen around the time of weaning. Can these squabbles be interpreted in terms other than those of conflicting fitness interests? A look at any weaning squabble suggests to us that it is an inevitable part of the process of behavioural maturation, during which the young are guided towards independence by their parents, sometimes against the youngsters' immediate and short-sighted wishes. Far from being an overt manifestation of a parent–offspring genetic conflict, we see weaning squabbles as a manifestation of the difficult processes of learning and unlearning. To illustrate why, we can use a classic example from Robert Trivers' book *Social Evolution*, in which he describes the work of Nicholas Davies on the interaction between parents and offspring in spotted flycatchers. These interactions are interpreted by Trivers as a clear-cut illustration of parent–offspring conflict, with parents 'hurrying' their offspring to independence. Trivers writes:

> For the first nine days that the spotted flycatcher chick is out of the nest, most parental feeding occurs when the chick is silent (instead of calling); but after ten days the parent brings food only when the offspring calls. At the same time there is a sharp increase in the rate at which chicks chase after their parents. Between the 10th and the 16th day the parents tend more and more to feed their chicks only after being chased by them and, at the same time, an increasing percentage of chases fail to result in food transfer [fig.]. There is even a decrease in the size of the food items transferred.
>
> The joint effect of these actions is a very sharp reduction in the amount of food transferred by the parent per unit time; by the 14th day the spotted flycatcher offspring does better by capturing prey itself than by begging food from its parents. This is exactly the time when it shifts from getting most of its food from its parents to getting the bulk by itself. In effect, the parent forces more and more of the burden of feeding onto the offspring. The offspring is forced first to call, then to chase, then to chase for longer and longer periods, all the while it is being provided with food, time, and – increasingly – the motivation to perfect its own prey-capture techniques. Most of the improvement in these techniques comes while the parent is still providing most of the food, so although the parent hurries the offspring to independence, it still seems to provide a cushion of safety for the offspring to develop its skills.
>
> (Trivers, 1985, p. 154)[15]

Any human parent reading this description will recognise the methods of positive and negative reinforcement that they themselves constantly use to encourage their young children to learn the new behaviours that are increasingly necessary for their future survival, and to dissuade them from behaving in ways that are either no longer adequate or are liable to be positively harmful. It is difficult to believe that this interpretation would not be accepted by Trivers and other biologists who see a reflection of evolutionary conflict in the sequence of events described for the spotted flycatchers. The 'conflict' that they see seems to stem not from the learning sequence itself, nor from the actual 'extorting' behaviour of the offspring, but from the assumption that parents *hurry* their offspring to independence. However, no indication of *hurrying* is evident in the behaviour. The training sequence in this and other cases seems to be carefully timed to ensure the most efficient learning through conditioning, with the level of stress being adjusted so that the youngster's motivation to learn is increased but its ability to learn is not hampered. There is no reason to think that, if the parents deferred the start of the learning session, or prolonged it, the fitness of the offspring would be enhanced. As any parent knows, timing is all important in teaching. This is particularly true when the breeding season is short and the 'education' of the youngsters has to be completed before the favourable season is over, as it is in flycatchers and many other birds and small mammals. It seems likely that the behavioural sequence and its timing is a result of selection for learning in the young and for the ability of parents to control this learning process of their offspring. Looked at this way, the squabbles between parents and young are not an outcome of evolutionary conflict, but are inevitable results of the learning process and the somewhat painful transition to the youngsters' independence.

It can be argued that it would have been much better from the evolutionary point of view if youngsters accomplished the transition to independence painlessly – if the transition were smooth and amicable. This, however, would be possible only if all aspects of foraging were completely innate, and if environments were fully predictable and free of competition. Only in such conditions is selection likely to favour an automatic switch from parental feeding to independent foraging. However, usually environments are neither free of competition nor entirely predictable. In almost all species of birds and mammals, young have to learn where, when and sometimes also how to forage. Mild hunger is probably a reinforcer of exploratory feeding behaviour, but offspring who have been accustomed to regard their parents as the source of food have expectations and demands that are directed towards their past benefactors. The behavioural clashes of parents and their maturing young reflect the learning process, in which both positive and negative reinforcement is involved. By definition, learning through negative reinforcement is neither pleasant nor amicable! But learning is to the advantage of both parents and young.

In monkeys, the disputes between mothers and offspring are often about the time at which suckling occurs, rather than about the amount of suckling. These disputes provide a good example of training by parents. The timing of suckling is very important, because, as the infant grows, its hold on the nipple can interfere with the mother's foraging. It is easier for both mother and offspring if suckling occurs when she is resting or socialising, rather than when she is out searching for food. Studies of yellow baboons, gelada baboons and other monkeys indicate that the time at which the infant suckles is controlled by the mother, and conflicts are the result of her training the infant to reach for milk at the right time rather than about the amount of suckling.[16] Learning the best time to suckle is to the benefit of both mother and her young.

Although the conflict theory view of parent–offspring squabbles assumes that hurrying their offspring to independence increases the fitness of parents, whereas remaining dependent increases the fitness of young, these assumptions are not substantiated by any data. It is at least as reasonable to assume that the fitness of the young will decrease, not increase, if the parents are too indulgent. If parents 'give in' to their offspring's demands, not only are the parents likely to suffer, but their offspring may fail to reach independence at the appropriate time and consequently be unable to compete successfully for food and status. Studies of primates and rodents have shown that when mothers are too protective, their offspring are excessively cautious and cowardly, and are reluctant to explore new objects, environments and situations.[17]

There is a detailed and long-term study of family relationship in domesticated budgerigars which has shown how important it is to take social learning into account before interpreting the behaviour of young and their parents in terms of genetic conflict.[18] At first it seemed that the behaviour of budgerigar parents and offspring would provide a particularly good illustration of evolutionary conflict. It was found that in nests in which both parents fed the young, offspring who begged more did indeed receive more food. In these asynchronously hatched broods, younger offspring begged more and received more than the older ones did when they were the same age. The parents, especially the father, seemed to be manipulated by the more noisy offspring, who seemed to get more than their fair share of food, just as evolutionary conflict theory predicts. However, contrary to the expectations of conflict theory, the increased begging and feeding did not result in an increase in an individual chick's size and weight at fledging. The investigators discovered that fathers tend to give more food to broods containing more daughters than sons (why this should be is not clear), and that, within a day or two of the father starting to feed them, the begging rate in these broods increased dramatically. This strongly suggested to the researchers that begging in such nests is positively reinforced by the behaviour of the feeding father.

The increased begging rate of these well-fed young budgerigar offspring may well be the result of social learning. Since budgerigars hatch asynchronously, there are substantial age differences among siblings. Those that hatch late are exposed throughout most of their chick life to the high begging rates of older sibs. It is therefore reasonable to assume that, as a consequence, the younger sibs, who have a longer exposure to begging, learn to display a particularly high begging rate.[19] In other words, younger nestlings may beg more frequently not because they need more food than their older sibs did, but simply because they have been better trained to do so.

Human beings provide another example of behaviour that is now interpreted in terms of social learning rather than conflict. These days, the tantrum fits of young children are no longer interpreted as the expression of weaning conflicts, with the child demanding more food and attention than a reluctant parent is ready to give. They are interpreted as expressions of frustration between two contradictory needs of the child: the need to get away and establish autonomy from the parent, and the need for parental security. Like young primates,[20] confident, well-attached children display fewer tantrums, because they are able to explore the world with a feeling of security, using their parents as a safe base. Temper tantrums in children are not a way of blackmailing the parent.[21]

Although the patterns of behaviour involved in so-called 'weaning conflicts' may have little to do with weaning or with evolutionary conflicts, we believe that some forms of begging and nagging are more than inevitable by-products of maturational learning. Any observer of human or animal behaviour has noticed that before, during, and after weaning, youngsters are masters of nagging their parents. This nagging takes many forms. Bird chicks, whether hungry or not, frequently beg from their parents, and even overgrown fledglings, who are already able to forage independently, often attract their parents' attention using begging displays. Mammal cubs are very good at pestering their mother. They may bite at her heels, pull tender parts of her anatomy such as ears or tail, or play-attack her when she is resting. Every dog or cat lover is familiar with the sight of a weary mother, dragging her feet along, closely followed by an excited gang of youngsters, each trying in its own way to attract her attention. Young children are particularly good at nagging, using every trick in the book, from intentionally dropping toys to crying loudly or blocking their mother's way, to attract her attention and ensure her close proximity.

Frequently these young are not after food. Sometimes they are looking for emotional security, but this is not the whole story, because they not only want the proximity of the parent, but also insist that the parent is active. We suggest that young birds and mammals, as well as children, are using active measures to obtain information and behavioural demonstrations from their parents. Youngsters are extremely curious, constantly seeking stimuli and information. Frequent nagging makes the parent play with the youngster, clean it once again, feed it a little more often or give it a comfort-suckle, thus repeating various behaviour patterns time and time again. Eventually the parent becomes impatient with the nagger and throws it off or withdraws out of reach, but parents are often remarkably patient, even when very tired. Nagging can be seen as an active measure employed by the young to extract the utmost of their parent's attention, and thus maximise the parental flow of information. Nagging, or even misbehaving, makes the parents aware of the developmental state of the offspring, and often induces the parents to add to or repeat their actions, thereby allowing the young to learn parental behaviour patterns more effectively. For example, we have seen that greylag goose parents utter action-inhibiting calls towards young who are prematurely attempting a solo flight. If the youngsters disregard the parental warning call and fly off, the parents quickly join them, assume the lead and land safely upwind with them, thus demonstrating how to avoid crash-landing. In mammals, a youngster who wanders too far is seized by its mother and brought back, but it tends to repeat its meanders despite parental exasperation and often apparent displeasure.

Overall, scrutiny of squabbles does not reveal convincing evidence for parent–offspring evolutionary conflict. Squabbles, nagging and rows seem to be either an inevitable part of a trying process of maturation through social learning, or honest signs signalling a need for food or information, or the outcome of premature attempts at independence that are suppressed by the parents. The apparent disagreements serve both the parents' and the offspring's long-term fitness interests.

Conflicts in the womb and beyond

At this point, we would like to digress to look at an aspect of parent–offspring conflict that is only indirectly related to overt behaviours and social learning. In the late 1980s, David Haig added an interesting molecular twist to evolutionary parent–offspring conflict theory. This twist is based on a phenomenon that was first discovered in the 1920s in insects, but has been found since in many different taxa. In many organisms, including mammals, there are genes that are expressed in a different way depending on whether they have been inherited from the male or female parent. For example, an allele derived from the mother may be active, and that from the father inactive (or vice versa), just because of the sex of the parent from which they have come. This parent-dependent difference in gene activity is the consequence of the different ways in which the gene is 'marked' during the processes of sperm and egg production. 'Marking' involves changes in the chemical groups attached to the gene's DNA or in the proteins bound to it. The differential marking of genes in males and females, and consequent differential gene expression in the embryo, is known as genomic imprinting.[22]

Haig suggested that, when embryos are nourished by maternal tissues, there might be a conflict of interests between the parents, which would be reflected in the way maternal and paternal genes are imprinted, or marked.[23] He argued that the conflict would be particularly acute when the relationship between the mates is not one of long-term monogamy. In such relationships, the father's interest is that his embryo exploits maternal resources as much as possible; since his next offspring will probably be with another female, the father's fitness will be unaffected if his embryos reduce their mother's ability to provision her future offspring. The mother, on the other hand, wants to distribute her resources equally to all her offspring, both present and future. The expected outcome is therefore an evolutionary arms-race between paternally and maternally derived genes that affect the embryo's growth. Paternal alleles are expected to be marked in ways that result in the expression in the embryo of growth-influencing genes that lead to the extraction of more resources than would be optimal for the mother; maternal alleles are expected to be marked in ways that lead to gene expression that allocates resources more equally and frugally, and counters and suppresses any overexploitation caused by paternal genes. The offspring, who are interested in their own reproductive success more than in the reproductive success of any future half-sibs, are expected to extract as much nourishment from the mother as they can. The interests of the offspring and their father therefore coincide.[24]

Haig's idea has been used to explain the pattern of expression of imprinted growth-affecting genes in the prenatal period, when the embryo is nourished by the maternal tissues. However, it should also apply to the way some genes are expressed after birth, before the young become independent. Studies of the regulation of lactase-I, the enzyme that breaks down the milk sugar lactose, might enable us to see whether offspring manipulate their mothers (via genomic imprinting) to wean them later than is optimal for her. As we discussed in chapter 1, with a few interesting exceptions, lactase-I activity is turned off during the weaning period, so adult mammals cannot make use of fresh milk as an energy source. Its consumption often causes them indigestion and diarrhoea. This benefits the mammalian mother, since it means that she will not have to rebuff the suckling attempts of a greedy grown-up offspring, and will have more resources for her next litter. Since once their lactase-I enzyme is turned off young mammals tend to get indigestion after drinking milk, grown-up offspring probably develop an aversion to it and this results in a more peaceful process of weaning. The evolutionary interest of the mother is therefore that the gene or the enzyme is turned off as soon as her offspring can look after themselves. The offspring, on the other hand, may have an interest in having the gene turned off later, so they can extract as much milk as possible from their mother, even if it is at the expense of their future sibs. This interest will be particularly strong if litters are unlikely to have the same father, and, consequently, future sibs are likely to share only 25 per cent of their genes. Maternal and offspring interests therefore differ, just as they do in the womb. The time during weaning when the lactase-I enzyme is turned off could therefore be the result of an evolutionary compromise between conflicting maternal and offspring interests. Alternatively, it could be to the benefit of both the mother and the offspring, because it leads to an efficient process of weaning. So how can we tell?

If the regulation of lactase-I activity is determined by an evolutionary conflict between mother and offspring, we expect the gene inherited from the father to be marked in a way that ensures that lactase-I activity continues for as long as possible, whereas the maternal gene will be marked in a way that will lead to a more limited activity. The imprints could be on the regulatory region of the lactase-I coding gene, or on any gene that regulates the lactase-I gene or enzyme. We also expect that maternal factors will re-mark paternal genes in ways that correspond with the mother's interests, partially or fully suppressing the paternally derived regulation. If, on the other hand, the regulation of the lactase-I gene was not driven by evolutionary conflict, no differential genomic imprinting and maternal suppression should be observed.

Following the same line of reasoning, if parent–offspring conflict is driving the evolution of parent and offspring interactions, we expect that, where there is polygamy, mothers will have additional ways of countering possible exploitation by their mates or offspring. For example, mammalian mothers could overcome the extorting strategies of their offspring through the evolution not only of ways of suppressing the effects of paternal marks, but also through a shorter pregnancy. This would minimise the opportunities for her embryos to exploit her excessively, and give the mothers earlier behavioural control over their offspring. Another way in which the mother could control weaning would be through changing the ingredients of the milk with time, introducing into it factors that lead to suppression of the activity of the lactase-I gene or enzyme in the offspring, or that cause an aversion to milk. Counter-strategies would be expected in the offspring, of course. If more of such maternal measures were found in polygamous and promiscuous species than in monogamous ones, it would indicate a role for evolutionary conflict in the evolution of this aspect of mother–offspring relationships. A lack of such evidence would point to more congruous mother–offspring interests. Comparing genomic imprinting of the lactase-I regulator/s, the length of pregnancy, and changes in milk composition and its effect on suckling, in species practising long-term monogamy and species with other mating systems, might help to distinguish between the two hypotheses about the evolution of the weaning process in mammals. Unfortunately, although molecular biology may provide us with potentially useful tests, at present relevant molecular data are non-existent.[25]

Parental control: 'From each according to [its] abilities, to each according to [its] needs'[26]

There is no doubt that bird and mammal parents are able to control some of the behaviour of their offspring. In some cases, by using overt physical harassment or other methods, they even stop their offspring from reaching reproductive independence. This may be against the immediate wishes of the offspring, but it does not reduce their inclusive fitness, because the young usually remain and help their parents raise another litter or brood. We shall discuss the evolution and evolutionary effects of this helping behaviour in some detail in the next chapter. For present purposes, it is sufficient to note that, as with the situations discussed earlier, the control a parent has over its offspring seems to be used for the benefit of both parties. There is no evidence that there is an evolutionary conflict of interests.

What about the converse situation – where offspring manipulate their parents? Are there cases in which offspring act in ways that increase their own chances of living and reproducing at the expense of the fitness of their parents? We know of no evidence of this in birds and mammals.[27] In these groups, it is the parents who are in command of resources and are able to use both active and passive measures to affect the number, well-being and behaviours of their young. Yet it is frequently argued that the seemingly selfish behaviour of some young is evidence that they are manipulating their parents in ways that potentially reduce the parents' fitnesses, while increasing their own. To see if this is so, we need to look more closely at what goes on in situations where the selfish behaviour of one offspring could reduce the parents' overall reproductive success. Many of the observations suggesting parent–offspring conflict have been made on birds, where sibling rivalry is particularly apparent. In extreme (but rare) cases, older birds kill their younger sibs. Usually, however, rivalry is expressed more subtly, as we saw with the great tit nestlings, where there was competitive begging and active manœuvring to find the best position, the one closest to the feeding parent.

Competition between nestlings is particularly likely when the young are of different ages. The hatching intervals between chicks are the result of a behavioural decision of the parents over when to start incubating. Typically, a hen bird can lay no more than one egg a day. The eggs, which need a high, well-regulated temperature for normal development, usually start to develop only after the initiation of incubation. If a bird does not start incubating until the last egg is laid, the young hatch synchronously and are of roughly the same size. This seems to be the situation with early broods of great tits in the Judaean hills. But, if incubation is initiated before the last egg is laid, the eggs hatch asynchronously – the first to be incubated is the first to hatch. The outcome in most cases is an age-related size difference between nestlings. In times of scarcity, this difference in size may lead to the ultimate death of the last-hatched nestlings, either because they are unable to compete with their larger siblings, or because parents tend to feed large nestlings. In many cases, it seems that, although the hens have the ability to control the hatching pattern, they choose a pattern that leads to the death of some offspring. The question is why.

There are at least seventeen different explanations for the evolution of hatching asynchrony.[28] As we emphasised when we discussed maternal licking behaviour and zebra stripes in chapter 2, there are commonly several factors influencing the evolution of any complex trait, and hatching asynchrony is no exception. Most evolutionary explanations of the asynchrony suggest that it represents the parents' way of maximising their own reproductive success. For example, a popular hypothesis, originally suggested by David Lack in 1951, posits that, when the environment is unpredictable, parents may start incubating a large brood early so that the chicks hatch asynchronously. If the breeding season happens to be particularly bountiful, the largest possible number of chicks will be raised. But, in a lean year, hatching asynchrony will lead to an automatic reduction in brood size, since the young and weak will die before much precious parental time and energy are invested in them.[29] Another hypothesis suggests that hatching asynchrony is a parental 'insurance policy' against unexpected losses, in case older nestlings die prematurely. The two asynchronously hatching eggs laid by Verreaux's eagle and many other large raptors seem like very good examples of this 'insurance' strategy, since the parents are rarely able to rear more than one offspring. Hatching asynchrony has also been interpreted as the parents' way of reducing competition among sibs, since competition is claimed to be greatest among individuals of equal size.

According to the class of explanations just outlined, parents are controlling brood size by manipulating the degree of sibling rivalry. Sibling rivalry is therefore not a reflection of the different evolutionary interests of parents and offspring – it is a mechanism through which parents directly maximise their own fitness. Others argue, however, that, in species with a short breeding season, hatching asynchrony is merely a side-effect of selection for early incubation. In such cases, the asynchrony can be exploited by selfish first-hatched offspring, who may harass or even kill their younger sibs, although this is not in their parents' interests. Consequently, parental measures that prevent older offspring from pursuing their selfish goals have evolved. It is significant that among many asynchronously hatching passerines, egg weight often increases from first to last-laid egg.[30] As these weight differences are translated into size differences in the nestlings, the extra investment in late-hatched eggs may be an evolved parental counter-measure that minimises the danger of early hatched sibs exploiting those hatched later. Canary mothers have a more sophisticated way of compensating for the differences between early and late-hatched offspring: it is not the size of the egg, but rather the temperament of the young that is controlled by the mother. The eggs that are laid last are supplied with higher titres of testosterone. Consequently, the late hatching and therefore smaller offspring are more aggressive than their older and larger sibs, compensating for their small size with extra belligerence.[31]

Another parental measure that helps the smaller and weaker offspring is discrimination in their favour during feeding. Parents usually feed the chicks closest to them, but, when they do discriminate, they commonly favour the smaller chicks. In horned grebes, pied flycatchers, budgerigars and blue tits, one or both parents preferentially feed the smaller offspring.[32] Similarly, in moorhens the parents are more aggressive towards larger and more competitive chicks, thus allowing the smaller and weaker more access to food.[33] Discrimination also occurs in American coots, where newly hatched chicks beg by exhibiting the ornamental orange plumage that covers part of their body until they are about three weeks old. The chicks are 'bald as a coot', and the naked patch on the top of their heads is also brightly coloured. Parents feed the young orange-plumaged offspring more often than older sibs who have lost these signs of immaturity. It is known that in other birds bright colours are indicators of health, so these visual signs may tell parents not only about a coot chick's age, but also about its well-being, and allow them preferentially to feed the youngest and most needy chicks in the clutch.[34] In pinyon jays, when the larger nestling continues to beg even though its bill is stuffed with food, the parent will remove food from this nestling's bill and feeds it to one of its less noisy sibs.[35] It seems that, in general, when parents actively intervene, they bias their investment towards smaller and weaker nestlings.[36]

The behaviour of the great tits we described earlier shows how parents can control the distribution of food to their young. It also illustrates something else that limits the ability of a 'bully' to grab everything – the demands of nest sanitation.[37] If damage to the nestlings' health from bacteria, viruses, lice and a host of other disease-causing parasites is to be minimised, excrement, which is frequently produced immediately after feeding, has to be removed from the nest. The faeces, which in songbirds are contained in a fairly tough gelatinous sac, are removed from the nest by one or both of two methods, both of which were known to Plutarch in the first century of the present era. The first, in which a parent actively solicits a dropping from a nestling and then either swallows it or carries it out of the nest in its bill, is more common for young nestlings. The second, in which the nestlings turn their tails out of the nest and then void their droppings without parental help, is more common for older nestlings. Either way, immediately after being fed, a nestling has to change its position and thus lose the prime feeding spot to another chick. Together with other parental behaviours, sanitation behaviour helps to ensure a relatively unbiased partitioning of the parental resource among nestmates.

In mammals, too, every detailed study of maternal care has shown that mothers carefully monitor the state of both their offspring and the environment, and regulate the number of young and the care they give to them. One rather extreme method of control is to kill the litter when the cost of raising it is too high.[38] Rodent mothers will usually kill and sometimes eat the whole litter if food is short or the mother is unable to find a safe place to nest. Females may control their litter size even earlier than this – in times of stress, mothers sometimes absorb some or all of their embryos. In times of plenty, on the other hand, mothers can afford to give extra care to frail offspring.[39] For example, goat and vervet monkey mothers who give birth to twins preferentially feed the weaker twin, and rats who are given pups younger than those they have been suckling, continue to lactate for a longer time. In vervet monkeys, conception of the next offspring is delayed if the present one is developing slowly, and white-tailed deer mothers extend their period of lactation in lean years. One spider monkey mother was seen to give extra care to her juvenile son after it lost its tail in an accident; although he was previously weaned, she nursed him and carried him around for five months after the accident.[40] It seems that, as with birds, mammalian mothers are in control of the allocation of resources.

Daughters and sons
Bird and mammal parents can affect the fate of their offspring through infanticide, through the regulation of feeding or through other forms

of behaviour. But do they have more subtle effects on their offspring's future? For example, is the care and education that they give the same for both sexes? We know that human parents both consciously and unconsciously treat and educate their male and female babies differently, according to the role in society they will be expected to play in the future. Yet, remarkably little is known about sex-specific transfer of parental information to the offspring in birds and mammals. In accounts of studies of monogamous species, which are mainly about birds, there is little to suggest that parents treat their sons and daughters differently.[41] At first this seems odd: after all, the future life styles of sons and daughters are going to be different, and we would expect parental care to have evolved in ways that tailor it to the future needs of the young. However, it is less odd than it seems, because the two sexes need much the same information, even though the way they use it will be different. For example, while a son needs to learn his father's song in order to sing in the right way later in life, a daughter needs to learn the very same male song, even though she herself will never sing it. She needs to know it for future diagnostic purposes – for selecting an appropriate mate. As we saw in the previous chapter, in some birds the local characteristics of the song are important criteria for mate choice. We know that female birds do learn the paternal song, because, if they are masculinised by treating them with testosterone, they start singing a version of the male song similar to the one they heard as nestlings. Similarly, although in birds the male is usually the philopatric sex, and therefore needs to know the local territory, we have seen that both male and female fledgling great tits, like many other bird species, join their father when he is foraging. The young sons learn from their father where and how to find and handle food in the local territory; they may also learn that this is how parents help their young to find food. Daughters also learn from their father the basic art of foraging; possibly they also learn the characteristics of a forager who in the future would be a compatible mate. Aspects of parental behaviour are therefore learnt by both sexes, even though males and females do not perform identical parental roles. As we see it, members of both sexes need to know their mate's behaviour in order to be able to achieve effective co-ordination. It is therefore not surprising that parents in monogamous species, where the lives of adult males and females are so intimately interwoven, transfer much the same information to sons and daughters. The same information and behaviour needs to be learnt by both, although its future use – whether for diagnostic purposes, for co-ordination with a mate or for actually performing some act – is likely to be sex-specific.

In polygamous species, things are different: there is evidence that sons and daughters are not treated in the same way, although, in the early stages of development, maternal care is often identical. Some of the differences between the behaviour of sons and daughters arise not because their mother treats them differently, but because the initial physiological and behavioural responses of male and female offspring are not the same. For example, although mammal mothers treat daughters and sons in the same way with respect to the nourishment they offer, the larger offspring (often the males) solicit or suckle more persistently. Therefore, at an early age, male African elephant calves, male lambs and the male young of many other species get more milk than their female sibs. However, when youngsters grow up, maternal behaviour does seem to be discriminatory, since the more frequent demands of the young males are rebuffed more often than those of young females. In sheep, maternal rejection is reflected not only in active rebuffal, but also in how close mothers stay to their male and female offspring. Mothers of male lambs keep a greater distance between themselves and their sons than do mothers of female lambs.[42]

Maternal discrimination over the amount of contact sons and daughters are allowed has also been found in monkeys. Mothers of pig-tail macaques and of captive rhesus monkeys rebuff contact initiated by sons at an earlier age than they do contact initiated by daughters. The result is that, relative to males of the same age, young females spend more time with their mother.[43] This makes sense because in most primate species females stay in the natal territory and lead complex social lives. They need to learn the idiosyncratic features of the local territory and the local social system; they need to form coalitions and other subtle relationships with the females, and to monitor the behaviour of males; and, of course, they will also be involved in long-term maternal care. Increased tolerance towards daughters gives them more opportunity to learn the local ecological and social peculiarities of the site, and allows them to become better socially integrated into the group. For males, who disperse during adolescence, it may be more advantageous to grow as fast as possible and acquire fighting skills, which will assist them to establish dominance status in their future groups. The longer exposure of females to the intricacies of family life may account for the observation that, in some polygamous primate societies, females have a more varied and richer repertoire of behaviours than males.[44]

How a primate mother behaves towards her infant depends not only on its sex and the likelihood of it dispersing, but also on her social rank. This can be seen in the extent to which rhesus monkey mothers let their offspring suckle.[45] The rhesus monkey social system, like that of most social mammals, is based on a group of females and their young, with adolescent males dispersing to other groups. The female lineages are organised in a hierarchical manner, and 'social mobility' is rather limited. Daughters inherit their mother's social rank, so daughters of high-ranking mothers become dominant, while daughters of low-ranking mothers remain low in social rank. Low-rank mothers are frequently in a poor condition, and are unable to produce much milk in a single bout of suckling. Their infants therefore demand and are given more frequent access to the nipples. Since in this species, as in so many others, frequent nipple stimulation inhibits reproduction, low-ranking mothers are usually unable to reproduce in the following year.

High-ranking rhesus monkey mothers are usually in good health, and rear their male and female offspring with equal success. Low-ranking females, on the other hand, not only start from an inferior position with regard to their nutritional condition, they also suffer more if they have a daughter than if they have a son. Low-ranking females with daughters are attacked more frequently than those with sons, and calm down their terror-stricken young (and maybe themselves, too) by letting them suckle. The attacks may be so stressful that they inhibit milk production, a situation that leads to even more nipple stimulation by the hungry daughter, and a consequent further reduction of their mother's fertility. Low-ranking mothers with daughters do not reproduce next year, and their bodily condition is particularly poor, whereas, if they have a son, they are much better off. The heightened attacks made on low-ranking mothers with daughters are understandable in this society, because daughters remain in the troop and may form coalitions with their mothers. The high-ranking matrilineages feel threatened by them. Low-ranking mothers with daughters are a potential challenge, to be oppressed and crushed by the female ruling class.

Rhesus monkeys are an excellent example of something we looked at in chapter 4 – the way in which a bird or mammal parent can affect not only the early development of their young, but also their adult life and behaviour as parents. The physical state and behaviour of a parent can influence its offspring in ways that lead to the perpetuation of the developmental cycle that is typical of the species, population or even lineage. Developmental systems are usually fairly stable, since the developing animals are able to compensate for variations in parental behaviour or environmental circumstances, but often there is more than one possible stable system of interactions, and these variant systems can be transferred from one generation to the next. This is the case with the rhesus monkeys, where female social rank is transmitted in families. Female offspring of high-ranking and low-ranking mothers are treated differently by their mothers and other members of the group, and this early experience may shape their future behaviours in ways that reinforce and perpetuate their social standing. Carol Berman, who studied free-ranging rhesus monkeys, has shown that the extent to which the mother rejects her offspring varies between matrilineages. Variations in the maternal style of parenting are transferred from mothers to daughters because daughters observe and participate in the rearing of their younger sibs. The daughters learn the idiosyncratic maternal style and, later, when they themselves become mothers, repeat the same style of behaviour towards their own offspring.[46] In vervet monkeys, the situation is slightly different, and the experience of the daughter while she herself was an infant has more influence on her future behaviour as a mother than does the maternal behaviour she observed as a juvenile, although this, too, is influential. The vervet mother's social rank, which is correlated with her maternal behaviour, is the most important thing determining her daughter's future style of parenting.[47]

There is clearly a very complex network of interactions shaping the behaviour of a mother and her offspring. The mother's social rank, the sex of the offspring, and the social and environmental conditions are all involved. Many aspects of behaviour are affected by the way the system develops, and variant social traditions may be constructed and passed on in the maternal line. One aspect of behaviour, the development of reproductive behaviour, has received special scrutiny by scientists, so it is worth looking at it more closely. It has been most thoroughly studied in rodents.[48]

Rodent mothers can influence the future sexual behaviour of their offspring through the extent to which they lick their pups' ano-genital region. Rat mothers, for example, devote a lot of their time to vigorously handling and licking their offspring. When the mother enters the nest and before she allows the pups to suckle, she moves around them and licks different part of their bodies. Often she holds one of the pups so that it is lying on its back, and licks its ano-genital region. The reaction of the pup is to extend its hind legs, become stiff and release urine. If the licking continues, it releases faeces too. Since very young pups do not control their excretions, this is an essential part of early maternal care, and each pup is licked sufficiently to ensure that it is void and clean. The licking is also to the benefit of the mother, who is attracted to the water and salt in the dilute urine of her offspring, which is a particularly desirable resource when there is a shortage of water. What is fascinating is that rat, gerbil and mouse mothers give their sons more ano-genital licking than their daughters. The sons seem to have some chemicals in their urine that make licking them in this area particularly attractive to their mother. Female offspring who are injected with the male hormone testosterone on the first day after birth get a son-like treatment – they get as much licking as their brothers, and significantly more licking than their non-treated sisters.[49] In addition to the initial bias, sons also release more urine than daughters, so the mother gets a larger dilute urine 'drink' from a son, and has more incentive to continue licking him. The mother's extensive licking of her young sons has quite important effects on their behaviour when they mature: sons who were licked more when very young have more intromissions during copulation. They are therefore more likely to fertilise a receptive female than are sons who were poorly licked during early life. The mother's behaviour thus bolsters the future reproductive behaviour and success of her sons. Through differential ano-genital licking, the mother reinforces the gender differences among her offspring.

Rodent mothers may have effects on their daughters that go beyond their reproductive behaviour as adults – the effects may sometimes be transmitted to subsequent generations. We mentioned this in chapter 4, where we alluded briefly to the studies of Mongolian gerbils by Mertice Clark and her associates. Maternal lines of Mongolian gerbils differ in a predictable and long-term manner in the sex ratio of their litters and in female reproductive behaviour. These heritable, persistent differences stem from acquired, non-genetic differences in the reproductive biology of the females. The important initiating factor is the level of testosterone to which females are exposed while they are in the uterus. Testosterone produced by male embryos diffuses away to affect neighbouring embryos. It affects both males and females, but the effect is more pronounced in females. The level of testosterone to which a developing embryo is exposed depends on its position in the uterus: a female who developed between two brothers, and was therefore exposed to high concentrations of testosterone, is likely to become sexually mature later, to have a larger territory and to be more aggressive than a female who has developed between two sisters. The late maturing, more 'masculine' females have a higher concentration of androgens and, for reasons that are not yet clear, they tend to produce litters with more sons than daughters. Because there are more males in their uterus, their daughters are also likely to develop between two brothers, and thus to perpetuate the cycle. Since Mongolian gerbil mothers lick daughters who developed between two brothers more generously than daughters who developed between two sisters, these more 'masculine' daughters receive a larger share of licking, and their more 'masculine' gender may be further reinforced. The opposite is true for early maturing 'feminine' females, who have developed in the uterus between two sisters. These females produce a female-biased litter and lick their daughters less, and the cycle perpetuates itself. In this way, two different reproductive patterns involving a different sex ratio and a different set of female behaviours are perpetuated via non-genetic means.[50]

What causes the initial difference in exposure to testosterone in the uterus? It is clear that chance plays a part in the process, since the position of an embryo in relation to its sibs' sex is unlikely to be rigidly determined. But other factors can also be important: environmental conditions, especially various types of stress before and during pregnancy, can alter the mother's hormonal balance and lead to biases in the sex ratio of the litter, which can be self-perpetuating. Stressful conditions during pregnancy can also lead to changes in the behaviours of the offspring, for example in the way they explore their environment, and these behavioural changes, too, can sometimes be carried over to subsequent generations.[51]

The development of gender roles in rodents shows that the routes to maturity in males and females are not the same. But, as we stressed earlier, this does not mean that the information that parents transfer to them must be different. Earlier we saw that, in most monogamous species, daughters and sons get similar (though differently utilised) information from their parents, and this identical information-kit serves them well in their future lives. In polygamous species, daughters and sons are indeed treated differently in some respects. Yet, even in these cases, we must bear in mind that, although some 'discriminating' patterns of behaviour (such as differential ano-genital licking) are clearly induced by the sex of the offspring and usually lead to sex-specific adaptive outcomes, not all differences in the way that parents behave towards sons and daughters need be related primarily to their sex. Some may be related to factors that are frequently associated with sex, but have an independent effect on behaviour. One such factor is the likelihood of dispersal. The dispersing sex – males in most mammals, females in most birds – is the one that suffers the greatest mortality, and may therefore need more parental investment.[52] The selection pressures leading to different treatment of male and female offspring are therefore probably varied, with differential treatment having evolved to ensure not only that daughters and sons have the appropriate reproductive behaviours, but also the survival of both dispersing and philopatric offspring.

A parent's effect on the way offspring develop and their behaviour as adults cannot be isolated from the way in which the offspring respond: the kind of care parents provide is influenced by their offspring's initial conditions and responses, which may, in turn, depend on the parent's physiological state. A feedback-loop is formed between parental caring behaviour and offspring responses, reinforcing and stabilising fundamental sex-specific behaviour patterns in the offspring, and leading to their reliable perpetuation.[53] The feedbacks between parents and their young mean that changes (whether genetic or not) in components affecting the behaviour of parents and offspring may have very different outcomes: some variations will completely demolish the network, others will be compensated for and have no long-term effect, while a few variations (like uterine position in Mongolian gerbils) can lead to a variant web of self-perpetuating interactions. Evolutionary explanations therefore have to take the structure and the regulatory properties of the network into consideration, for this structure constrains the type of variation that can be of evolutionary significance. The nature of the interaction system and the mutual interdependence of the behaviours of parent and offspring make it difficult to see them as conflicting. For example, as Celia Moore has pointed out, the extra licking that male rat pups get from their mothers is not energetically expensive for her, and she even has a direct fitness benefit, because she gets more urine than she would from licking her female pups. Similarly, in many birds and mammals, the offspring's nagging and the parents' measured responses to it look like mutually dependent and mutually beneficial behaviours. The web of interactions between parents and offspring, though very intricate, appears like a well-balanced and stable system that cannot be readily decomposed, and that is often optimal for both.

The family unit

Animal families are clearly immensely complex systems – merely trying to describe the relationships between mates, between parents and offspring, and among sibs requires the literary talent of a Tolstoy. The behaviour of family members towards each other is sometimes aggressive and quarrelsome, while at other times we are struck by remarkable acts of co-operation. Interactions within the family are always intricate, sensitive to alterations in social relations and full of what looks like subtle calculations. We have tried to show that squabbles and aggressive encounters between parents and offspring are frequently not symptoms of underlying genetic conflicts, and that interpreting the relationship between parents and offspring in such terms is often inadequate. Even dramatic fights between siblings, which do indeed reflect the sibs' differing fitness interests, are usually perfectly compatible with the parents' genetic interests.

The evolutionary way of looking at family disputes and family organisation has undoubtedly stimulated new interpretations of the behaviour of family members. No aspect of human family relationships – from what seemed like straightforward 'weaning conflicts' to Freudian insights (or mis-sights) – has escaped conflict-based and gene-centred evolutionary interpretations. In this book, we are not going to devote a lot of space to criticising these gene-centred interpretations of specifically human and profoundly cultural psychological notions, but we do want to consider just one aspect of human relationships that may also be important in other animals. Frank Sulloway recently explored how birth-order in humans affects personality.[54] Sulloway collected an unusually large amount of data about scientists and artists, which showed that first-born children tend to be conservative and to take on traditional societal values, while late-born children tend to be adventurous, supporting revolutionary theories and rebelling against the accepted order of things. He believes that the different psychological profiles of first- and late-born individuals are due to sib competition within what he has aptly called the 'family-niche'. The first-born 'occupies' the most apparent psychological and social position within the family niche, identifying itself most strongly with the parents and with the parents' expectations, and demanding their undivided attention. A first-born child, who at least for a while is the sole focus of its parents' attention, tries to retain its favoured position when a younger sib comes along. It therefore tends to be jealous and status conscious. Late-born children, on the other hand, have to construct a niche that is different from that already occupied by older sibs, yet one that endears them to their parents. They are therefore better natured, more relaxed and outgoing. These psychological characteristics, as well as the position of the first-born in the family, make late-born children more likely to take physical and intellectual risks.

Sulloway looked at the evolution of the different behaviour patterns associated with birth-order from the child's point of view, and argued that the behavioural strategies of first and late-born sibs are the evolutionary results of direct and indirect competition for parental care and love. However, there may be an alternative explanation. The correlation between psychological profiles and birth-order need not be an evolved result of sib competition, but rather an outcome of changes in parental caring skills. First-borns are the offspring of inexperienced, overanxious parents, whereas with late-born offspring parents are much more relaxed and skilful. Most studies of primates show that mothers are more cautious, nervous and overprotective with their first infant than with later ones (especially when their own mother, the grandmother of the infant, is not around).[55] As we have seen, the maternal style of care has a lasting influence on the behaviour of the offspring. First-born offspring of vervet and rhesus monkeys are cautious, and much less enterprising than late-born offspring of the same mother,[56] just as Sulloway describes for human children. As in non-human primates, the different psychological profiles of first and late-born human beings may therefore stem from differences in the parental care they get. Nevertheless, it is possible that the differences between first and late-born offspring, although initially a side-effect of the caring of the parents, are reinforced to form a more stable (though not necessarily more peaceful!) organisation of sibling relationships within the family niche.

Summary

Tolstoy's great family novel *Anna Karenina* starts with the well-known statement: 'All happy families resemble one another, but each unhappy family is unhappy in its own way.' We all understand what Tolstoy meant, because we have all experienced various idiosyncratic versions of family troubles, and we all seem to share the same ideal of family bliss – a conflict-free, harmonious family. But a family with no clashes between parents and offspring and no quarrels among sibs is almost unheard of. Dissonance and struggle are integral parts of the family life of humans and any other mammal or bird. However, conflicts among family members do not necessarily represent underlying genetic conflicts. Conflicts between parents and offspring are often routes towards fitness-enhancing co-ordination. Hence, weaning squabbles are associated with youngsters learning how to forage and live independently. Other squabbles between parents and offspring may be honest signals of need – not only for food, warmth and security, but also for information. The youngsters' tiresome demands may also inform their parents of their maturational state. Conflicts between sibs, which reflect genuine competition, are often the outcome of parental strategies that enhance the parents' fitness but do not reduce the offspring's inclusive fitness. Similarly, differential treatment of offspring according to their sex, birth-order or life style may contribute to the long-term fitness of all family members. Quarrels, rows and treating offspring differently are therefore not just inevitable consequences of conflicting fitness interests, they are also adaptive processes through which offspring mature and learn the skills necessary for adulthood.

Notes

[1] The account of the family life of great tits is based on the personal observations of E. A. and the following sources: Bengtsson & Rydén, 1981, 1983; Gosler, 1993; Perrins, 1979; Royama, 1970.

[2] Gilbert, 1995.

[3] Mock, Drummond & Stinson, 1990.

[4] See Hamilton, 1964. According to Hamilton, altruism will spread when $C < rB$, where C is the cost to the individual who performs the altruistic deed, B is the benefit to kin resulting from the altruist's behaviour, and r is the coefficient of relatedness. The coefficient of relatedness is a measure of the probability that an allele in two genetically related individuals is common by descent. It is 0.5 between full sibs, and between parents and offspring, and 0.25 between an individual and its grandchild, or niece or nephew.

[5] A discussion of Trivers' ideas can be found in Trivers, 1985. See also, Clutton-Brock, 1991; Godfray, 1995a,b.

[6] Burke & Brown, 1970.

[7] Godfray, 1995a,b; Parker & MacNair, 1978.

[8] For an excellent, non-mathematical review of models of parent–offspring conflict that takes into account asymmetries in offspring size, see Godfray & Parker, 1991.

[9] For discussion of the additional complications in parent–offspring relations, and especially the argument that signalling by offspring is honest and therefore costly, see Eshel & Feldman, 1991; Godfray, 1995a,b; Johnstone, 1996. It must also be remembered that the proper interpretation of signs as honest clues about need may be problematic, because the strong and the healthy offspring may be those most able to scream loudly and extract most from the parents (see Kilner & Johnstone, 1997).

[10] For evidence that begging is cheap see McCarty, 1996. Bergstrom & Lachmann (1998) have shown theoretically that honest cheap signals may be evolutionarily stable.

[11] Mock & Forbes, 1992.

[12] Mock & Parker, 1997, p. 255. For further discussion, see Bateson, 1994; Clutton-Brock, 1991; Mock & Forbes, 1995.

[13] Bateson, 1994.

[14] For a discussion, see Fox Keller, 1996, pp. 154–72. She has shown how the use in evolutionary biology and ecological studies of 'competition' to describe both fitness differences and actual competitive interactions leads to the mixing of levels of explanation and to limited and limiting interpretations of data.

[15] The work that Trivers discusses can be found in Davies, 1976.

[16] Altmann, 1980; Barrett, Dunbar & Dunbar, 1995.

[17] Fairbanks, 1996.

[18] Stamps *et al.*, 1989.

[19] Muller & Smith (1978) found that when tape-recorded begging was played in the nests of zebra finches, it elicited both elevated feeding by parents and elevated begging by nestlings. The nestlings started begging immediately after the recorded begging was heard and before the parents started feeding, suggesting that nestlings respond to the sound of begging in addition to stimuli from parents.

[20] See Fairbanks, 1996.

[21] Bhatia *et al.*, 1990; Lieberman, 1993.

[22] For a review of work on genomic imprinting see Jablonka & Lamb, 1995, chapter 5.

[23] See Haig, 1992, for a brief review of his ideas.

[24] Several mammalian genes that affect embryonic growth have been found to be imprinted in ways that correspond to the expected pattern, but others are not. Haig's theory also predicts that the war of attrition between maternally and paternally marked genes should lead to high rates of molecular evolution, but this prediction is not supported by the data, which show that imprinted genes have low rates of molecular evolution (see Hurst, 1997). Although male–female conflict may have something to do with some cases of genomic imprinting, imprinting has many other functions (see Jablonka & Lamb, 1995).

[25] Lactase-I activity and the level of behavioural conflict have not been compared between species. Some comparisons of imprinted genes in mammals indicate that paternally derived growth-promoting factors are more active in non-monogamous species, but other results point in the opposite direction. See Pennisi, 1998.

[26] Karl Marx's phrase was actually 'From each according to *his* abilities, to each according to *his* needs' (*Critique of the Gotha Programme*, 1875). In the second half of the nineteenth century, the abilities and needs of females did not receive much attention. Biologists today have to be less prejudiced.

[27] The only evidence for offspring manipulating their parents comes from the study of eusocial insects such as some bee and ant species (see Mock & Parker, 1997). The workers in these species form an absolute majority in the nest and are in full control of resources. They seem to manipulate the sex ratio of their mother's reproductive offspring in accordance with their own genetic interest, biasing it towards overproduction of reproductive females. In birds and mammals, there is evidence for the opposite situation: mothers can affect the sex ratio of their offspring, and the biased sex ratio may enhance the mother's fitness (Trivers, 1985).

[28] Stoleson & Beissinger (1995) reviewed 16 different hypotheses about hatching asynchrony. They fell into 4 major categories: 2 hypotheses assumed that hatching asynchrony is a non-adaptive consequence of physiological, energetic or phylogenetic constraints; 4 hypotheses suggested that it is a by-product of selection for early laying and incubation; 8 hypotheses suggested that it is due to selection for early hatching patterns; 2 hypotheses suggesting that it is the result of selection for the optimal allocation of resources during the different phases of the nesting cycle. A 17th hypothesis, suggested by Ricklefs (1993), is that hatching asynchrony is a by-product of selection for increased longevity in altricial birds. For a short recent review of hatching asynchrony, see Stenning, 1996.

[29] David Magrath (1989) has shown that Lack's theory may be valid for blackbirds. He experimentally manipulated hatching synchrony in broods and the supply of food in some territories. Asynchronous broods were more productive than synchronous broods when food was scarce, but when food was abundant, the survival of offspring in synchronous and asynchronous broods did not differ.

[30] Ehrlich *et al.*, 1994.

[31] Schwabl, 1993.

[32] Ferguson & Sealy, 1983; Gottlander, 1987; Slagsvold, Amundsen & Dale, 1994; Stamps *et al.*, 1985.

[33] Leonard, Horn & Eden, 1988.

[34] Lyon, Eadie & Hamilton, 1994; Pagel, 1994.

[35] Marzluff & Balda, 1992.

[36] Clutton-Brock, 1991.

[37] For information about nest sanitation, see Skutch, 1976, pp. 281–5.

[38] Mock & Forbes, 1995.

[39] Bateson, 1994; Klopfer & Klopfer, 1977; Wiesner & Sheard, 1933.

[40] Chapman & Chapman, 1987.

[41] The two exceptions that we are aware of are enhanced feeding of female-based broods by fathers in budgerigars and Eastern bluebirds.

[42] Redondo, Gomendio & Medina, 1992.

[43] Fedigan, 1982, pp. 181–6.

[44] Rowell, 1988.

[45] Redondo, Gomendio & Medina, 1992.

[46] Berman, 1990.

[47] Fairbanks, 1996.

[48] Our account of the work on sex and gender in rodents is based on the excellent review of Moore, 1995.

[49] Moore, 1982.

[50] For further details, see Clark & Galef, 1995; Clark, Karpiuk & Galef, 1993.

[51] See Kinsley & Bridges, 1988; Kinsley & Svare, 1986. The timing of maturation is also related to female size, which may either reinforce or counter the other developmental legacies.

[52] Sex and the likelihood of dispersal are closely associated. In birds, females are usually the dispersing sex, while in most mammals the males disperse. In birds, male mortality when juvenile is greater when the male is larger than the female, but, in most species, female mortality is higher than male mortality during all stages of the life cycle. There is no convincing evidence that the greater mortality of female birds is principally due to their larger investment in parental care. Rather, it seems that their high mortality is associated with their greater dispersal. In mammals, too, there is a correlation between high adult mortality and dispersal – males disperse more and have a higher mortality. (In polygynous mammals, as in some polygynous birds, high male mortality is also correlated with competition for access to females.) The dispersing sex may need extra energy, to prepare it for the precarious life away from the natal area. This could be the evolutionary reason for the behaviour of budgerigars and bluebird fathers who feed female-biased broods more generously than male broods, and may thus enhance the survival of their dispersing daughters. For a discussion of mortality patterns and parental investment in monogamous birds, see Breitwisch, 1989.

[53] Excellent reviews on the interaction between parents and offspring and the construction of relationships can be found in the volume edited by Rosenblatt & Snowdon, 1996. For details of the parent–offspring behaviour in rats see Fleming, Morgan & Walsh, 1996; for rabbits, see González-Mariscal & Rosenblatt, 1996; for voles, see Wang & Insel, 1996; for sheep, see Lévy et al., 1996; for primates, see Maestripieri & Call, 1996; Pryce, 1996.

[54] Sulloway, 1996.

[55] See, for example, Hiraiwa, 1981.

[56] Fairbanks, 1996.

7   Alloparental care – an additional channel of information transfer

According to the Bible, the Lord commanded Moses to tell his people 'Thou shalt love thy neighbour as thyself.'[1] Regrettably, most of us fall short of this high moral standard: the interests of friends and neighbours are usually not as close to our heart as our own interests. Although human beings often co-operate with each other, strikingly altruistic acts are far from being the rule. When we do encounter them, we tend to regard them with surprise, admiration and sometimes even with contempt, indicating that these acts are seen as something exceptional. Impressively altruistic acts, especially those that are not directed towards close relatives, are often thought of as biologically 'unnatural' – the result of ideals imposed on us by custom, law or God, or else the unfortunate outcome of some miscalculation. Biologists have therefore been extremely puzzled by the observation that many birds, mammals and even insects perform what seem like acts of self-sacrifice. They take risks by warning others of lurking predators; they fight, sometimes to the death, to protect other individuals; and they take upon themselves the onerous chore of caring for the young of others. In several hundred species of birds and mammals, from bee-eaters and kingfishers to jays and woodpeckers, from voles and mongooses to bats and marmosets, parents are helped to rear their offspring by other individuals who seem to surrender, at least temporarily, their own reproductive rights and opportunities, and become 'helpers'.[2] In a few hundred other species, individuals actually adopt unrelated young and take full parental responsibility for them.[3]

Why should an animal invest time and energy in raising, or helping to raise, someone else's offspring, instead of looking after its own? How can we explain such seemingly altruistic behaviour in evolutionary terms? Who are these helpers and adopters and what is in it for them? This issue has been one of the most active areas of behavioural ecology since the 1960s, and there are many excellent reviews and analyses of helping behaviour. We are not going to attempt to review this vast literature here, but rather will concentrate on two related aspects of extra-parental care – the transfer of information through social learning during helping and adopting, and the role of this learnt information in the evolution of these behaviours. But, before going any further, we want to describe the lives and the typical behaviours of some of the helpers and adopters. The most detailed studies of extra-parental (or 'alloparental') care in vertebrates have been those on birds, where individuals who are currently non-breeding (the 'helpers'[4]) assist others to rear their young, but do not take full parental responsibility for these young. So, we will first take a fairly close look at the helping behaviour in a colony of European bee-eaters in the upper Jordan valley of northern Israel.[5]

Here, in early spring, the vegetation along the Jordan river is composed of a rich variety of trees, bushes and climbers, which in places form a dense thicket. The place teems with bird and insect life. The frequent chatter of a flock of yellow-vented bulbuls blends with the constant humming of a host of honey-bees and the sharp occasional buzz of a carpenter bee. More than a dozen species of colourful dragonflies and damselflies, dressed in various combinations of red, blue, yellow and black, alternate between perching and hunting flying insects.

The traditional breeding site of the bee-eater colony is the plant-bare part of a steep bank, a hundred metres long and four metres high, which contains many dozens of horizontal nesting burrows, about a metre and a half deep. Each of these currently unused burrows houses just spider webs and piles of old insect skeletons. But they will not remain desolate for long. Pleasant, liquid voices pour down from the clear sky. These voices and the colourful flashes that follow announce that the bee-eaters are back from their African wintering grounds, and are carrying out their intricate insect-hunting manoeuvres high up in the sky. By late afternoon, the bee-eaters have gracefully descended and settled noisily. They perch, mostly in pairs, on the dry outer branches of a large fig tree overhanging the river.

Next morning presents us with a family portrait of the future breeding colony: forty-three colourful bee-eaters, with gold, green-blue, chestnut and black plumage, are perched along the branches of the fig tree. On arrival from Africa, 34 birds are already paired, 2 new pairs are currently forming, and 5 birds, all of them young males, probably in their first year, are left unpaired.[6] The fig-tree perching sites are good vantage points for the bee-eaters. They offer them a wide-angle view of the surrounding area, enabling them to spot lurking predators as well as unsuspecting prey.

The buzzing, undulating flight of a couple of large carpenter bees attracts the attention of two paired males. Both immediately fall into the air and each quickly intercepts his chosen victim, grabbing it cautiously in the middle with the tip of his long beak, holding it crosswise to avoid being stung. Each male returns to his perching post beside his mate, cautiously holding his trophy, and starts a courtship feeding display, calling loudly and violently vibrating his tail. In order to turn his beakful of dangerous prey into a courtship gift, he has to render the bee harmless by first battering its head against the perch, and then crushing its abdomen and sting by violently rubbing them on the perch. Once made harmless, the prey is presented to the eagerly awaiting female, who devours it, emits a high-pitched call, and assumes a horizontal position against the perch, inviting her mate to copulate. This scene, taking place between paired males and females, is repeated time and again throughout the day.

But later something different happens. A young unpaired male hunts and finally catches a damselfly, and brings it back to a female whose rightful mate is out insect hunting. The young male performs a full-scale courtship feeding display and the female eagerly accepts his insect gift, but she denies him copulation. During the next two days, the same event reoccurs, with each of the young unpaired males being 'faithful' to a particular paired female, courting and feeding her on a regular basis, without copulating with her. The mates of the paired females seem indifferent to the young males who courtship feed their females, and aggressively chase them away only when the young males signal their sexual intentions. Yet, the same paired males are very aggressive towards almost all other birds that come too close to their perches, driving them away with wide open beaks and raised feathers. So why do the paired males tolerate the social, but not the sexual, advances of the young males towards their mates? Why do the young males help to feed the females? Why are they allowed to help? Why have they chosen to help and remain with particular females?

Using their defended perches as a temporary home base, the mates fly out, calling excitedly, to hunt insects and test the suitability of various places on the earthen bank as nest sites. They thoroughly examine old, disused nest burrows, and hover in front of the bank face, striking it at various points with a half-open beak. After several false starts, members of each pair finally find a suitable nest site, a metre or two away from any neighbouring pair, and start a two week digging marathon, using their beaks to dislodge earth and their feet to kick it backwards. During these two weeks, each couple scrapes, digs and removes ten kilograms of earth, making the bank area look like a quarry. When the digging is over, each couple owns a safe, metre-and-a-half long nest burrow, ready for egg laying.

By mid May each female has laid, at one- or two-day intervals, a clutch of 4 to 6 eggs. As the energy demands of a laying female are high but her ability to satisfy them is low, her mate must now work harder. Each mated male makes rapid shuttle flights between the insect-rich aerial hunting zones and the laying hen. A returning male loudly calls his laying hen to the nest entrance, feeds her with the largest insect he has been able to find and often copulates with her before hurriedly flying away to hunt again. Each of the five unpaired males is adjunct to a particular mated pair. He hangs about the nest, escorts the 'rightful' male on his insect-hunting forays, and frequently hunts large insects and brings them back to the laying hen. However, the young helper usually shows less dedication and success in bringing in food and feeding the female than does her rightful mate.

During the day, both mates incubate the eggs, changing shifts every 30 to 90 minutes, but during the night the female alone incubates. In captivity, helpers are known to share incubation duties with the mated pair, but in this species this has not been reliably observed in natural conditions. In early June, when the chicks hatch, the helpers start to assist the parents to feed them, and continue to do so for the following month or so, until the young bee-eaters can forage independently. The average rate at which the young are fed by the parents alone goes up from 12 feeds per hour for a newly hatched brood to around 35 during the second week of life. At this stage, a single helper can usually add 15 to 25 feedings per hour to the parental score, and sometimes a lot more. The magnitude of a helper's contribution varies widely among individuals, with some helpers bringing food to the young only when they beg loudly or approach the nest entrance.

By three weeks of age, the fledglings have become curious, alert, mobile, highly vocal and quite quarrelsome, especially around feeding time. Some of them try to come to the entrance of the nesting burrow, a vantage point for securing food, and wait there for the next delivery, meanwhile blocking their brothers' and sisters' access to meals. Since the other kin are no dupes, these passage-blockers are often forcefully tugged by their tails back into the depths of the nest and out of the passage.

When the chicks are about three and a half weeks old, both the parents and the helper gradually encourage them to leave the nest by first bringing less food, and then, a few days later, refusing to feed them inside the nest, thus making them all fledge within a day or two. For the first few days out of the nest the young are completely dependent on their parents and the helper for food. Over the next three weeks, feeding by both parents and helper steadily decreases, so the inexperienced young have no choice but to learn to hunt insects and generally help themselves. By the end of this period, begging young bee-eaters are either ignored or driven away by their former benefactors. Now they must be independent and fend for themselves.

The helping behaviour of the European bee-eaters in Israel is typical of many other species of birds. The benefits both to the helped young and to their parents seem obvious: additional food and care should translate into better survival for the young and greater reproductive success for their parents. In European bee-eaters, brood size at fledging is known to be larger for helped broods, and a higher proportion of helped (compared with non-helped) individuals return to the same breeding site next season, suggesting that the survival of both parents and chicks is probably enhanced by helping.[7] However, we cannot take it for granted that helping enhances the fitness of parents and chicks in this or any other colony or species. The correlation between the presence of helpers and the reproductive success of the helped pair may be a consequence of the quality of the territory, rather than of the assistance of helpers. A rich, plentiful territory will increase the parents' chances of surviving, and also the chances that helpers will be allowed to remain, because they will not significantly deplete food resources. The better survival of the 'helped' parents may therefore have more to do with the good territory than with the assistance of the helpers. Inexperienced helpers may, in fact, hinder rather than help. They may act clumsily as food providers and harm the young, or be too noisy and attract predators to the vicinity of the nest. Helpers may even gorge themselves and overexploit the feeding territory, or try to cuckold or even displace the breeding males. Therefore, we cannot automatically assume that helping always benefits breeders and their young. It needs careful studies to show that helpers really help.

What about the helpers? What do they gain or lose? Helpers seem to put a lot of time and energy into helping to raise the offspring of other birds, when their efforts might have been devoted to raising offspring of their own. At first sight this situation seems to be biologically absurd, although not nearly as absurd as the behaviour of adopters who, unlike helpers, take full parental responsibility for rearing foreign young. So, why do helpers help, and adopters adopt? Before we can answer these questions, we need to know the extent to which the parental roles played by helpers and adopters are an addition to or substitute for their normal roles as parents to their own young. For adopters, alloparenting is frequently additional, while for helpers, who are almost always currently non-breeding individuals, it seems to be a substitute for raising their own young. But is it? Are helpers really able to become breeders, yet refrain from it in order to become helpers? The answer is not straightforward. First, breeding may not be possible for a helper, either because there are no vacant breeding territories, or because there is a shortage of mates, or because they have inadequate parenting skills.[8] In these cases, refraining from rearing their own young is certainly not a matter of choice. Second, a non-breeding individual does not have to become a helper. It can, for example, stay on someone else's territory, feed itself, refrain from helping and patiently wait for a future breeding opportunity. Similarly, a breeder is not compelled to become an adopter. It can drive the potential adoptee away or, better still, as sometimes happens, turn it into a nourishing meal.[9] The point is that, whether or not breeding is possible, we still need to find evolutionary reasons that can explain why it is worthwhile for these seemingly altruistic helpers and adopters to help rather than not to help, to adopt rather than not to adopt.

So what explanations have been offered for the widespread occurrence of helping and adoption? They could be non-selected by-products of parental behaviour, ineluctably occurring under dense colony conditions, where adult animals are constantly exposed to the presence of young and cannot help but express parental behaviour towards them.[10] Alternatively, these behaviours could be beneficial for both helpers and adopters. The adaptive explanations of alloparenting fall into three categories. First, the care that adopters and helpers provide may benefit them, even when it does not benefit those they help. Caring may enable a young helper to stay in the parental territory, or to learn parenting and foraging skills. It could also be a way of advertising the caregiver's qualities to potential mates. Second, caring can benefit the relatives of caregivers, thereby enhancing the caregiver's own inclusive fitness through kin selection. Third, caring can benefit everybody involved, with caregivers gaining the advantages that living in a group can provide, including greater protection and the possibility that they themselves will receive help later in life. All three types of adaptive explanations may be valid, although their relative significance will vary according to ecological and social conditions, the type of care given and the sex of the caregiver.[11]

In the following sections, we will look at the different categories of adaptive explanations, focusing especially on the processes and benefits of the non-genetic transfer of information that takes place during extra-parental care. Although adopters and helpers have many things in common, and the adaptive explanations for their care-giving behaviour overlap, there are important differences between the two types of allo-parental care, so it is easier to treat them separately. We shall therefore look first at the evolutionary interpretations of helping behaviour, and then examine adoption.

Self-interested helping

One of the direct and most obvious benefits of helping is that a young helper acquires information that will make it a better forager and parent. This hypothesis has been called the 'skill hypothesis'. More than forty years ago, David Lack suggested that helping in birds is associated with the benefits of staying longer in the natal territory, and that young birds should not leave home unless they have learnt the skills necessary for future survival and breeding.[12] Prolonged association with either their parents or other breeding adults increases the foraging proficiency and parenting skills of many young birds and mammals.[13] This increase is most pronounced when individuals need special foraging skills, as they do in various birds of prey, oyster-catchers, tropical insect-eating birds and vampire bats,[14] and in omnivores such as meerkats and dwarf mongooses, who have to learn where, when, and how to find and handle various sometimes elusive and dangerous food items.[15] In the European bee-eaters, we saw that handling stinging insects requires considerable skill. The young helpers are probably less adept at this than the breeders, but their skills are likely to improve while courtship feeding the females and helping to feed the young. By practising on the broods of others, helpers build up their parenting skills, and become better providers and caregivers.

Good evidence that helping is important for practising foraging and parenting skills has been found in Seychelles warblers, where birds that have had prior helping experience are better breeders than inexperienced birds.[16] Even better evidence comes from white-winged choughs, for whom a period of helping is an integral part of the life cycle. These Australian crow-like songbirds live in permanent, closely knit co-operative groups of up to twenty individuals, consisting of parents and young of both sexes. Group members wander over large areas and forage for invertebrates, which are often hidden in the soil or amongst leaf litter. Their food supply fluctuates widely with the local climate. In these circumstances, young birds are likely to die from starvation, but for several months, while they are acquiring the necessary foraging skills, they stay in their natal group and depend on food provided by helpers, who are usually their older siblings. The young, in turn, then spend more than three years as helpers, developing their foraging and parenting skills both by observing the breeders and by practising on their younger sibs. With both age and experience, the young individual becomes a more skilled forager and provisioner, bringing in more food items for itself and for the youngsters it feeds and helps to raise.[17]

It may take a long time for birds and mammals to perfect their skills. In many habitats, predators and prey change with the seasons or between years, so if youngsters of a long-lived species stay with their parents for just the short breeding season, they will lack the foraging skills and knowledge of predators that are necessary at other times. They will be only partly educated. One way for them to complete their education is to stay with their parents as helpers, learn the 'extra' skills, and only then disperse and attempt to breed. Young adults of many species, including the European bee-eater, tend to delay dispersal and stay with their parents when conditions become harsh. In such lean years, their chances of breeding are slim anyway, but, by staying as helpers, the young are likely to learn from the experienced breeders the skills that will be useful when a lean year reoccurs. It would be interesting to compare the survival during lean years of adults (and their offspring) who have received a lean-year education as young helpers, with those who lack this education either because they have dispersed, or were helpers only in a good year. Acquiring specific 'lean-year skills' from experienced parents may give an adaptive edge to the former helpers when they mature and re-experience such conditions. From the point of view of the parents, the prolonged stay of their offspring can be seen as an extension of their parental care, which benefits these offspring but need not affect their own survival or that of their current brood.

A youngster may sometimes attempt to stay with its parents and 'help' when its physical state is such that it needs another season of parental care. Such a 'helper' (usually a weakling) may be unable to offer its parents any significant assistance, but it is also unable to become independent. The reason why parents allow such burdensome, unhelpful offspring to stay and 'help' may be that by allowing the weaklings to spend one more season in the protective environment of the parental territory, they are given a better chance of gaining strength and experience, and hence of becoming breeders next season. This seems to be the case in pinyon jays, where sons who do not manage to find mates sometimes stay with their parents. Although these sons go through the motions of helping, bringing some food to their younger sibs, they do not really contribute to the rearing of additional sibs, nor do they prolong the life of their parents. John Marzluff and Russell Balda, who have carried out an extensive long-term study of pinyon jays, believe that this extended parental care gives these weakling sons a better chance to survive and reproduce.[18]

The need to acquire a variety of information and competence by learning from experienced adults may be one of the reasons for the evolutionary maintenance of delayed dispersal and helping, but it is difficult to know whether skill acquisition was the original function. Although the need to acquire skills could be the cause of late dispersal and thus encourage helping, acquiring skills could also be the result, rather than the cause, of delayed dispersal and helping. If helping has been adaptive for other reasons – perhaps because it benefits relatives or because help will be reciprocated later – the incidental benefits of learning from experienced adults will increase the value of the behaviour even more. Helping may, in fact, have originated as a by-product of parental care – as some initially non-selected, neutral patterns of misplaced parental behaviour, which were inevitable in dense breeding colonies, and which only later acquired functions and started to be selected.[19] But, whatever the origin of helping, and whatever its original adaptive function, there is no doubt that the skills that helpers acquire contribute to their present reproductive success and the persistence of helping.

A very different kind of explanation for helping, although one that again stresses the self-interested aspect of this behaviour, was suggested by Amotz Zahavi more than twenty years ago.[20] Zahavi claimed that, by engaging in costly, self-handicapping, altruistic acts like helping and adopting, individuals are advertising their high quality to those around them, and thus are consolidating their present social status and improving their future breeding prospects. In other words, by helping and adopting, helpers and adopters are actually investing in quality advertising, which, sooner or later, pays off. Zahavi's application of his handicap principle to co-operative breeding was inspired mainly by his study of Arabian babblers.[21] These are group–territorial and communally breeding birds, in which each group usually consists of 1 breeding pair and 2 to 11 non-breeding males and females. The breeding mates are unrelated to each other, but in many groups the non-breeders are different-aged offspring of the breeding pair. According to Zahavi, the extent of helping reflects the quality of the helper – the more an individual helps, the higher the status it advertises and establishes, although helping may also be affected by the size of the group and its composition.[22]

Zahavi's explanation of helping may apply to other species, such as the pied kingfisher. In this colonial, fish-catching species, helpers are currently non-breeding individuals who are either sons of the breeding pair or unrelated males. Unrelated males are accepted as helpers only if they regularly bring in fish at a time when the breeding couple is experiencing difficulties in providing enough food for their chicks. These unrelated helper males sometimes displace the original 'husbands' in later seasons, and pair with their females. It is possible that, in this case, consistent and successful helping not only makes the breeding female more familiar with the helper, but also indicates to her that, compared with her 'husband', he is a superior provider and therefore a higher quality male, a good future mate.[23]

Sometimes helpers may improve their chances of breeding not because they displace the previous 'husband', but because, by helping, they increase their chances of later pairing with the young they helped. This type of benefit is seen in riflemen (small New Zealand birds), where breeding pairs can have regular and irregular helpers. The irregular helpers assist in several nests, whereas the regular ones help one single pair and feed the chicks as much as the parents do. It is these regular helpers who later, when the chicks mature, have a very high chance of pairing with their previous 'helpees'.[24]

Skill acquisition, advertising and other direct but delayed benefits are all explanations of helping that stress the advantages to the helper, whether or not the help provided contributes to the well-being of the helped young or the breeding pair. It is, of course, likely that the helpers' care will often be of value to the young or their parents, and it may even be essential, as it is in white-winged choughs. Nevertheless, according to this type of explanation, the major function of helping and the main reason for its maintenance during evolution is the direct benefit to the helper itself. When there are benefits to other individuals, they are incidental and inevitable by-products of skill and status acquisition. One cannot acquire a parenting skill without practising! However, since in many species, including European bee-eaters, Arabian babblers and white-winged choughs, the helpers may be genetically related to the breeders and their offspring, the inclusive fitness benefits resulting from enhancing the survival and reproductive success of kin also have to be considered.

Help thy kin and spread thy genes

Helping is found in many different species and in different ecological and social situations, but it seems to be most common in dense breeding colonies. Often, but not always, helpers are young individuals, offspring of one or both members of the breeding pair, that associate with and are tolerated by their own parents at a time when they might be expected to be independent. In the European bee-eaters, the helpers are either young, currently non-breeding offspring, or neighbouring individuals who have recently made an unsuccessful breeding attempt. The neighbours usually seem to be related. Brother bee-eaters often disperse together, and males who have relatives nesting nearby are more likely to provide help if their own nest has failed.[25] Here, as in many other species of birds and mammals, individuals seem to prefer to help their kin rather than unrelated individuals.[26] Unfortunately, there are many problems in determining the actual genetic relatedness between individuals in the family. As with humans, the offspring that parents care for are often not their genetic offspring! Extra-pair copulations are surprisingly common in socially monogamous birds and mammals, and intra-specific nest parasitism (dumping eggs in someone else's nest) is also much more prevalent in birds than was previously assumed. The only reliable method of estimating genetic relatedness is DNA fingerprinting, and this has not been extensively used in most studies of helping and adoption, including those of European bee-eaters.

Currently, we have much more information on the genetic relationships in a related bee-eater species, the white-fronted bee-eater. This species inhabits the southern part of the African rift valley, and populations at Lake Nakuru National Park have been thoroughly studied over many years by Stephen Emlen and his collaborators.[27] The site of the colony is a nearly vertical cliff-face, free of vegetation, which is densely perforated with burrows. These serve as both roosting and nesting sites for the couple of hundred colony members, who belong to about a dozen extended families, or clans. When they are not breeding, clan members occupy adjacent burrows and defend them against individuals from other clans. Each clan consists of up to 17 members: 2 or 3 breeding pairs, and several unpaired, single or widowed individuals. They belong to 3 or 4 overlapping generations of kin, and each is able to individually recognise at least everyone else in its own clan. Clan members forage together in a private, insect-rich, foraging area, showing tolerance to each other but attacking non-clan members. In the evening, when they return from their daily insect hunts, members of the clan congregate at the colony site and frequently greet each other and visit each other's burrows.

Like their European counterparts, white-fronted bee-eaters have helpers during the breeding season. Helping is more common, however, with about half of the breeding pairs having helpers. It is also more extensive: helpers participate in digging the nest chambers, feed the breeding females during the egg-production period, take turns with the parents in incubating eggs, and they defend the young as well as constantly provide them with food. Helpers continue to help to feed and care for the young throughout the six-week period that it usually takes to bring the youngsters from fledglings to self-sufficient birds. Their help is enormously beneficial, particularly since the supply of flying insects, the major source of food for the young, often varies unpredictably during the breeding season. It is estimated that just a single helper can cut by half the chances that nestlings will die from starvation, the most common cause of death.

Helpers are always clan members, and are therefore usually genetically related to one or both members of the breeding pair and their offspring. Most are unpaired individuals, frequently young adult sons of the couple they are attached to, although some come from the ranks of mated pairs that currently are not attempting to breed, or have recently failed in their breeding attempts. Among mated helpers, those that were born within the clan have a much higher tendency to become helpers than do their mates, who are probably unrelated to the local breeders. Fathers sometimes force their newly paired adult sons to become their helpers by interfering with their courtship feeding, blocking the access to their nest chambers, or annoying them by frequently visiting their nest chambers before the eggs are laid. This parental coercion does not have any significant effect on the sons' long-term success: both sons who left the nest and those who stayed to help have been found to have similar inclusive fitnesses. As Emlen has pointed out, in the white-fronted bee-eaters, the intricate and subtle behaviours of family members towards each other seem like the outcome of a precise and cunning calculation based on inclusive fitness considerations![28]

Both silver-backed and golden jackals also help family members.[29] Sexually mature offspring often stay with their parents and help them rear their younger siblings. This help may considerably enhance the sibs' survival, and thus contribute to the inclusive fitness of both parents and helpers. As with bee-eaters, 'deciding' whether to stay and help, or to disperse and breed, depends on the breeding opportunities open to the young adults. When all local territories are occupied and breeding is therefore impossible, jackals stay with their parents and help them to rear younger sibs. Kin selection, as well as learning foraging and parental skills, have probably been very important for the evolution of helping behaviour within families in jackals and many other mammal and bird species.

Helping is good for all – delayed benefits, reciprocity and mutualism

In the numerous cases where helpers are known to be unrelated or only distantly related to the breeders, kin selection is irrelevant and other adaptive explanations for the observed co-operation have to be invoked.[30] The major explanations for co-operation among non-related individuals are mutualism, where co-operators benefit immediately and suffer no temporary reduction in their fitness, and reciprocity, where the fitness of a co-operator is temporarily reduced, but this reduction is later compensated for by an overall increase in fitness. In the early 1970s, Robert Trivers suggested that reciprocal altruism – 'trading' in altruistic acts (the 'you scratch my back and I'll scratch yours' principle) – can be important in social groups in which members recognise each other as individuals and have many opportunities to interact.[31] The 'trading' of favours can increase the long-term fitness of all co-operating parties. The classic examples of co-operation that is assumed to have evolved via reciprocal altruism are green woodhoopoes among the birds, and vampire bats among the mammals.

Vampire bats are small, nocturnal mammals that feed exclusively on blood, usually that of domestic stock such as horses and cattle, occasionally that of wild mammals, and rarely that of humans.[32] In Costa Rica they live in groups of between 8 and 12 adult females and their young. Members of a group share day-roosts in either hollow trees or caves, often for several years. Juvenile females remain in their natal group, while their male counterparts disperse before they reach sexual maturity. Since the females mate with several males and sometimes change groups, the degree of genetic relatedness within a group is rather low.

If they are unable to get a blood meal for three consecutive days, vampire bats usually die of starvation. However, bats that fail to acquire a meal (and this can happen occasionally to even experienced foragers) are usually saved because well-fed bats within the group regurgitate some blood for them. Even bats that are genetically unrelated regurgitate for each other if they have had a lengthy roost association. The donations are reciprocal – those who get blood later give blood preferentially to their former benefactors. Cheaters can be detected because roost-associates groom each other continuously, especially before and after food-sharing, so information about an individual's foraging success cannot be hidden. Vampire bats have good eyesight and a well-developed sense of smell, and they are known to learn fast and can easily be trained to change their behaviour using simple conditioning techniques. This suggests that learning plays an important role in their daily lives.[33] Female bats also nurse each other's babies, and this kind of help may also be reciprocal, although relevant field data about this are as yet insufficient.

Very strong evidence of reciprocity that involves 'trading' in helping behaviours has been found by David and Sandra Ligon in the green woodhoopoe in central Kenya.[34] In this territorial, co-operatively breeding bird, the social unit is the flock, which consists of a single breeding pair and up to fourteen other individuals. Non-breeding flock members are often, but not always, genetically related to the breeding pair, and serve as helpers. They courtship feed the female, feed nestlings and help to defend the territory against intruders from neighbouring flocks. Newly formed flocks, or established flocks that get new members and reorganise, often include individuals who are related neither to the breeding pair nor to the nestlings, but nevertheless help them enthusiastically, sometimes providing more than 80 per cent of the nestlings' food. This help is far from trivial, because the local supply of moth larvae, their major food source, depends on rainfall that is both variable and unpredictable, and competition over food can sometimes be fierce. Helpers, whether related or not, vie for the opportunity to feed nestlings directly. They beg or steal food from one another, and actively avoid the food-soliciting breeding female in order to be able to feed the nestlings themselves. Why is direct feeding so important to them?

Evidence gathered over the years by the Ligons shows clearly that by forming close relationships with the nestlings – by grooming them, vocalising to them and feeding them as much as they can – the helpers not only invest in the nestlings, they actually recruit them as future allies and helpers. In due course, the helped nestlings will become the helper's own helpers! There is a high rate of mortality among the wood-hoopoes, mostly from predation, and this creates a high turnover of breeding vacancies. By using their coalitions with the young they have helped to rear, surviving helpers have a good chance of attaining breeding status and successfully raising young of their own. The aid the helper gives to the young is later repaid by the recipient of the favour, so both are better off than they would be if they did not engage in these altruistic acts. The young would probably die of starvation if helpers did not provide food, and the helpers would have little chance of rearing their own offspring when they become breeders unless they have helpers. Although the benefits for the helpers are delayed, the way the system works makes it cheating-free and hence stable. First of all, to reproduce successfully, the helped young will need their own helpers, so it does not pay them to act selfishly and try not to return favours which are going to be to their future benefit. Secondly, there are immediate benefits for young helpers. By associating with a familiar, vigorous former helper – one who gave them food, grooming and attention – a young helper is associating with an obviously successful individual, a devoted caregiver and good food-gatherer, from whom it can acquire useful skills and information.

Co-operation is sometimes enforced.[35] We have already mentioned the enforced co-operation in male white-fronted bee-eaters, who may use active measures to force their young adult sons to become their helpers, in spite of the usually impotent resistance of the sons. Enforced co-operation can also occur among unrelated individuals. In superb fairy-wrens, because the female emigration and death rates are high and extra-pair copulation is frequent, the young adults who are helpers are often unrelated to the breeders they help.[36] If these helpers are experimentally removed from the group while the breeding members are rearing young, a period when the helpers normally play a crucial role in feeding the youngsters, then on their return they are attacked and harassed by the breeding males. Yet, helpers are never attacked if removed and returned at other, non-breeding times, when their help is not needed. The helpers seem to be expected to pay for the right to stay in and use the breeders' territory by helping the territory owners, and they are punished if they fail to pay the 'rent'. It is a situation not unlike the old human feudal system. Something similar is seen with rhesus monkeys, where a group member who does not co-operate is often punished. For example, if an individual finds a favourite food source but fails to emit food calls attracting the attention of others in its group to the food, it becomes, if discovered, a target of aggression.[37] Punishment (in the biological jargon sometimes called 'negative reciprocity') seems to be an effective way of maintaining relatively cheating-free co-operation.[38] In the case of punishment by rhesus monkeys, the cheater–punisher relationship is not restricted to the members of a single pair who do favours for each other and expect to receive them in return. Punishing can involve everyone in the group, some of whom may have had no particular alliance with the cheater.

Apparently altruistic acts can be quite subtle. In many bird species, it has been observed that occasionally adults feed the young of unrelated pairs.[39] For example, during the last stages of nestling life, shortly before fledging, groups of pinyon jays fly slowly through the colony, peering into nests, sometimes preening and feeding the resident, unrelated young. It is possible, as Marzluff and Balda have suggested,[40] that the adults are gathering information about the colony, but, in addition, such seemingly benevolent visits may make the adults and nearly fledged young familiar with each other, and thereby decrease future aggression within the colony. Remarkably, adult birds sometimes feed nestlings that belong to another species. Over the years, this behaviour has been observed in many species, regions and circumstances, but it has generally been dismissed as being the non-adaptive outcome of the overeagerness of parents to feed any begging nestling.[41] We believe, however, that, whatever the origin of this behaviour, it may now have adaptive functions. By occasionally feeding neighbouring nestlings of another species, the feeders make themselves familiar to their future neighbours and will probably eventually enjoy increased tolerance from them, which may save wasting energy on futile interspecific strife. In other words, a modest investment in tolerance-promoting familiarity may pay rich dividends. If the familiar neighbours also emit alarm calls in the presence of their past benefactors and vice versa, the resulting benefits may be even greater. Such good neighbourly relations are probably mutualistic, a form of no-cost co-operation.

Co-operation and reciprocation may take even more circuitous forms. Human experience shows that an individual who performs many altruistic acts and is also showing other signs of high quality, such as good health and vigour, attracts 'friends' – individuals who vie for his or her company and alliance, and are ready to offer more help than they are ever willing to offer other, less illustrious individuals. The Old Testament saying, 'Many will intreat the favour of the prince: and every man is a friend to him that giveth gifts',[42] reflects this utilitarian tendency to make friends and co-operate preferentially with the successful and the generous. Rich philanthropists and generous altruists are frequently considered as valuable members of the community. As Alexander argued some years ago, and as has recently been demonstrated by mathematical modelling, co-operation will spread if those who perform a lot of altruistic acts, and through them receive high 'image scores' in the eyes of others, later become the preferred recipients of acts of altruism.[43] Since altruistic acts increase the 'reputation' of the altruist, altruism spreads through a mechanism of positive feedback. In this case, as in the case of the punishment of selfishly acting individuals that we discussed earlier, those who favour a generous individual need not be direct allies of that particular individual. The return of the favour is not necessarily carried out by the recipient of the altruistic act, but by other members of the same group. Information about the social behaviour and the status of others, and hence about their potential utility for oneself, is probably constantly exchanged among group members.

Although among birds and mammals there is ample evidence that group members are frequently attracted to socially dominant individuals, preferentially grooming them and associating with them, as yet there is no substantial evidence that this attraction has anything to do with reciprocal altruistic acts involving other members of the group. However, an intriguing behaviour displayed by sexually immature white-winged choughs does indicate that the good opinion of other group members is of importance for their future. Juvenile birds help the breeding pair by feeding the young, but, when food is short and other birds are not at the nest to witness exactly what the juveniles are doing, they consume the food themselves, sometimes even placing the food in the chicks' gaping beaks and then quickly swallowing it themselves. Such 'cheaters' show special enthusiasm for preening the young, and are apparently trying to impress their group members with their devotion. For a white-winged chough, 'reputation' may be important for establishing the new social coalitions that form among group members when a breeder dies and the members disperse.[44]

The adaptive explanations for helping are of course not mutually exclusive. Kin selection, self-interested helping, and exchange of benefits may all contribute to the evolutionary maintenance of helping in social groups.[45] The relative importance of the various benefits may differ in different species and between males and females. Andrew Cockburn has recently highlighted the different fitness interests that male and female helpers have in birds, where males are usually the philopatric sex. By helping, males may increase their chances of inheriting the natal territory, and may later be able to court the females they have helped. For them, the advantage in helping is likely to be associated more with providing direct access to territory, prestige and females, than with fitness gains from helping to rear non-descendant kin. Females, on the other hand, disperse, so they are unlikely to inherit territory, and are less likely to encounter the males they have helped to rear. When females do act as helpers, the major advantage to them may be the increase in their inclusive fitness resulting from helping to rear their kin.[46]

The helping behaviour of white-fronted bee-eaters is probably maintained by a combination of all the three types of benefit. Since many helpers are the offspring of the pair they help, kin selection may be of major importance in this species. However, because the young males are likely to remain for a long time in their natal clan and know their clan members as individuals, the 'trading' of co-operative acts is possible. The inheritance of territory by young males and an enhanced access to breeding females are also likely to be important. Finally, the benefits resulting from acquiring skills – from the perfecting of 'ordinary' skills to the acquisition of season-specific or lean-year skills – may also contribute to the fitness of helpers.

The benefits of adoption
Just like helping, adoption is now acknowledged to be much more common than was previously thought. It has been found in hundreds of species of birds and mammals, and for many of them it seems to be a

normal and customary part of life. Although it is not a unitary phenomenon, and it is sometimes difficult to distinguish adoption from helping, adopting behaviour has two special characteristics. First, adopters often take unrelated young; therefore kin selection cannot provide an explanation for the adaptive evolution of this behaviour. Second, unlike helpers, adopters often take exclusive responsibility for rearing the young they have fostered. These two features of adoption have made this behaviour particularly difficult to explain in adaptive terms.[47] It is therefore not surprising that adoption is often regarded as an incidental by-product of parental care – a mistake due to abnormal personal, social or ecological conditions. The mistake might be in the recognition of one's own young in crowded conditions, or a by-product of an unusually high hormone titre in a failed breeder. Adoption has also been explained in terms of evolutionary conflict. Hence, when it is the young who initiate adoption, as is the case in several species of gulls and terns, adoption is seen as the outcome of parent–offspring conflict, with the adoptee trying to manipulate the prospective parent to adopt it, while the parent is doing its best to avoid being so manipulated.

There are also adaptive explanations for some forms of adoption. In ostriches and many species of waterfowl, where the adoption of whole broods is common, adopters benefit by increasing their family size and therefore their ability to dominate food resources. In other species, adoption is related to essential aspects of life like helping or mate selection. As we will show in what follows, these cases suggest that adoption can have advantages for adopters, including the advantages that result from information that is socially learnt in adoptive conditions.

The origins of helping behaviour in white-winged choughs are lost in the annals of evolutionary history, but now it is often associated with adoption. In this species, helping is more than helpful: it has become obligatory. Without helpers the whole group perishes. Groups with less than four members are not viable because they do not succeed in raising young. Juveniles are therefore very valuable to adult birds, who go to great lengths to acquire them and recruit them as future helpers. Remarkably, still-dependent, unrelated fledglings are actively kidnapped during battles between groups! The kidnapped young are adopted by the kidnappers, and receive substantial, devoted parental care. They end up as helpers in their new group, contributing to the group's size and therefore to its success in rearing young. Since in this species there is no individual dispersal, and new groups become established only when a large group breaks up, inbreeding is a potential problem. The adoption of unrelated young not only strengthens the group, but also probably solves some of the inbreeding problems, for, when the adoptees gain breeding status, they pair with unrelated individuals.[48] Adopting is clearly adaptive in this species.

Forced adoption is also known in mammals. A group of hamadryas baboons is a complex hierarchical social unit made up of families that are organised into clans, with several clans forming a band and several bands forming a troop. A family unit consists of a single mature male and several 'wives', who are guarded by this male, and are respected as belonging to him by other mature males in his clan. 'Wives' are acquired by abducting mature females from other clans, or by a young male who joined the family and followed it for years displacing an old male and acquiring his 'wives'. Alternatively, a young male may kidnap and adopt a young, two- or three-year-old female. In this latter case, the young female is taken away from her mother, often screaming and protesting, to become the protégée and future 'wife' of the young male. The kidnappers are young adult males, eight to ten years old, and for several years the relationship between the young male and female is non-sexual. Hans Kummer describes the relationship between the kidnapper and his future wife in this way:

> these are not real marriages yet. There is no sex in them at all, because the female is too young to have swellings and the size difference is absurd. Instead, the male replaces his little companion's mother whom she would otherwise still need . . . He makes allowances for her deficient climbing skills and carries her on his back over difficult passages, whereas adult females are never carried by their spouses. When there is some disturbance in the group, the young male shelters the little female in his arms, where she may almost disappear behind the curtain of his mantle, and in the evening she usually goes to sleep nestled against his breast.
>
> (Kummer, 1995, p. 217)

By becoming a substitute mother, the young male is securing a wife for himself, a demanding task in hamadryas baboon society. A young female is much easier to acquire than a mature one, and the relationship that develops between the prospective mates creates a bond of familiarity and affection. When the female is sexually mature, the older males in the clan will regard her as belonging to the adopting male. In this way, by adopting a 'child-bride', bonding with her and defending her, a young male acquires a 'wife' who is very familiar, maximally attuned to and highly compatible with his personality and habits, and more difficult to abduct than a less devoted 'wife'.

We believe that 'arranged-marriage', which ensures a high degree of compatibility between mates without the problems of inbreeding, may help to explain other cases of adoption too. In birds such as ostriches, ducks and shorebirds, whose offspring are born largely independent, adoption is quite common.[49] It usually involves brood amalgamation: an entire brood, or part of a brood, joins or is taken to join another brood, and remains with the adopting parent(s). Since the adopting parents are successful reproducers and not failed breeders, their behaviours cannot be attributed to misplaced parental care, although brood amalgamation sometimes results from broods mixing accidentally following the confusion accompanying territorial disputes.[50] There is no evidence that adopting reduces the reproductive success of adopters. In fact, it may sometimes be beneficial to the adopters as well as to the adoptees. In these precocial birds, parental care is largely a matter of vigilance and food-selection demonstrations, so the cost of care does not increase very much with increasing brood size. But a larger brood size has several potential benefits. One is that the risk of predation is decreased,[51] both because the chances that the parent's own offspring will be taken are reduced, and because early detection of predators is more likely with a greater number of vigilant juveniles in the family. Another advantage is that larger families are often dominant over smaller ones, because they can acquire and defend better foraging areas. Consequently, chicks grow faster and survive better in larger families.[52]

A further benefit of this form of adoption may be that it allows the parents to match their own offspring with non-related but compatible potential mates.[53] As we argued in chapter 5, a certain degree of similarity in habits and experience increases the efficiency of co-operative activities such as nestling provisioning and defence against predators or neighbours. A high degree of compatibility between mates is particularly important in monogamous species in which both mates contribute to parental care and the pair bond is stable and long-lived.[54] By adopting unrelated chicks and rearing them with their own, parents may be 'choosing' compatible future mates for their own young. The adoptees and the adopters' own young will be familiar with each other, and behaviourally and ecologically compatible. Consequently, when the adopters' offspring mature, they have available highly suitable mates – their adopted sibs. This form of 'match-making' is expected to be most common in species in which the mates pair for life, or in any other circumstances in which a high level of compatibility between mates is required. It may not be a coincidence that in North American species of geese, which have long-term pair bonding and where both parents care for the young, brood amalgamation is common.[55] Although to the best of our knowledge such data have not yet been reported, we expect that preferential mating between non-sibs belonging to the same amalgamated brood will be found.[56]

Biasing mate preference may also play a role in the evolution of another form of adoption, adoption by replacement males. In many bird species, when a female becomes widowed, a male attaches himself to her, and in some cases adopts and cares for her current nestlings, which are genetically unrelated to him. This form of adoption may be similar to the helping by unrelated pied kingfishers that we described in the previous section: by caring for the young of a prospective mate, the male may be demonstrating to her his capabilities as a provider and parent, thus increasing the chances that she will take him for a mate in the next breeding season.[57] Viewed in this way, the male's adopting behaviour is functionally a more demanding variant of courtship feeding.

There may be benefits from this behaviour that are even more delayed. The unrelated foster-daughters may become sexually imprinted on characters that are unique to their foster-father, such as his song or the particular colour–pattern of his plumage. In this way, a foster-father may inadvertently bias the future mate choice of the fostered daughters towards those males with features similar to his own. If a fostered female chooses an individual who is similar to her father, the fostering male may improve his own future mating chances, as well as those of his sons and male sibs, who are similar to him. In cases where such biased mate preference is important, we expect foster-fathers to be more 'generous' towards foster-daughters than towards foster-sons. Of course, this prediction will not hold when foster-fathers need to recruit males as future helpers, as they do in green woodhoopoes.

What we know about sexual imprinting suggests that young may become sexually imprinted not only on their foster-parents, but also on helpers. We know of no studies that have looked at the effect of helpers on the future sexual preferences of the helped young, but there are some indications that helpers could have such an effect. There are factors that can influence and sometimes even override initial imprinted preferences. For example, in the snow goose, mallard and zebra finch, the characteristics of other siblings affect the future sexual preferences of the young.[58] Sexual imprinting is also influenced by social interactions and in particular by feeding: in experimental mixed-species pairs, where one of the mates was a zebra finch and the other a Bengalese finch, the parent who had most interactions with the young, or the one who fed them most often, was the target of their imprinting. Ten Cate has suggested that early learning, such as sexual imprinting and song learning, are preceded by a period of familiarisation, when an initial attachment is formed between the caregivers and the tended young. Imprinting and other forms of early learning occur on the basis of this attachment.[59] Young certainly become familiar with their helpers during the period in which imprinting is first established: helpers interact with the young, feed them, vocalise to them and sometimes groom them. It would be surprising if helpers had no influence on the future sexual preferences of those they help.

If helpers do bias the future sexual preferences of the helped young towards their own phenotype, the benefits of imprinting, especially imprinting on male helpers, may have contributed to the evolution of helping. Since, in birds, females are usually the dispersing sex, their mortality is often higher than that of males, so mature females may be in short supply. If female youngsters become sexually imprinted on a helper male, this may increase the chances that the helper will be selected by and mate with one of these females as they mature. His male kin, who are likely to be similar to him because of their shared genes and shared early environment, will probably enjoy a similar increase in their chances of mating. The imprinted females can also benefit: if helping leads to familiarity between helper and helped, and if familiarity facilitates pair formation, the time spent searching for a compatible male may be shortened, aggression curbed and the success of the pair bond enhanced. Male helpers with characteristics that make them good targets for imprinting, and females who readily become imprinted on such males, would have an advantage, so imprinting on helpers as well as helping would spread. We mentioned earlier that 'regular' helpers in riflemen have a very high chance of later pairing with the chicks that they helped to rear.[60] We suggest that the benefits here are not only to the helpers who secure mates for themselves, but also to the young and parents who, in addition to the obvious energetic benefits of help, acquire mates with whom they are already familiar and hence behaviourally compatible. Sexual imprinting and the effects of early familiarity may be the psychological mechanisms underlying this biased mate choice.

Although some of the ideas in this section are speculative, there is little doubt that the socially learnt behaviours and preferences that are acquired and transmitted by helpers and adopters can contribute to their reproductive success and to the evolution of seemingly altruistic behaviours. Helpers gain important skills by learning from experienced individuals, and both helpers and adopters transmit information which could benefit them in the future by biasing the mate preferences of the helped or adopted young. These ideas can be tested experimentally. However, we expect to find differences between the effects of such information transfer in helpers and adopters. When adopters assume full parental responsibility, they are very much like genetic parents, major transmitters of learnt information. Compared to such adopters, what helpers are able to transmit is generally of secondary importance. Most helpers spend less time and effort feeding the helped young than the parents do, so the information they are able to transfer is likely to be less effectively learnt and remembered than parentally transferred information. Nevertheless, the information supplied by helpers can contribute to the youngsters' store of knowledge in two general ways. First, if the information is the same as that which the parents transmit, it consolidates behaviour learnt from the parents, because repetition and positive reinforcement encourage habit formation. Second, new behaviour patterns or preferences that are introduced by the helpers can be added to those acquired from the parents and thus enrich the potential repertoire of socially learnt behaviours in the helped young. In these ways, young raised by both parents and helpers probably gain 'informational' benefits in addition to energetic benefits, and may consequently show a wider range of behaviours than young raised solely by their parents.

Phenotypic cloning through alloparenting

Alloparenting, whether through adoption or helping, may result in the 'phenotypic cloning' of some of the behavioural phenotypes of foster-parents and helpers.[61] Helpers and adopters can transfer information to

non-descendent young at the time when the young learn most effectively. Like genetic parents, they can transmit information about their
food preferences, foraging techniques, mate-preferences, ways of avoiding predators, mobbing techniques and styles of parenting. When the
young who acquired behaviour-influencing information from their alloparents mature, they can transmit it either to their own offspring
through conventional parental care, or to the offspring of others
through alloparenting. In this way, lineages with somewhat different
patterns of behaviour may develop, and, through the differential survival and breeding success of the individuals in these lineages, the behavioural repertoire may evolve. Essentially, this is an extension of the skill
acquisition hypothesis: if adaptive variations in the skills acquired by
the young are transmitted across generations, they will spread and may
evolve further. The process is one of natural 'cultural' selection of variant socially learnt and transmitted preferences and skills.

Even if a behaviour has no adaptive value, it can nevertheless spread
through adoption and helping. Imagine a population where non-
adopters are the majority, and there are only a few adopters. Now
assume that, by chance, adopters have a new socially transmissible trait,
such as a special food preference, that is neither harmful nor beneficial
compared with the traditional ones. In such a case, adoption may lead
to the spread of this rare trait, because of the inherently asymmetric
nature of information transfer. Adopters adopt the offspring of both
non-adopters and adopters, and transmit information to them, as well
as to their own offspring. When adoptees become parents, they transmit their socially acquired patterns of behaviour to their offspring
through conventional parental care. If the practice of adoption does not
lead to a reduction in the adopter's fitness, a socially acquired behaviour, which was present by chance in just the adopter fraction of the
population, may spread in the population at large, even if it confers no
benefits.[62] As Cavalli-Sforza and Feldman have emphasised, the spread
of any 'cultural' trait depends on both the relative success of 'cultural
parents' in transmitting learnt behaviours to their offspring and other
'students', and its effects on biological fitness.[63] The explanation of the
distribution of behavioural phenotypes in a population consisting of
both adopters and non-adopters, or helpers and non-helpers, therefore
has to take into account the asymmetrical mode of transfer of behavioural information by adoption, as well as the effects of the transmitted behaviours on fitness.

What is fascinating about the transmission of information in this manner is that it can explain the transmission and even the spread of alloparental care itself! If helping or adoptive behaviours are associated with a particular style of parental care, and that style of care is transmitted through social learning, the transmission of the tendency for alloparenting may follow. Imagine that a rare adopter or helper appears in a population. Its alloparental behaviours may spread because it will transmit its parenting style not only to its own offspring (vertical transmission), but also to the offspring of non-adopters or non-helpers that it adopts or takes as helpers (oblique transmission). In contrast, individuals with non-adopter and non-helper parenting styles can transmit their parenting style only vertically, to their own biological offspring. Simple mathematical models have been used to investigate how the 'altruistic' parenting styles of adopters and helpers can spread in a population. Although adopting and helping usually have immediate and delayed benefits for the caregivers and the care receivers, which contribute to the persistence of these behaviours, the models show that the peculiar dynamics of the spread of alloparental care can sometimes be a sufficient reason for its presence in a population.[64] The spread of the adopting (or helping) parenting style depends on its transmissibility from parents to young, and on the chances that a parent will encounter and take an offspring who is not its own as an adoptee or as a helper. With helpers and with adopters who are not themselves parents, the spread of the parenting style also depends on the proportion of such non-reproducers in the population. Even adopters with no family of their own can transfer patterns of behaviour, so their influence on the frequency of behavioural phenotypes in the population may not be zero.

We have already seen that styles of parenting can be transmitted within lineages.[65] For example, the way rhesus monkey mothers treat their offspring is passed from one generation of females to the next.[66] Similarly, infant-abuse runs in pigtail macaque families, where offspring who observe the abusive behaviour of their mothers towards their sibs are likely to repeat it when they become parents. As in some human families, in which there are repeated cycles of child abuse, once abusive behaviour is precipitated (by stress, or for any other reasons), it seems to perpetuate itself through early social learning.[67]

The transmission of parental neglect is not a peculiarity of primates: female rats who have been prematurely separated from their mothers during early infancy spend significantly less time with their offspring.[68]

But the most dramatic evidence that parenting styles can be transmitted by social learning comes from voles.[69] Meadow voles have a promiscuous mating system in which the female mates with several males. These males do not share her nest or display any paternal behaviour. The basic social unit in meadow voles is thus the same as in most mammals: a mother and her current offspring. Prairie voles, on the other hand, are monogamous, and the male bonds with the female and helps in nest building and food hoarding. After the birth of the young, he huddles over the infants and grooms them. The prairie vole mothers also provide more care and nursing effort during the lactation period than do the females of the promiscuous meadow vole. Moreover, most juvenile prairie voles, both male and female, stay in the nest beyond the weaning period and help their parents rear subsequent litters. The basic social unit in the prairie vole is, therefore, the extended family. The differences in family life style and style of parenting between these related species can hardly be more profound. Yet, the prairie vole's caring paternal style can largely be transferred to meadow voles if the latter grow up in prairie vole families. Cross-fostering experiments in which newborn meadow vole babies are transferred to a prairie vole nest have shown that, when the fostered males grow up and mate, they cohabit with the female inside the nest, and after the birth of the young they frequently huddle over them and groom them. The foster-sons copy their foster-father's set of parental behaviours! Female meadow voles who are cross-fostered and reared in prairie vole families also show more grooming, huddling and nursing than meadow vole females who grew up with meadow vole parents.

Even in normal prairie vole families, the effects of upbringing on the future parenting style are very apparent. Living in an extended family with helper sibs influences parental behaviour. When individuals who as juveniles cared for their younger siblings become first-time parents, they show more parental care than those who were not helpers. Presumably their experience with younger siblings allowed them to practice their parenting skills, and later enhanced their tendency to care. Surprisingly, the extent of helping is also influenced by the father's behaviour. The length of time that a juvenile prairie vole spends in the natal nest, helping its younger siblings, is correlated with the length of time that the father of this juvenile spent caring for his offspring. The father's example therefore affects not only the paternal behaviour of his male offspring when they become breeders, but also their helping behaviour when still young. In this species, the behaviours of the male parent seem to be transmitted to its offspring through social learning, thus profoundly influencing the social organisation of the family.

Cross-fostering experiments are not carried out as often with birds as with mammals, so there is only anecdotal evidence that their helping behaviour can be transmitted through social learning. However, some of the observations of pinyon jays suggest that helping may be associated with a socially transmitted style of parental care. In this species, sons who do not manage to disperse are frequently given a second chance by their parents, and are allowed to stay for another year in the parental territory as 'helpers', although the value of their help is doubtful. But helping by young pinyon jays and parental tolerance towards grown-up offspring are family-specific affairs – they are not seen in every lineage. Some lineages consistently include helpers, while others never do. The difference between the two types of families is interesting, and its causes are not entirely clear. However, the fact that helping runs in lineages may be the result of social learning and transgenerational transmission of parenting styles. An indication that the difference between the lineages could be due to socially induced and socially transmitted behaviours is suggested by a nest transfer experiment. When a young male nestling from a non-helping lineage was transferred to a helping-lineage nest, it grew up to become a helper to its foster-parents.[70]

There are probably several different kinds of parenting style associated with helping and adopting behaviours. In colonial species such as gulls or terns, where straying young initiate adoption, the extent of parental tolerance towards youngsters who stray is likely to influence whether and how much parents adopt. Individuals are known to vary in their degree of tolerance or aggression towards both their own young and other individuals.[71] As they themselves become parents, young adults who have experienced parental tolerance as juveniles will probably be more inclined to act tolerantly towards non-descendant young, and hence become adopters or allow their offspring to become helpers. The helping behaviour of the cross-fostered pinyon jay that was 'allowed' to remain with its foster-parents may be a result of acquiring the foster-parents' tolerant parenting style. If so, when a young male becomes a parent, its behaviour towards its own offspring is expected to reproduce the tolerant parenting style of its parents. A tolerant parenting style may spread because it is advantageous, but it may also spread if unrelated young occasionally happen to or are allowed to stay with helper–lineage pairs. If this occurs, both a tolerant parenting style and helping will be introduced into non-helper lineages, and the relative number of helper-lineages will consequently increase.[72]

Adoption is not always initiated by the young. It can be initiated or even enforced by the adopters, as it is in white-winged choughs and ostriches. In these cases, it is not parental tolerance that is socially transmitted, but probably a set of recruiting, coercive or kidnapping behaviours. Since these behaviours are observed by the 'adopted' and other young, either as participants or onlookers, they may be learnt by them. Later, as they become parents, they may display the behaviours they learnt as youngsters, and thus socially transmit them to their own biological and cultural offspring.

We see adopting and helping behaviours as effects of particular styles of parenting, but we still need to know how such parenting styles originated. It is reasonable to assume that occasional caring for non-off-spring may be the result of a slightly generalised or extended form of normal parental behaviour. For example, under favourable conditions, such as when food is abundant, particularly successful individuals may show increased tolerance towards unrelated young, and allow them to help or to become adoptees. More active adoption probably occurs when parents have unusually high hormone levels, or when they have cur-rently lost their own young but their hormonal state still primes them to care for young. Whatever the original reason for tending unrelated young, however, and irrespective of whether it is a mistake or adaptive, once present the parenting style may spread in the population through social learning. Its subsequent perpetuation may become dissociated from its origin. We can speculate that the differences between the family organisations of meadow voles and prairie voles may have originated in this way. If we assume that prairie voles evolved from meadow voles, then the initiating event could have been an ecological change that limited suitable space and cover, and led male meadow voles to try to cohabit with the females with whom they mated and who tolerated their presence. While cohabiting, these males inevitably became more familiar with mother and young, and consequently showed a limited amount of parental care, as indeed meadow voles do in the crowded and often exposed conditions of captive husbandry. The young voles born into the altered social structure may have perpetuated the new life style, even if the circumstances that led to it changed in subsequent generations. The effects of the transient ecological conditions may have become integrated into the family social system, initiating a self-perpetuating change in its social organisation. The new habits would be further stabilised by any genetic changes that were congruous with the new family organisation.

A particular style of parenting will not lead to adoption and helping under all circumstances, of course. It is unlikely that even very tolerant parents would let unrelated young join their family when food is scarce. But we expect that, in similar ecological conditions, young reared by adopting parents will have a higher tendency to adopt than young reared by non-adopting parents, and young reared with helpers will be more likely to become helpers themselves and, later, become parents who encourage helping. It is important to realise that, even when adults with the help-inducing or the 'adoptive' parenting style do not actually take helpers or adopt, the general parenting style that their genetic off-spring experience and learn from them may lead these offspring to adopt or to induce helping whenever ecological conditions allow it.

In species that breed in large colonies, the spatial structure of the colony and the relative positions and distances of adopting or helping families may influence the future tendency of adoptees to adopt, and of helped young to become helpers. Young may learn not only from parents, but from neighbours too. For example, if adoption is common in a particular part of the colony, young may learn by observation from both parents and neighbours, and in time the tendency to adopt may diffuse out of this locality. The more instances of adoption that are observed, the greater the efficiency with which the adopters' parenting style will be learnt.

The spread of adoption and helping through social learning illustrates the point we made in chapter 2 – that it is necessary to consider not only the adaptive functions of learnt behaviours, but also the mechanisms underlying their production and transmission. The psychological basis for caring for non-offspring (for example, parental tolerance) is intimately tied up with the circumstances in which this trait is socially learnt, remembered and transmitted. Consequently, we should always identify not only the adaptive functions leading to the evolution of a learnt behaviour, but also the underlying psychological 'substrate' of evolution, the 'proximate' causes of the behaviour we study. We have to understand the mechanisms of social learning, the role of familiarity in inducing and encouraging early learning, and the manner in which behaviour patterns and preferences are learnt by both adults, young and juveniles. The cultural transmission of patterns of behaviour is part of their development.

Our focus on the transmission of alloparenting through social learning does not mean that alloparental care has not been selected at the genetic level as well as at the 'cultural' level. Since parental behaviours, like most other behaviours, are quite plastic, the adaptations to new ecological conditions or social situations that promote alloparental care are initially likely to be behavioural rather than genetic. But, if such alloparental care is adaptive over long periods, selection for the stabilisation of the new set of behaviours may result in the genotypes that confer or favourably affect such stability becoming common. For example, there may be selection for genotypes that predispose the young to learn parenting styles earlier, or predispose adults to be more tolerant (or aggressive!) towards offspring and non-offspring.

Summary

Looking at alloparental care as a route for the transfer of learnt information makes it much easier to understand the persistence and evolution of adoption and helping behaviours. The time and energy that caregivers invest in non-descendent young is often more than compensated for by the benefits these caregivers gain through the exchange of information. First, alloparents who act as helpers may acquire and perfect useful skills by staying as 'assistants' to the breeding pair. Second, by engaging in costly caring behaviour, helpers and adopters may honestly advertise their high quality. Third, by transmitting preferences and skills, helpers and adopters may bias the future social and sexual behaviour of those they help, so that the tended young will favour them or their close kin as mates or social partners.

Like parental behaviour, alloparental behaviour may often lead to the 'phenotypic cloning' of acquired patterns of behaviour. Through helping and adopting, useful behavioural variations, such as a more efficient method of foraging or a more memorable variant of an alarm call, may be transmitted between members of different lineages, and spread in the population as a whole. Furthermore, the styles of parenting may themselves spread from members of one lineage to another. When a particular style of parenting increases the tendency to adopt or allow helping, this parenting style will itself be transmitted and possibly evolve. In other words, the transmission mechanism itself can undergo cultural evolution. The transmission of the parental behaviour that leads to helping or adoption is inherently asymmetrical, because alloparents transmit their behaviour to both genetic and cultural offspring, whereas non-alloparents transmit only to genetic offspring. Alloparenting may therefore spread even if it is selectively neutral or slightly deleterious. The enhanced spread of alloparenting through the cultural transmission of the parenting style that promotes it, and the advantages that the information exchanged during alloparenting confer, may allow rapid cultural evolution. When these aspects of alloparenting are considered, it is easier to understand the persistence and evolutionary elaboration of the apparently paradoxical, 'altruistic', patterns of behaviour seen in adoption and helping.

Notes

[1] Leviticus 19, verse 18. In the original Hebrew version, the word used for what in the King James's version is translated as 'neighbour' is 'rhea', which means 'friend'. It is certainly easier and biologically more sensible to love a friend than a neighbour!

[2] For general information on helpers, see Emlen, 1991; Skutch, 1961; Stacey & Koenig, 1990.

[3] Discussions of adoption can be found in Boness, 1990; Packer, Lewis & Pusey, 1992; Riedman, 1982.

[4] Skutch, 1935.

[5] This account is based on E. A.'s observations of breeding colonies at two sites along the upper Jordan river during the 1996 and 1997 breeding seasons, and the following sources: Fry, 1984; Lessells, 1990; Lessells & Krebs, 1989; Lessells, Rowe & McGregor, 1995; Lessells et al. 1991.

[6] This particular colony was observed during 1996 at a breeding site a kilometre and a half from Kibbutz Ashdot Yaakov.

[7] Lessells, 1990.

[8] Emlen, 1991; Stacey & Koenig, 1990.

[9] Pierotti, 1991.

[10] For discussions of the 'unselected' hypothesis, see Jamieson, 1986, 1989, 1991; Jamieson & Craig, 1987.

[11] The different adaptive explanations of helping are discussed in Brown, 1987; Cockburn, 1998; Emlen, 1991; Hamilton, 1964; Heinsohn, 1991a; Ligon, 1997; Ligon & Ligon, 1983; Trivers, 1971; Zahavi & Zahavi, 1997.

[12] Lack, 1954.

[13] Wunderle, 1991.

[14] For birds see Ashmole & Tovar, 1968; Gill, 1995; Norton-Griffiths, 1969; Skutch, 1976; for bats, see Altringham, 1996.

[15] Brown, 1987; Doolan & Macdonald, 1996a; Johnston, 1982; Lack, 1966.

[16] Komdeur, 1996.

[17] Heinsohn, 1991a; Heinsohn, Cockburn & Cunningham, 1988.

[18] Marzluff & Balda, 1992.

[19] See note 10.

[20] Zahavi, 1975; Zahavi & Zahavi, 1997.

[21] Zahavi, 1990.

[22] According to Wright (1997, 1998a,b) all group members, irrespective of their sex or rank, provision the young at similar rates and make similar adjustments in feeding effort, presumably in accordance with the changing needs of nestlings. He argues that kin selection and group augmentation have been important factors in the social evolution of Arabian babblers. However, Zahavi claims that, by averaging over different groups and sub-groups, and ignoring subtle, yet highly important, manifestations of competitive superiority, Wright has obliterated the significant status-related behavioural differences between individuals (Zahavi, 1999, and personal communication).

[23] See Reyer, 1990. Fessl, Kleindorfer & Hoi (1996) found that in the moustached warbler the mate and male helpers compete directly for the female. Helpers assist breeding females in incubating the eggs and in feeding and defending the chicks. Females sometimes switch mates to pair with the helper for the second breeding attempt in the same season, but never switch mates if they were not helped.

[24] Sherley, 1990.

[25] Lessells, Avery & Krebs, 1994.

[26] Summarised by Emlen, 1997a,b.

[27] For details of the life of white-fronted bee-eaters and the role of helping in this species, see Emlen, 1990; Emlen & Wrege, 1988, 1989, 1992.

[28] Emlen, Wrege & Demong, 1995.

[29] Moehlman, 1979, 1983, 1986.

[30] The evolution of alloparental care of unrelated young is discussed by Dunn, Cockburn & Mulder, 1995; Pusey & Packer, 1994; Riedman, 1982; Woodroffe & Vincent, 1994.

[31] Trivers, 1971.

[32] Wilkinson, 1984, 1988.

[33] See Altringham, 1996; Kunz, 1982; Racey & Swift, 1995; Turner, 1984. The excellent learning abilities of vampire bats may explain the almost complete switch they have made from their former wild hosts to domestic herds, a switch that occurred in central and south America during the last four hundred years, following human hunting and cattle ranching.

[34] Ligon & Ligon, 1978a,b, 1982, 1983.

[35] See page 219 and Clutton-Brock & Parker, 1995a; Emlen, 1995, 1996; Emlen & Wrege, 1992; Emlen, Wrege & Demong, 1995. Coercion does not always require brute force. In mammals, coercion through a combination of chemical, pheromonal and subtle psychological means is often quite sufficient to

suppress the reproduction of offspring, but suppression is usually combined with psychological stress.

[36] Dunn, Cockburn & Mulder, 1995.

[37] Hauser, 1992; Hauser & Marler, 1993.

[38] Clutton-Brock & Parker, 1995a,b; Mulder & Langmore, 1993.

[39] Summarised by Skutch, 1976.

[40] Marzluff & Balda, 1992.

[41] But see Skutch, 1976.

[42] Proverbs 19, verse 6. In the original Hebrew version, the word for what is translated as 'prince' in the King James's version is 'nadiv', which means 'generous', without implying anything about the social class or status of the generous individual.

[43] Alexander, 1987; Ferrière, 1998; Nowak & Sigmund, 1998.

[44] Heinsohn & Legge, 1999.

[45] An additional reason for the persistence of co-operation has been suggested by Sella & Lachmann (personal communication). They argue that, when new groups are formed by very few founders (e.g. a pair of individuals), and when only co-operative pairs can establish an effective group, most offspring will be co-operators. However, selfishly acting individuals may appear, and, because of their higher fitness, they can become relatively more abundant within the group. They are therefore liable to lead to the destruction of their group – it will eventually 'die'. However, if some co-operative individuals disperse to found new groups before the group dies out, groups composed of co-operators will always exist, and co-operation can be maintained over time. According to this model there is no competition among groups, and all groups are equal in productivity. Co-operation is maintained by the dynamics of the birth, growth and death of groups.

[46] Cockburn, 1998.

[47] For a review of ideas on the evolution of adoption, see Avital, Jablonka & Lachmann, 1998.

[48] See Heinsohn, 1991b, for an account of kidnapping in white-winged choughs. Selection between groups, favouring groups with devoted helpers and effective kidnappers, may also contribute to the evolution of helping and adoption in this species. We discuss group selection in the next chapter.

[49] Beauchamp, 1997; Bertram, 1992; Eadie, Kehoe & Nudds, 1988.

[50] Savard, 1987.

[51] Bertram, 1980.

[52] Beauchamp, 1997; Choudhury et al., 1993; Eadie, Kehoe & Nudds, 1988; Riedman, 1982. According to Beauchamp, brood amalgamation of chicks is positively correlated with the brood amalgamation that occurs pre-hatching, when one female lays her eggs in the nest of another. The correlation between the two types of brood amalgamation supports the idea that an increase in family size is adaptive in these species.

[53] Avital, Jablonka & Lachmann, 1998.

[54] Black, 1996; Black & Owen, 1988; Choudhury & Black, 1994.

[55] Eadie, Kehoe & Nudds, 1988.

[56] When match-making is an important function of adoption, the timing of sexual imprinting and the pattern of dispersal are expected to determine the age at which young are adopted, and possibly also the preferred sex of the adoptee. We expect (i) a correlation between the timing of sexual imprinting and of adoption, and (ii) that adoptees will mate more often than at random with their foster sibs.

[57] Reyer, 1990.

[58] Cooke, 1978; Klint, 1978; Kruijt, ten Cate & Meeuwissen, 1983.

[59] ten Cate, 1987; ten Cate, Kruijt & Meeuwissen, 1989.

[60] Sherley, 1990.

[61] Avital & Jablonka, 1994; Avital, Jablonka & Lachmann, 1998.

[62] Avital & Jablonka, 1996; Avital, Jablonka & Lachmann, 1998; Hansen, 1996.

[63] Cavalli-Sforza & Feldman, 1981; see also Boyd & Richerson, 1985.

[64] In some cases of adoption, for example in the common eider, white-winged scoter, barnacle goose, Hawaiian monk seals and many other species, there is no apparent advantage or disadvantage for the adopting parent. For references see Avital, Jablonka & Lachmann (1998), where two simple models showing the invasion of rare, neutral, socially learnt adoptive behaviours are described. Similar models can be developed for helping. It is easy to show that even if adoption and helping are slightly deleterious, under some conditions they can still spread.

[65] See chapter 4 (pp. 130–2); chapter 6 (pp. 198–200); Berman, 1997; Fairbanks, 1996.

[66] Berman, 1990.

[67] Maestripieri & Wallen, 1997.

[68] Skolnick *et al.*, 1980.

[69] Wang & Insel, 1996.

[70] Marzluff & Balda, 1992, pp. 219–29.

[71] Benus & Röndigs, 1996; Rohwer, 1986; Scott, 1975; Southwick, 1968.

[72] An increase in the proportion of individuals belonging to helper lineages has been observed in a study of a town flock of pinyon jays in North Carolina, but the reasons for this have not yet been found.

8   The origins and persistence of group legacies

Up to this point, we have concentrated on social learning and its consequences in nuclear and extended families, where information is transferred between mates, between biological or adoptive parents and their offspring, between helpers and those they help, and among sibs. We now want to widen our discussion to see what goes on in those species of birds and mammals that are highly social, living in more or less permanent groups composed of both related and unrelated individuals. Our aim in this chapter is not to carry out an extensive review of the social group-life of birds and mammals. Rather, we want to look at some aspects of behaviour and psychology that throw light on the formation and maintenance of group traditions, and see how these group traditions themselves influence, directly or indirectly, the evolutionary development of social behaviour. We shall show how the various psychological mechanisms that serve to organise and co-ordinate the activities of a group depend on a constant flow of information among its members. This flow of information is mediated through social learning and maintained by frequent social interactions.

Before starting this discussion, we want to take a close look at the real-life intricacies of a group-living social mammal. So, imagine a cloudless day in January, in the Kalahari desert of south-west Africa, where we are watching the activities of meerkats, the social mongooses that live in small groups on the dry, open plain along the Nossob river.[1]

The sandy plain stretches along the dry river bed for as far as the eye can see. It is covered mainly with small greyish-green bushes, with much open sand between them. Several large acacia trees add a vertical dimension to the otherwise flat scenery. The early afternoon sun is merciless. Sources of drinking water are nowhere to be seen. The dazzling sun and the scorching heat at first make it difficult to spot the meerkats, but rhythmic spurts of flying sand and noises that sound almost like human murmurs finally disclose their presence. A dozen slender, grizzled-grey, 30-centimetre-long animals, spaced about 3 metres apart on the vegetation-free exposed sand flat, are all excavating and shifting sand. The meerkats in this band, each within viewing, smelling and hearing distance of the others, are searching for sand-dwelling insect, arachnid and reptile prey.

A closer look on one member of the foraging party finds it sniffing at a certain seemingly indistinctive location, and then digging frantically with its long clawed forefeet. It unearths a gecko, pulls it to pieces with claws and mouth, and devours it. A fat beetle grub comes next; unlike its predecessor, it is not dismembered, but is grabbed eagerly and eaten. Another bout of digging and sand shifting brings up the next victim, a skink. The meerkat, behaving as if unsure of how to deal with this relatively uncommon item of prey, sniffs repeatedly at the motionless skink, cautiously knocks it about with its foreclaws and test-nips it. The skink now tries to wriggle back into the sand, but the meerkat finally gives it a hard bite and shakes it to death.

A loud screech sends this and the other meerkats running at top speed for their burrows, and sends the approaching large, black-and-white Bateleur eagle floating silently away to try its luck elsewhere. The screech, a life-saving alarm call, was sounded by a lone meerkat sentry. This sentry, standing erect at a vantage point a few metres up a dead acacia tree overlooking the meerkat territory, constantly scans the surrounding sky and ground for potential predators. During its hour-long shift other band members are free to relax their vigilance, lower their heads and devote their attention exclusively to foraging.

After a few more minutes the aerial menace has gone. The meerkats peer out from 3 of the 10 entrances to their extensive burrow system, which they both dig and amicably share with a colony of social, vegetarian African ground squirrels. One by one the meerkats cautiously sneak out of the burrows to resume foraging at another location, a hundred metres or so from the first. It will be at least a week before they return to forage at the first site.

High on the acacia tree, a change of sentry takes place. It is late afternoon, but the meerkat foragers do not seem to tire. They eagerly continue to dig and hunt a wide variety of insect and arachnid sand-dwellers. The new sentry is now on guard and, as it detects a passing silver-backed jackal, it emits a warning call that is gruffer and more abrupt than the previous eagle-induced one. This call immediately brings the whole meerkat band to tense, erect attention. They watch the jackal until it trots out of sight, and then resume their digging. It does not take long to realise that, by using a variety of alarm calls, a sentry is able to communicate to his band members different kinds and degrees of danger.

A neighbouring meerkat band is approaching. The sentry shrills. Twelve members of the group scurry towards the trespassing band, calling loudly. They stand erect, tails pointing upwards, and face the eight opponents. For a few uneasy moments both groups do nothing but survey one another. Finally the trespassing, smaller band retreats. Fierce territorial battles, lasting until one of the bands retreats, do occur; but not this time.

Back home, in front of one of the entrances to the burrow system, three six-week-old, half-weaned youngsters, closely guarded by a currently non-breeding male baby-sitter, spend their time play-fighting. They ambush, chase and grasp each other, falling over and pestering their baby-sitter. The baby-sitter occasionally participates in these play-fights, but usually he seems to be very patient with the young. Five non-breeding individuals share the baby-sitting duties, each in turn acting as the babies' guardian. The baby-sitters include two adult females who gave birth at the same time as the dominant female, the babies' mother. These two females lost their entire litters shortly after producing them, probably through infanticide by the dominant female. Yet they guard her young and seem to care for them just as devotedly as they would care for their own offspring, and there are even signs that they suckle them.

Early evening brings all the band members, the babies' mother included, back from their foraging grounds to the safety of their burrows. The babies rush to their mother and in a few moments are eagerly suckling from her, purring softly, as she lies on her back. She licks and picks fleas off her babies, but looks very weary. During the eighteen weeks of gestation and lactation, she has sustained herself and her milk supply by foraging intensively, consuming a greater quantity and diversity of food than ever before. The babies, however, do not allow her much rest. After suckling, they pay attention to a large wolf spider, which they discover by one of the entrances, but they do not treat it as food. The mother bites the spider hard and grasps it in her mouth, but does not swallow it; instead, she runs back and forth with it in front of her babies. All three of them quickly respond by snatching parts of the dismembered spider from her mouth and eagerly devouring them. Mother certainly knows what is edible, but other members of the band also help with the weaning process. A non-breeding female is trying to eat a large locust. One of the babies, already satiated, runs over to her, sniffs her mouth, and tries to snatch the already crushed locust that protrudes from it. The female relinquishes it. Babies and youngsters always come first. Other band members are clustered around the entrance, either huddling together grooming each other, or else sprawled on their backs or sides, absorbing the last rays of the setting sun. Eventually, they all go to sleep, piled tightly together in the depth of their burrows.

A few weeks later, the three youngsters join the adults in their insect forays on the dry plain. Nearby, several African ground squirrels forage for edible plant parts. Each meerkat youngster is attached to and closely supervised by a personal, non-parent adult male or female tutor. By following one of the tutors, a mature male, we see a relatively uneventful series of encounters with edible beetle grubs, which he hands over almost immediately to his ever-hungry young apprentice; it is prey that can be devoured without further processing. In the late morning, however, routine sand-shifting suddenly produces a large, fat, but highly venomous, scorpion. The experienced tutor circles the scorpion and constantly teases it, using fast, forward and backward movements of its forefeet, always keeping a short but safe distance from the dangerous sting. The scorpion tries to direct and thrust its sting at the meerkat, but it meets nothing but air. After a few minutes of being teased, the scorpion tires, and its stinging movements become slower and less directed. The meerkat tutor now cautiously approaches the scorpion and quickly delivers a sharp, accurate bite that cuts the section containing the sting and venom glands off the scorpion's abdomen. The detoxified and practically harmless scorpion is now handed over to the eagerly awaiting youngster, who has been intently watching the whole show. In the following weeks, this educational ritual will frequently reoccur, until eventually the tutor allows the eager young apprentice, who has by then gained experience in handling de-venomised scorpions, to tackle a dangerous intact scorpion all by itself.

The food tutorial continues. The next item of prey, a black, 12-centimetre long millipede, found by the tutor among the leaf litter under an acacia tree, requires the use of a different handling technique to make it edible. Time and again the tutor rolls the millipede quickly between his paws, making it secrete most of its semi-volatile, cyanide-based, defence chemicals. Once detoxicated, the millipede is handed over to the young apprentice to practice millipede rolling and, finally, to eat.

The tutorial is suddenly cut short. A martial eagle, perched on a large, distant acacia tree, starts to fly into the air space over the foraging ground. This time it is a ground squirrel who is first to emit a loud alarm screech, and everyone, meerkats as well as ground squirrels, is quick to dash into the safety of the shared quarters. As the meerkats sneak out of their burrows again and start moving towards a new foraging ground, one of the adults suddenly halts, bristles and emits a series of short, loud calls, which immediately brings the whole band to its side. This individual has alarmed and summoned its group members because it has come face to face with a metre-long yellow cobra. This quick-moving, highly venomous, bright-yellow snake is one of the most dreaded of the meerkats' foes, not least because, unlike most of the meerkats' predators, it is able to slide unnoticed into the meerkats' burrows and prey on their young. When a yellow cobra is around, no meerkat is safe. This foe must be driven out of the meerkat territory at all costs.

A dozen bristling and nervously calling meerkats surround the snake from three sides, cautiously keeping themselves outside its striking range. The three young-sters are kept well back, but watch the action from a safe distance. The yellow cobra faces the meerkat band with body raised and neck-hood expanded, ready to strike. Several meerkats, facing the snake's tail, quickly move forward, and each in turn half-heartedly tries to bite the tail. The yellow cobra turns around anxiously. The attackers retreat, nervously performing displacement digging movements. Other band members try the same trick. The snake turns round and strikes at the fast-retreating attackers. Too late! The attackers are already out of range. Twenty minutes of continuous feinting, circling and futile strikes finally exhaust the cobra's patience, and it quickly leaves. The winning meerkats stand together, intently watching the defeated snake as it slides out of their territory.

The meerkats are exemplary group-living social carnivores. Individuals live most of their lives deeply immersed in a social group milieu. They excavate their burrows together, fight enemies together, play together, sleep together, raise young together and change quarters together. Together they form a highly successful, well-co-ordinated co-operative unit, breeding and raising young, and making the best of their meagre, enemy-infested and often unpredictable desert environment. The func-tioning of such a group relies on a system of communication that pro-motes the rapid spread of information among individuals and the co-ordination of activities. When communication is efficient, every indi-vidual can benefit from the experiences of the others. Being in close proximity most of the time helps to make the flow of information fast and easy. Through its behaviour, a foraging meerkat who finds an unusu-al item of prey will probably attract the inquisitive attention of those nearby who, by studying both the discoverer and the prey, may learn how to identify and deal with a new type of food. Similarly, an indi-vidual who suddenly halts and bristles, because it has almost stumbled on a well-concealed snake, will immediately attract nervous attention from its neighbours, who will congregate and learn, practice or refresh their snake mobbing techniques. In this way, the close proximity of group members enhances social learning. However, efficient communi-cation requires additional and more deliberate mechanisms of co-ordination among individuals – mechanisms that can be applied over long as well as short ranges. The summoning call emitted by a meerkat when it encounters a snake, the alarm screech of a sentry when it feels a neighbouring raptor has become a menace to the foragers, and the shrill call a sentry makes when it notices a foreign, trespassing band of meerkats are typical examples of vocal signals that enable co-ordination. In each case, the message is highly specific and causes all group members to act in concert in an adaptive way.

Information transfer and social learning go on continuously in social groups, holding the group together and making it a functional unit. But not everything that could be learnt is learnt, nor is every learnt behaviour likely to become a part of the life style of the group. In order to understand how a group functions, we need to know how an individual becomes a group member, how its social behaviour is constructed, and how this behaviour then fits with the behaviour of others and helps to maintain group cohesion. When we focus on social learning, it becomes clear that the relationship between individual psychological development and group structure and function is essentially cyclical. As the social psychologist Solomon Asch stressed, to understand social psychology it is necessary to constantly switch between two perspectives: 'To understand the individual we must study him in his group setting; to understand the group we must study the individuals whose interrelated actions constitute it.'[2] This is true both when we study the maintenance and functioning of groups on the time-scale of an individual life, and when we study the evolution of groups over generations of individual lives. Neither perspective is better than the other. Which is more useful and relevant depends on the questions being asked, the trait being studied and the time-scale being considered. Here we will begin by looking at the psychological mechanisms that direct and organise social learning within groups, and then look at evolutionary explanations for group-life. Studying the immediate and the evolutionary causes of group behaviours will help us understand how and when variations in socially learnt patterns of behaviour can support cultural evolution.

Social cohesion and social death

In well-integrated social groups like those of meerkats, many of the adult behaviours resemble those that we have already encountered in families. In the wider social context of the group, these behaviours have often acquired new functions, and sometimes also take new forms. It may seem, therefore, that all we need to do in order to understand the mechanisms that bind individuals into groups is to look again at the basic psychological and behavioural processes already found in the family, and see how they have been extended and modified. However, it is not as simple as that: social life in groups goes well beyond that of the family. In a meerkat group, for example, there is a division of labour among members, and group-wide activities such as mobbing predators and attacks on rival groups have become highly organised. Moreover, not all of the behaviours seen in groups can readily be traced back to behaviours within the family. For example, all females within a group may reproduce synchronously, (although different groups reproduce at different times). Such reproductive synchrony must involve some type of group co-ordinating process, which has no obvious parallel in nuclear families. Social groups therefore seem to be bound together by new, group-emergent mechanisms, as well as family derived behaviours.

Mechanisms of cohesion: transfer, rituals and rites
We will look first at those behaviours that are obviously derived from family living. In the meerkats, a male baby-sitter is a perfect mother-substitute in all ways except one – he does not lactate. The evolution-ary origins of the baby-sitter's caring behaviour clearly lie in the realm of parental behaviour, with the co-ordinated alternation of baby-sitting shifts probably originating from the co-ordinated shifts of duties between monogamous mates. The way the parental and between-mates behaviours are combined in a baby-sitter may either be innate or learnt by experience. From an evolutionary point of view, baby-sitting behav-iours, whether genetic, cultural or both, are a set of slightly modified parental behaviours that are displayed by currently non-breeding adults towards non-descendent young. The two sets of behaviours – caring for the young and co-ordinating shifts – originate from different sources, but are now linked together in one functional baby-sitting behavioural set. Although based on established family behaviours, the sequence and context of the behaviours are different – whereas mates alternate shifts only with each other, a meerkat baby-sitter alternates with all caregivers in the group (except the parents!). The 'tutorials' given by tutors to the young are also slightly modified parental behaviours applied in a dif-ferent context. The tutor's instruction is very similar but not identical to a parent's: whereas a parent has to divide its attention among all its offspring and manage their squabbles over food and information, a tutor is devoted to a single youngster. Consequently, the pupil enjoys its tutor's undivided attention without competition from siblings, which no doubt enhances the quality of its learning processes and the rate with which it acquires information.

The derived behaviours of meerkat baby-sitters and tutors are very similar to the original parental behaviours, but in other cases family derived behaviours have acquired new meanings, lost old ones and been combined with other behaviours to form group behaviours with new functions. The social behaviour of wolves provides many good examples. Wolves live in packs of several adults and young, centred on a dominant, alpha male and female pair. All pack members care for the youngsters born to the alpha pair, regurgitating food for them and alternating baby-sitting duties. The adult wolves hunt co-operatively, often catching and killing prey that a single wolf could not. The emotional bonds between the pack members are very strong, and rely on intricate communication that includes rituals. In his famous and beautiful book on the life of wolves in North America, David Mech describes some of the rituals in a pack of wolves he observed for many years on Isle Royale.[3] When the wolves of a pack reassemble after sleep or regroup after a chase, and sometimes when preparing for a hunt, they perform a ritual directed towards the leader, the alpha male. They surround him and nose-push, all trying to nuzzle up to him and lick his face, seizing his muzzle gently while wagging their tails sideways. As Mech points out, this ritual is very similar to the food-begging behaviour of wolf pups, who, by licking the mouths of adults, stimulate them to regurgitate food. It may well be that, when they wish to hunt, adult wolves beg the pack leader in an almost symbolic way to provide his leadership for the hunt, and hence for food.

The ritualised submission of a subordinate wolf to a dominant one, usually a response to the dominant wolf's exploration of the subordinate's genital region, is very different. Whereas the greeting ritual of subordinates conveys the pup-like emotional state of happy expectation, submission entails expressions of inferiority and helplessness. The subordinate wolf lies, tail wagging very slowly, with the ventral side of its chest and sometimes its abdomen exposed. According to Schenkel,[4] this posture is similar to that of very young pups who expose their abdomens and genital regions to their carers in order to have them cleaned. Therefore, like the greeting and pre-hunt ritual, adult submission behaviours seem to have been derived from pup–adult relationships. In all these adult rituals, the original pup behaviours have been modified, and are directed towards different recipients and carried out in different circumstances.

Other striking examples of family based behaviours being transferred and modified for group settings are seen in primates. In many species of monkeys, subordinate members of the society use presenting and permit mounting, which are sexual activities, as appeasement signals directed towards dominant, potentially aggressive individuals. Presenting is normally a gesture of a female in oestrus, who presents her swollen and often colourful behind to a male, inviting him to copulate with her; the male's typical response is to mount the female from behind, grasp the back of her knees and thrust his pelvis back and forth. These actions, in somewhat abbreviated and modified forms, are also used for appeasement purposes in various non-sexual encounters between both different and same-sex individuals of different status. Similarly, allogrooming – using the hands and mouth to rid someone's body surfaces of parasites, pathogens and dirt – is used for social as well as hygienic purposes in mother–infant relations and among close relatives, but often it is also transferred to a more general social domain.[5] Between non-relatives, grooming functions as a friendly gesture that is far more binding than presenting and mounting. Both reduce tension, but grooming also helps to establish alliances between the grooming partners. In hamadryas baboons, there is yet another transferred pattern of behaviour, the embrace, which is the most intimate and affectionate gesture between any two individuals. The embrace is also derived from mother–infant behaviours, and among adults is used only between very close friends, usually mates who are in danger of separation.[6]

In all social mammals, it seems that courtship and parent–offspring behaviours are 'borrowed' and transferred into the social domain, where they are transformed into socially relevant messages, forming and cementing the emotional and instrumental relationships among individuals within the group. However, not all such behaviours can be traced back to a family origin, and, even when they can be, they are often so fundamentally transformed that their original structures and functions are not easy to identify. The famous howling of wolves is a good example. Wolves howl in groups, and seem to enjoy it; communal howling is accompanied by general excitement and friendly tail-wagging. Before it starts to howl, a wolf wags its tail and whines; it then lifts its muzzle upwards and forwards and starts howling. Howling is contagious. Once one wolf in a pack starts howling, it is usually not long before others join in. Each wolf starts with lengthy low-frequency howls, then utters shorter and higher-pitched ones, tending to howl slightly out of synchrony with the others, so the wolf chorus does not sing in unison.

The wolf howl seems to serve several basic emotional–social group functions: assembling the group, advertising the extent of the territory to neighbouring groups, and reinforcing friendly ties among group members. It is difficult to derive the wolf howl from any non-group behaviour. It has probably emerged in a social, rather than a strictly family, context.

Activities normally performed by single individuals can also be transformed into communal activities. Scent-marking with anal-gland secretions is typical of many solitary and social species of mongooses. Single individuals mark territorial boundaries with their individual scent tag.[7] In the more social mongooses, territorial scent-marking has turned into group behaviour. Dwarf mongooses mark their territories communally, each individual smearing the same smell-post (usually a branch or exposed root of a bush near their sleeping quarters) with secretions from both its anal and cheek glands. The anal-gland secretion decays at a slow rate, and conveys information to other bands, as well as to the marking-band itself when it returns to the same area. From anal scent marks, bands can tell when the marking-band last visited the place, how long it stayed and how many members it then had; additional information about both the sex and the age of members is probably also available. Marks from cheek-gland secretions, unlike anal marks, last for a short time and carry information about the emotional state of the individual who marked, so they seem to be a within-band signal. Scent marks have additional functions when there is war between dwarf mongoose bands. Frantic and intense communal marking precedes fights, with all members of a band marking each other. Individual scent differences are probably masked, and for a while all are united through their joint, communal, olfactory 'badge'. Why is this needed? Does the communal badge carry a simpler and more distinct message in the turmoil of the battle, so that members of fighting bands are not confused? Is a temporary overruling of individual differences and the creation of a group identity necessary to increase the fighting ability of the group? Even during the intervals between fights, the intense marking of smell-posts and each other continues. Once one of the bands retreats, members of the winning band furiously mark the smell-post of their opponents, as well as each other.[8]

Belonging to such a cohesive social group is clearly not a passive state. When a newcomer joins a group, it adds its secretions to the group's smell-post, and although this does not secure its acceptance, because acceptance is a process that can last many weeks, it signals its own self-categorisation as a member of the group. It is not enough to just physically join a group – if an individual is not born into a group and naturally socialised within it, it has to make active efforts to become part of it. It may try to become a leading member by force, displacing a previous leader. Or a would-be member, who is unable to force its way in, may make positive efforts to identify itself with the group and gradually be accepted. While doing so, it faces both the dangers and the excessive stress of living on the physical and the social periphery of the group.[9] The slow process of being accepted into a group could be seen when an individual American crow was introduced into a group living in captivity, and began to imitate song elements specific to that group. As its imitation improved, aggression towards it decreased and social contact increased.[10] Any new member joining a group has to participate in the group's activities. Howling by wolves, their pre-hunt rituals, and the scent-marking of mongooses before a fight, are all active behaviours that reinforce a member's group identity. With humans, experiments have shown that imposing a group identity on people who are complete strangers, even by doing something as *ad hoc* and trivial as giving them a common badge, leads to intra-group favouritism.[11]

In all social mammals, familiarity, which is reinforced by things like local dialects, group smells and communal rituals, allows an individual to discriminate between members of its own group and those of other groups, and consequently to be more tolerant and overtly altruistic towards its own group members. In previous chapters, we have discussed how the ability to recognise each other as individuals is important for promoting co-operation and reciprocity. But an individual's identity as a member of the group may sometimes be of even greater importance than its identity as an individual. During conflicts between groups, the ability to recognise fellow group members may be the most relevant factor for co-ordinating the group's fight against its rivals. The idiosyncratic personality of an individual and its personal relationships with other members of the group are temporarily of no importance.

More about group cohesion: contagion and reproductive suppression
Behaviours such as greeting in wolves and scent-marking in social mongooses can be traced back to parent–offspring or territorial behaviours, and to the powerful emotional motivations inherent in them, but there are aspects of these communal behaviours that show something rather

different. Consider the pre-hunt ritual of hunting dogs or of wolves. As the ritual proceeds, excitement builds up: all pack members crowd around their leader, sniff each other and, finally, through their behaviours towards each other, come to share much the same emotions and motivations. It seems that some process of emotional contagion is at work. Similarly, the communal scent-marking of social mongooses before a band-fight seems to enhance and co-ordinate the excitement and motivation in the group. Since processes of contagion seem to be very important for group cohesion, we will look at some of them more closely.

The mechanisms of emotional contagion result in the triggering, synchronisation and sometimes mutual enhancement of emotional states, vocalisations, postures and other behaviours of the interacting individuals in a group.[12] Aristotle was the first to seek an explanation for contagious phenomena, wondering about the mechanisms that bring them about. In the 'Problems Book' he asks: 'Why do men generally themselves yawn when they see others yawn? . . . Why is yawning caused by the sight of other yawning and so also the passing of urine, particularly in beasts of burden? . . . Why is it that when we see anyone cut or burned or tortured or undergoing any other painful suffering, we share mentally in his pain?'[13] As well as yawning, we know that laughter, crying, scratching and vomiting are all highly contagious in humans. We also know that, in group situations, seemingly more complex emotional contagion in the form of panic, joy, aggression or mourning can occur. Mass hysteria has appeared in all cultures and at all historical times.[14] At the physiological level, laughing and yawning have been studied most intensely, but even here our understanding of the development and social dynamics of the contagious effect is still very poor.[15] Our lack of knowledge is surprising in view of the potential importance of the subject. The causes of crowd behaviour, such as the murderous and destructive behaviour of individuals in Nazi crowds, have been a subject of great concern but little empirical study. The nineteenth-century sociologist Gustav Le Bon, who discussed the bloodthirsty behaviour of members of crowds during the French revolution, likened the intense contagious power that sentiments, emotions and ideas have in crowds to the contagious effect of microbes.[16] He believed contagion to be the basis of crowd behaviour that leads to a 'hypnotic' state in which the feeling of individual responsibility is obliterated.

The contagious effects of fear and distress have been studied in some species of animals, especially in interactions between monkeys. These studies have shown that, just as Aristotle described for humans, the expression of fear in one monkey (the 'demonstrator') elicits an anticipation of fearful and stressful events in others who observe it. Seeing the stressful behaviour primes the observers to behave in a way that suggests that the sight of distress influences their own emotions, that the expression of pain causes them to anticipate pain. In fact, the observers were found to learn to alleviate the demonstrator's pain as rapidly as they learnt to alleviate pain caused directly to themselves.[17] At the group level, emotional contagion often leads to more or less uniform group behaviour, as during mobbing or panic reactions, where all members behave in an essentially similar manner. But contagion in small and well-organised groups may also lead to members taking on non-identical yet typical group roles, as with organised hunting or fighting.[18] In both crowds and small, organised groups, contagion tends to lead to stereotyped behaviours that make group members behave as a cohesive unit.

Perhaps the best-understood case of contagion in animals is the panic reaction in a school of fish following attempted predation.[19] In 1938, von Frisch discovered that mechanical damage to European minnows results in the release of a chemical that, at very low concentrations, causes fright responses in receiver fish. This chemical, which von Frisch called 'alarm stuff', is a pheromone that is released from special cells in the epidermis. It is perceived by the olfactory system of receiver fish, who typically dash away with rapid darting movements and form tight schools. The effect of this alarm pheromone is often associated with the visual stimulus of seeing the darting, frightened movements of fish. Indeed, merely seeing the panic responses of fish in a nearby aquarium induces panic reactions in observer fish.[20]

Alarm signals causing synchronised group behaviours are well known in all predator-vulnerable social mammals and birds. They do not always result in communal flight – often they lead to mobbing or predator harassment. Mobbing starts when an individual who discovers a potentially dangerous predator emits a loud alarm call and begins to perform certain stereotyped movements. These attract other members of the group, who join in, call and follow or attack the predator. The behaviour of the meerkats towards the yellow cobra is a typical mobbing response, while their response to the martial eagle and other aerial predators is a typical flight response. In both cases, the alarm signal elicits a response that is enhanced and escalates as group members respond to each other. Naïve individuals learn the relationship between the predator, the alarm call and the proper response by observing the behaviours of experienced individuals towards the threatening object. The information about the particular object that forms the target of the mobbing or flight reaction is transmitted by social learning both within and between generations.[21]

Many of the co-ordinated behaviours of members of groups seem to involve contagion. Group migration may be another example, since the migratory behaviour of individuals – particularly young wanderers or stragglers – seems to be enhanced by seeing the migratory behaviour of others in the group.[22] The reproductive synchrony that is found within some groups of meerkats, hamadryas baboons and catta lemurs[23] may also be socially induced and contagious. The mechanisms leading to reproductive synchrony are not always clear, but they are likely to be based on a combination of pheromonal and behavioural signals. Reproductive synchrony has strong effects on the behaviours of group members, altering levels of vigilance, aggression and co-operation. It leads to marked differences between groups at different phases of their reproductive 'life cycle'.[24]

One of the things that can greatly enhance the functioning of a group as a cohesive unit is allowing only a single pair to reproduce.[25] Generally, a group can function efficiently only if competition between its members for food, mates and space is regulated and curbed and, at least for a while, the interests of the group override the interests of the individuals in it. If all except a single pair have zero individual fitnesses because only the dominant pair reproduces, group interests become paramount. The reproducing alpha pair is equivalent to the 'germ-line' in a sexually reproducing multicellular organism. Psychologically, the reproducing male and female tend to assume parental authority and have the status of actual or symbolic parents for all other individuals in the group. Through reproductive suppression, the individual fitness-promoting behaviours associated with producing and caring for their own young are eliminated, so group members behave more as a unit. Reproductive suppression is thus a powerful means of inducing group cohesion. The almost complete sterility of the worker caste in ant colonies, and the presence of a dominant reproductive pair in dwarf mongoose bands, wolf packs, groups of Arabian babblers and naked mole rats, suggest that reproductive suppression enhances group integrity. In some meerkat bands, reproductive suppression is achieved by pheromonal means, but in others, such as the band we portrayed at the beginning of this chapter, mature females reproduce synchronously and then all young except those of the alpha female are killed. Only one litter, that of the dominant female, is reared.[26] Such infanticide is functionally the same as reproductive suppression.

The various behavioural processes that bind group members together are not isolated – they can combine and reinforce one another. Emotional contagion, for example, can be reinforced by authority: a leader can precipitate, enhance and maintain the contagious state.[27] Through the psychological and physiological mechanisms that allow dominant individuals to impose parent-like authority, that produce reproductive suppression, and that culminate in various types of contagious behaviour, a group is made to behave as a cohesive unit. These mechanisms help us to uncover the basic rules of social interactions.[28] In the same way that there are fundamental rules that lead to the organisation of perception and allow efficient and clear categorisation of stimuli and responses, there are rules that lead to the organisation of social interactions and underlie the ability to form clear-cut social categories. For example, establishing criteria that allow a sharp and clear distinction between one's own group members and individuals belonging to other groups is an important social schema. Learning, recollecting and transmitting information about social identity must be simple, easy and reliable. It must be grounded in general and trustworthy psychological mechanisms, such as the generalisation of family derived behaviours and their transformations into group rituals that create group identity.

The benefits of sharing information

We have seen that a meerkat band is a closely knit system of interdependent individuals who, through their influences on each other's behaviours, develop collective emotions, attitudes and skills. The survival of a naïve individual depends on parents, tutors and the other band members transferring to it reliable and up-to-date information about enemies, guarding, the location of food and methods of foraging. It may seem obvious that individuals in the group must benefit from such information-sharing, but we need to look at this more closely, because acquiring and sharing information are not without costs. First, acquiring information about the world around can be costly, because it takes time and energy, and may expose the individual to danger; it may sometimes be more worthwhile to try to 'steal' information from others, rather than acquire it oneself. Second, sharing already acquired information can be costly; for example, warning others about the presence of a predator may endanger the informer. For information-sharing to evolve, the costs of sharing must be offset by the benefits. What, then, are the benefits of information sharing?

The rotation of guard duties in a meerkat band illustrates some of the benefits of sharing information. Assuming that guard duties are more or less equally shared among all adult group members, then, in a group of ten adults, each will guard for just a tenth of the time available, and can forage for the other nine-tenths; in a group of only three adults, each would have to guard for a third of the time and could forage for only two-thirds. Consequently, a member of the larger group has more time to feed. In this case, the advantages of belonging to a large group in which information is shared are obvious: individuals acquire good-quality information about predators at a reduced cost in time to themselves.

Even when all members of a group are doing the same thing, such as foraging near one another or guarding together,[29] there are still benefits from sharing information. The chances of detecting predators or food increase with the number of eyes, ears and noses, and the probability of responding to false alarms, which wastes time and energy, decreases. Sharing synchronously gathered information can compensate for individual mistakes and increase accuracy, thereby improving the reliability of environmental monitoring. Moreover, as we have stressed before, unlike energy, information that is given away is not lost. Sometimes, of course, sharing information can be costly: an individual who shares information about a source of food may then lose some of it. However, sharing information is frequently either cost-free or almost cost-free, as it is, for example, when an abundant food resource is found, or when alarm calls do not endanger the signaller. But, even when sharing information does have costs, the average cost per individual of getting information about predators or a source of food decreases as more individuals acquire and share it.[30]

The problem with sharing information is that there is always some danger that a selfish individual, a 'social parasite', will acquire information from others but not share its own. Since this may improve its chances of surviving and transmitting the selfish behaviour, such conduct could eventually lead to the collapse of co-operation within the group. However, the effect of selfish behaviour on the fitness of others may ultimately act as a boomerang: poor group performance through the failure of co-operation could greatly decrease the chances that the selfish individuals themselves survive and reproduce. As we have seen in previous chapters, in some groups 'parasites' are punished. In a small group, where members know each other personally and frequently act as a team, an individual who repeatedly acts selfishly may be harassed or denied essential help, and may finally leave the group. Another individual, living at the periphery of the group, eager to join it, may then be admitted. But, in a large group, individuals may not know each other personally, so, unless there is some collective 'rule' or a collectively accepted 'norm' of behaviour, cheaters may not be easy to detect. In the previous chapter, we saw how vervet monkeys punish group members who find a rich food source but do not give the food call. Punishing such 'criminal' behaviour may be an example of a simple 'social norm'. The evolution of co-operation in animal societies may often be associated with the evolution of punishment for social parasites, and possibly also the evolution of special rewards for co-operators.[31]

There are many examples of information-sharing and trading in animal groups. As Ward and Zahavi argued, a group can function as a food-related 'information centre' even when there are no specific or deliberate sharing relationships among its members.[32] The sight of an individual feeding, or the sight or smell of food or food residues on a particular individual, are reliable and difficult-to-hide clues to the identity of a successful forager. Individuals who return from a successful foraging trip are likely to be followed on their next trip by those who were less successful. In habitats where the food supply is patchy and varies over space or time, the flow of information within the group may be beneficial to all, because today's successful forager will often be tomorrow's less successful one, and vice versa. As followers and those that they follow change roles, a kind of mutualism emerges.

There are several well-studied examples of information centres in birds and mammals that live in stable social groups.[33] One of the most spectacular is found in ospreys, fish-eating raptors that sometimes nest colonially in coastal habitats. These birds not only share information about foraging, they can also assess the quality of the information they get.[34] The ospreys are more likely to go fishing after the return of an individual who caught a schooling fish than after the return of one who caught a non-schooling fish. When a successful individual arrives, the birds fly off in the direction from which it came, and fish there. An osprey who has found a rich school of fish may actively signal the find by calling loudly and flying in an undulating manner towards the colony, upon which all the non-fishing birds take off and start hunting near the site where a fish was caught.

The information that is shared among members of a group is not always about sources of food and predators. Through communal nursing, information about food preferences and pathogens may also be shared. Communal nursing, in which young are suckled by more than one female, is known for a number of mammals, from mice and rats to lions and cats.[35] Typically, it occurs among familiar, often related individuals. As we discussed in chapter 4, a mother's milk contains a high concentration of antibodies. Milk from different mothers will probably have slightly different types of antibodies, because of their differing medical histories. Communal nursing by mothers who have experienced different diseases may therefore confer resistance to a broader spectrum of pathogens than suckling by single mothers, because the young will share the immunological knowledge of several females.[36]

It would be wrong to assume that information is transferred and shared only among close acquaintances. The common raven shows that information may be shared with those who are both unrelated and unfamiliar.[37] Young ravens leave their parents and join other juveniles to roost, feed together and find mates. They wander over large areas, frequently changing roost sites. A young bird who finds a carcass takes active measures to inform others about it, often calling and attracting those nearby to the carcass. It probably also advertises its find either through vocalisation while on the roost tree, or through 'leading' the others to it by being the first to leave the roost. Ravens who are unaware of the location of the food follow their knowledgeable roost mate. When a carcass has been almost completely cleaned up, the ravens move to a new roost site, closer to a new carcass. The move is preceded by spectacular soaring displays, which may last for two hours, with an ever-increasing number of ravens ascending high up into the air and then diving and tumbling. The display seems to mobilise the ravens to change sites, since, when the aerobatics are over, all ravens take off and fly in a long line to the new roost. Since the young ravens in a roost are not related to each other and do not stay together for long, kin selection and classical reciprocal altruism can be ruled out as explanations for their seemingly altruistic carcass-sharing behaviour. So why do young ravens share?

The long-term study by Heinrich and his colleagues of ravens in the forested mountains of Maine shows that there are good reasons for young ravens to share. First, common ravens choose a mate for life, which may last up to forty years, so it is important that they find the best mate possible. A young raven who finds a carcass and informs other roost mates about it shows itself to be an expert carcass finder, a potentially good provider and hence a desirable mate. Second, by ganging up together, young birds can feed on a carcass that would otherwise be unavailable to them. Carcasses are jealously guarded by mated pairs of ravens, and a single juvenile has no chance of joining them and feeding from it, but a gang of nine or more youngsters is able to chase away an aggressive, well-coordinated pair of adults. Third, carcasses are a scarce food resource, and the more pairs of searching eyes the better the chances of finding one. In addition to all these benefits, for a single raven the cost of sharing is not very high. Large carcasses are rare, and in harsh winter conditions they are attractive to other scavengers, such as wolves and coyotes. They must therefore be consumed quickly, before the competitors arrive. But carcasses are usually frozen, so are not easy to dismember and eat. With other scavengers around, it is unlikely that an individual raven could make full use of a carcass, so sharing it with other ravens is not really very costly to the individual. The small cost and many benefits make sharing information with roost mates a good investment for a young common raven.

Which information is shared, how knowledge is disseminated and how new behaviours are learnt depend on the composition and organisation of the group, including the age and sex of its members. The role of old, post-reproductive animals usually receives little attention from biologists, since it tends to be assumed that individuals rarely reach post-reproductive age in natural conditions. When post-reproductive survival is thought about in evolutionary terms, it is usually seen either as a non-adaptive result of senescence, or as an adaptive response to slowly maturing offspring.[38] Old individuals, it is argued, preferentially care for their last-born young or the young of grown-up daughters, and through doing so increase their own inclusive fitness. However, there is probably more to it than this: the survival of individuals to post-reproductive age can be of advantage to *all* members of the group, whether related or not. The superior experience and knowledge of old individuals can benefit everyone in a hamadryas baboon troop or in a herd of African elephants. Old individuals can lead the group to distant and rarely visited water holes,[39] incidentally transmitting this valuable information to the younger members of the group. The old-timers may also be better at coping with rarely encountered predators, which are familiar to them but not to younger members of the group. It seems as if the unique information-store of older, non-reproductive individuals is 'traded' for co-operation, for example for the defence that is provided by stronger, younger members. The problem with this view is that, although the fitness of the young is increased by the old-timers' knowledge, for post-reproductives there can be neither direct fitness benefits nor costs. So, why do they share information with younger members of the group? Of course, their co-operative behaviour could be driven by kin selection, since, when the group thrives, relatives of the elderly thrive as well. But, in addition, the co-operative behaviour of the elderly may be a by-product of fitness-enhancing aspects of their earlier life, before they lost their reproductive value. In fact, co-operation with and by the elderly may be driven by tradition. By adopting the attitudes of their parents towards the elderly and learning to pay particular attention to older individuals, co-operating with them and learning from them, the young perpetuate the behaviours of the previous generation. If when they mature they then transmit their behaviours to youngsters, the cycle will continue, and, when they become old, they themselves are likely to be regarded as a source of information and someone with whom to co-operate.[40] Tolerance and regard for the post-reproductives may thus be perpetuated through social learning. If so, it would be interesting to compare the age structure of differently organised social groups. We would expect cohesive well-organised groups to contain more post-reproductive individuals than less organised, more diffuse groups.

Some groups consist of just males or just females, and this may have interesting effects on the type of information shared. In group-living mammals, where males often disperse out of their natal group and females stay behind, males sometimes spend much of their time in all-male groups, leaving them only for the relatively short breeding season. Females remain with their adult female relatives and the young. This pattern is typical for many hoofed mammals such as mountain sheep, red deer, Nubian ibex and mountain gazelles.[41] A similar pattern is seen in African elephants, where females and their offspring live in tightly knit family units, while males lead more or less solitary lives.[42] Sex-specific or sex-biased groups also form among young humans and young dolphins.[43] Initially the formation of sex-biased groups may be a by-product of dispersal patterns, but, once such groups are formed, the behaviour of the individuals within them is likely to be affected. Since individuals are surrounded by, and exposed to, the behaviours of other individuals of the same sex and frequently also the same or a similar age-group, sex-specific behavioural 'conventions' may become established. As a result, differences between the behaviours of males and females may be enhanced. For most species, it is still not known to what extent sex-specific behaviour is really influenced by either sex-specific or sex-biased social associations, but it would not be surprising to find that typical gender behaviour is standardised through social learning and information transfer in sex-specific groups.

Social death

We have argued that the integration of a social group depends on the diffusion of useful information among its members through various mechanisms of social learning. Sometimes learning is guided by emotions, motivations and behaviours derived from those found in parent–offspring or mate relationships, and sometimes from emergent group interactions. If the functioning of a social group does indeed depend on the transmission of information among its members, then we should be able to learn quite a lot about the cohesive and organising functions of social learning from cases where groups disintegrate – from 'social death'.

For several decades, the American behavioural scientist John Calhoun conducted the most detailed and thorough long-term studies on social organisation and social death in domestic mouse and Norway rat colonies.[44] In a now classical study, he constructed a perfect universe for the domestic mouse – a universe with food, drink and nesting boxes in abundance, free from predators and external sources of disease. The only snag in this 'paradise' was that, since emigration was not possible, the density of the colony grew far beyond that which is normal in natural conditions. This high density did not affect the physical requirements of these mice, however – they still had food, drink and nesting boxes in excess, and were free from predators and external sources of disease. Nevertheless, after reaching a certain population density, reproduction slowed down. Young were often prematurely rejected by their mothers, and the males aggregated in large numbers near the centre of the floor, inactive and apathetic for most of the time, except for occasional vicious attacks on neighbours. Females withdrew, or became aggressive as their nest boxes were frequently invaded. They often wounded and killed their offspring, and the incidence of conception dropped while that of resorption of embryos increased. Population growth eventually came to a complete halt. The mice began to die and, eventually, the whole population died out. Even when an individual was taken out of this 'universe' and placed in non-crowded conditions with an individual of the opposite sex who had been raised in normal conditions, it still showed abnormal reproductive behaviour. Calhoun's analysis showed that death was caused by socially induced stress. He summarised his results as follows:

> The results obtained in this study should be obtained when customary causes of mortality become markedly reduced in any species of mammal whose members form social groups. Reduction of bodily death (i.e. 'the 'second death') culminates in survival of an excessive number of individuals that have developed the potentiality for occupying the social roles characteristic of the species. Within a few generations all such roles in all physical space available to the species are filled. At this time, the continuing high survival of many individuals to sexual and behavioural maturity culminates in the presence of many young adults capable of involvement in appropriate species-specific activities. However, there are few opportunities for fulfilling these potentialities. In seeking such fulfillment they compete for social role occupancy with the older established members of the community. This competition is so severe that it simultaneously leads to the nearly total breakdown of all normal behaviour by both the contestors and the established adults of both sexes. Normal social organization (i.e. 'the establishment') breaks down, it 'dies'.
>
> Young born during such social dissolution are rejected by their mothers and other adult associates. This early failure of social bonding becomes compounded by interruption of action cycles due to the mechanical interference resulting from the high contact rate among individuals living in a high density population. High contact rate further fragments

behaviour as a result of the stochastics of social interactions which demand that, in order to maximize gratification from social interaction, intensity and duration of social interaction must be reduced in proportion to the degree that the group size exceeds the optimum. Autistic-like creatures, capable only of the most simple behaviours compatible with physiological survival, emerge out of this process. Their spirit has died ('the first death'). They are no longer capable of executing the more complex behaviours compatible with species survival. The species in such settings die.

For an animal so simple as a mouse, the most complex behaviours involve the interrelated set of courtship, maternal care, territorial defence and hierarchical intragroup and intergroup social organization. When behaviours related to these functions fail to mature, there is no development of social organization and no reproduction.

(Calhoun, 1973, p. 86)

Calhoun's domestic mouse population simulates some of the stages seen in natural population cycles of small rodents such as mice, voles and lemmings.[45] Fairly regular fluctuations in density are followed every few years by a sudden and often manyfold increase in density, terminating with a no less spectacular population crash. Whatever the reasons for the increase, be it temporary abundance of food or temporary absence of predators, the dramatic increase in density invariably brings with it greatly increased stress, which leads to social disintegration. As Calhoun and many others since have shown, overstressed mothers do not conceive, they resorb foetuses, readily abort or, when they do conceive and give birth, they reject or otherwise mishandle their infants.[46] The few infants who survive are further abused by the aggressive surrounding adults. In such circumstances, the overstressed youngsters learn very little and find it very difficult or even impossible to cope with predators, diseases and dwindling resources. The intergenerational chain of acquired practical knowledge about the habitat has been broken.[47]

Any breakdown of social structures can lead to social death. For example, the domestication of mammals and birds often leads to the dissolution of their original social structure and their original social behaviour. Man selects, unintentionally as well as intentionally, individuals who are least disturbed by transportation and human presence, least exacting and discriminating in their choice of mate or food, least interested in territorial defence, least aggressive towards members of their own species and towards humans, and least dependent on social learning from other members of their species.[48] The perceptual, emotional and motivational world of the domesticated animal is altered, and usually severely impoverished. Domestication often leads to diminished perceptual acuity, and it is not surprising that it is often associated with a reduction in brain size.[49] Engulfed as they are within the social world of man, objects of ownership, subject to human whims and to human criteria of utility, the social structure of many domesticated mammals and birds has largely ceased to be functional and self-sustaining, having become increasingly dependent on human presence and provision.[50] The prime example is, perhaps, the domesticated wolf – the domestic dog.

Man's best friend has been socially mutilated by domestication. This is revealed when stray dogs, frequently former human pets, live together in packs at the periphery of human habitation. They cannot reconstruct their ancestral social structure, which is a pack of currently non-breeding individuals who form a dominance hierarchy and are often led by a dominant breeding pair; nor do they manage to form an alternative sustainable social organisation. When we look at the social structure of packs of stray, feral dogs, we find that these dog packs are just aggregates of several monogamous pairs and their pups, without the higher-level pack structure typical of wolves. The feral dogs do not care communally for the young as wolves do, and, although they defend their food resources as a group, their foraging and barely organised defence is still very far from the effective, well-coordinated communal hunting that is carried out by wolves. Unlike wolves, all females in a feral dog group reproduce, and there is no indication of paternal care. Domesticated dogs are, in fact, the only members of the dog family in which the father does not help to care for the young! The provision of care by man has apparently eliminated paternal care, as well as selected for females with more oestrus cycles per year and a larger number of pups per litter. All these factors combine to make groups of feral dogs incapable of maintaining themselves over time, because they are unable to limit reproduction or raise their many young.[51] Pups die because mothers cannot provide enough care without help from others. The large number of oestrus cycles and litters all too frequently lead to high pup mortality, because the mother tends to care less for her offspring as she enters, too early for the previous litter's sake, a new oestrus cycle. Consequently, less than 5 per cent of the pups reach one year of age. The pack is maintained only through the ongoing recruitment of new stray dogs. Without recruitment, the pack goes extinct.[52]

There is no doubt that in domestic dogs selection by man has led to both physiological and behavioural changes which do not fit into the original life style of the wolf. However, it is not clear what exactly was selected during domestication. Genes, ecology and traditions combine to affect behaviour, and it is not easy to tease them apart. Was it only the selection of genotypes that led to the behaviour of the domestic dog? Or was behavioural–cultural selection also important? To what extent, for example, does the lack of paternal care in domestic dogs reflect changes in genes affecting parental behaviour? Is it not possible that it was cultural selection, rather than gene selection, that led to the break-up of social traditions and played a decisive role in dissolving the wolf group-binding mechanisms, thus altering social behaviour? Can the remaining, genetically based, behavioural plasticity still provide a way back to a wolf-like life style? For example, would male dog pups raised in a wolf pack show paternal care when they mature? Would female dogs raised in a wolf pack increase the intervals between oestrus cycles and become susceptible to reproductive inhibition by a dominant female? In the previous chapter, we described how rearing conditions could have a profound effect on paternal and helping behaviour in voles, and the same may be true for dogs. Cross-fostering experiments, which follow the development and behaviour of dog pups that are introduced into wolf packs, would help sort out genetic and cultural effects.[53] Even if the potential to adopt an effective pack life style does still exist in feral dogs, it remains unfulfilled because they have no suitable social model during early life – dogs and wolves copy whatever other pack members do. Thus, feral dogs could not easily revert to pack-life. The belief that the observed inability of domestic dogs to adopt a pack life style is due exclusively to genetic rather than cultural selection is just an assumption, and it is probably wrong.

Normal social structures, behavioural traditions and emotional motivations break down when numbers in a wild population decrease drastically, so that normal information acquisition and transmission are severely hampered.[54] A breakdown also occurs when animals caught in the wild are brought into zoos. Young members of the next generation are born into a new physical and social milieu that often bears only a partial resemblance to the original wild one.[55] When endangered species of birds and mammals that have been kept in zoos for just a few generations are reintroduced into the habitat occupied by their immediate ancestors, they often fail to survive, or find it difficult to adjust to the local, natural conditions.[56] Part of the reason for this failure could well be that the intergenerational chain of socially learnt practical knowledge, vital for success under natural conditions, was cut off in captivity. The first generation of youngsters born in a zoo is already behaviourally and socially impoverished. What can their offspring possibly learn from them that is of real value under wild conditions? Obviously nothing concerning food, predators and normal social behaviour. This means that in order to reintroduce individuals of endangered group-living species back to the wild, they need to have the right social and ecological know-how. Both social and ecological skills are at least partially acquired through social learning in natural conditions, so the conditions enabling the acquisition and transfer of useful knowledge have to be reconstructed. A reintroduction scheme based on such considerations has helped to save the Arabian oryx from extinction. As a result of hunting, this group-living desert antelope, which once used to inhabit much of the Arabian Peninsula, almost became extinct in the 1970s. It now persists in Oman, thanks to a successful reintroduction project.[57] One of the most important lessons from this ongoing project is that optimal diet selection in this species requires social learning within the social group. The rearing and reintroduction of these group-living herbivores into the wild therefore needs to be implemented in ways that enhance the transfer of knowledge from experienced to naïve individuals. The researchers involved in the project concluded that, for reintroduction to succeed, cohesive social units have to be released.

Studies of social dissolution show us how tight the package of ecology, genetics and behavioural tradition is in animal societies. All social vertebrates respond to ecological conditions by learning and constructing sophisticated adaptive life styles within the bounds of their genetic constitution. The breakdown of social structures through drastic changes in ecology, genes and traditions, either alone or in some combination, often leads to social death. However, in normal conditions, the overall life style, though flexible, is quite stable, much more stable than any of the single behaviour patterns that constitute it. This stability is the result of both the firm basis provided by fundamental social categories and psychological group-binding mechanisms, and the reinforcing interactions and feedback between the flexible patterns of behaviour that together form the overall life style.

The existence of cohesive groups consisting of individuals who exchange information and are involved in other co-operative activities is a problem for evolutionary biologists, who want to understand how groups and their social-binding mechanisms evolved. In this and previous sections, we have discussed some of the psychological processes that underlie the social behaviours of individuals in groups, but we have not yet discussed the evolution of group-life itself. We want now to turn to the functional explanations of group evolution, and look at the circumstances in which group living and variations in the mechanisms that bind individuals into social groups are actually selected.

The evolutionary origin and maintenance of groups

What are the evolutionary origins of group living? Did solitary individuals come together and, for the benefit of all concerned, form a group? There is a tendency to assume that something like this happened, and that communities invariably started as aggregates of solitary individuals that had evolved as separate beings before they joined together and began evolving as a group. Communication between individuals is envisaged as evolving from a state of non-communication, sociality from a state of non-sociality. This view, which gives priority to the individual, probably stems from an unconscious parallel with non-living objects: since a man-made object (a brick house, for example) is constructed from units, communities are also thought to be formed by putting together separately produced units. To see why this way of thinking may be misleading when the evolution of biological entities is being considered, we need to look at something rather basic – at the outcome of reproduction.

The result of reproduction is an increase in the number of contiguous, related individuals. When, to what extent, and in what manner these individuals disperse or remain in close proximity is an empirical question. The answer frequently depends on the distribution of resources in the surrounding environment. If the nature of the environment and the interactions between individuals are such that individuals tend to stay together or to disperse as a group, the evolution of the properties of both individuals and the group will be closely tied together from the outset. In such circumstances, it may be impossible to understand the one without the other, the individual without the group and vice versa. There would be no historical priority to 'the individual' or 'the group' level of organisation; many of the group-dependent properties of individuals would be as ancient as their group-independent properties. The historical synchrony of individual and group development calls into question the assumption that the properties of the group stem from and are caused primarily by the properties of individuals. The way that being in a group causes individuals to behave in a certain way, which then directs evolutionary changes in the properties of individuals, has to be considered too. For example, rather than assuming that the advantages of co-operation through reciprocal altruism have led to the evolution of mechanisms that enable group behaviours based on familiarity, it can be argued that familiarity with other individuals in a primitive group setting is the basis for the development of reciprocity. Familiarity is the consequence of learning and habit. Animals learn about all aspects of their environment, including the commonly encountered individuals. Early social familiarity frequently leads to tolerance of familiar individuals; this familiarity-based tolerance already contains the seeds of reciprocity, and may be the basis for the evolution of further co-operation through reciprocity.

Whether, and for how long, group members stay together is likely to influence the evolution of groups. When the environment can support more than a narrow family group, an extended family group may be formed. The extended family group can be the basis for the development of larger groups, which often consist of several family groups, as in troops of hamadryas baboons, colonies of white-fronted bee-eaters, or coteries of prairie dogs. Kin selection will lead to co-operation among related individuals, even when the reproductive success of some members of the family is temporarily or permanently diminished. In addition, mutual advantages that result from the interactions within groups will be reinforced by selection. The same psychological mechanisms that are recruited and modified to support co-operative behaviours within the extended family may be transferred to the more extended group. Even groups of non-related individuals, such as the roosts of young common ravens, are made up of individuals who once lived in some kind of family group.

It is not sufficient to posit a general kin-based or group-emergent advantage to explain why groups form and are maintained. We need to understand the particular selective advantages that occur through living

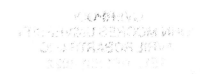

in social groups. We looked at many of these when we discussed the benefits of information-sharing. One of the most obvious advantages is improved protection from predators. This applies not only to highly cohesive groups such as those of meerkats, with their elaborate system of alarm calls, but also to species such as ostriches and wood-pigeons, whose groups are less cohesive and use simpler alarm systems. Individuals benefit because the chances of spotting a predator increase with the number of scanning pairs of eyes, because they have more time to respond to danger, and because the chances they will be singled out as prey are lower.[58] Being part of a large group also enhances the ability to deter predators. Mobbing, such as the meerkats' response to the yellow cobra, is common in group-living species.[59] Work with fieldfares that breed colonially in Scandinavian forests has shown that group size affects the deterrent effects of mobbing – the larger the mobbing party, the less likely the predator will return to the same area.[60] Predators may also be confused by the reactions of a large group. For example, a cheetah approaching a herd of antelopes often has difficulty in picking out a potential victim from among the many animals fleeing in all directions. Its attack is sometimes delayed and less directed, so each fleeing individual has a better chance to escape than it would have had had it been alone.[61]

The importance of group-life for protection from predation is shown by comparative studies relating the life style of animals with the openness of the habitat. Looking at the social organisations of a large number of African antelope species showed that most species that live in open areas, where hiding is difficult, lead a group-life, whereas those that live in forests are solitary.[62] Similar conclusions have been reached from studies of avian social organisation: grassland and ground-living are associated with group breeding, whereas arboreal life is associated with pair breeding.[63] Even within a species, group size is related to the nature of the habitat. The size of groups of black-tailed deer in forest and scrub is much smaller than in open habitats.[64] The same is true for British red deer, and for African oribi and klipspringer antelopes.[65] The open nature of typical group habitats not only exposes animals to more danger, which gives group living an advantage, it also makes the transfer of information among group members easier, thus enabling more elaborate patterns of group co-ordination. Visual, vocal and scent communication can be utilised much more efficiently in open areas.

The other important advantages of group-life are food-related. Group living can improve the chances of finding or processing food. Either with or without their consent and help, animals frequently learn about food from others in their group. Sometimes a group is essential for getting food: social carnivores such as lions, wolves, spotted hyenas and hunting dogs hunt in groups. Communal hunting requires good communication, precise co-ordination, and sometimes a division of labour. In a lion hunting-party in Namibia there are individuals who habitually stalk to the left, others who habitually stalk to the right, and others who always face the potential prey.[66] This specialisation presumably makes communal hunting more effective. Groups are also better able to protect their food from scavengers and competitors, and can probably scavenge more efficiently.

There are clearly many advantages in living in a social group. In closely knit groups such as a wolf pack, a meerkat band or a hamadryas baboon clan, group members forage or hunt together, share or divide chores, groom each other and frequently achieve goals that cannot be achieved individually, such as overcoming large prey, defending a large territory or driving away dangerous predators.[67] However, in order to understand the evolution of groups it is not sufficient to list the many obvious (and less obvious) advantages of group-life. Whereas for a solitary animal a change such as an increase in skill or strength, which allows better access to a limited resource, can be assumed to have straightforward fitness benefits, in a group things may be rather different. A better-endowed individual who increases the well-being of other group members can be exploited. Even if 'generous' individuals improve the well-being of others inadvertently and without any cost to themselves, the *relative* fitness of the beneficiaries, including non-co-operators, may become greater, so that in time non-cooperative individuals become more numerous and potentially capable of taking over the population. So, in order to explain how groups can persist over evolutionary time and how they can become more organised and cohesive, it is necessary to show that the advantages of group-life for any individual member are sufficient to make the group immune from destruction by free riders or social parasites. We need to show that the short-term advantages to selfish individuals cannot undermine the long-term existence of the group. The interplay between the two types of interest – the individual's short-term interest and the group's (as well as the individual's) long-term interest – have to be considered.

In many cases, co-operation is to the benefit of the individual – that is, the fitness of a co-operator is always higher than that of a selfish individual. Such co-operation is evolutionarily stable, because over evolutionary time it will not be disrupted by individuals who behave selfishly. For example, a selfish young common raven who fails to recruit its roost mates to a carcass will probably not attract a mate. In addition to the individual benefits of co-operation, kin selection is often important in the evolution of co-operation in small kin-groups. Reciprocal altruism may sometimes build on the psychological mechanisms established by both self-interested co-operation and kin selection, and extend 'altruistic' co-operation to non-relatives. As many experiments on human interactions have shown, altruistic co-operation is greatly enhanced when the group is small, when individuals are familiar with each other, when they share common goals or a common leader, and when they communicate frequently with each other.[68] In fact, some social psychologists argue that being aware that one belongs to a group, even when the group is temporary and based on very superficial characteristics, is a necessary pre-condition for effective co-operative activity within groups.[69] When trying to understand how individuals within groups interact and how group characteristics evolve, we need to consider both the interests of individuals and the effects of collective behaviour and emotions.

Group selection and selection for interactions within groups
Selection among individuals and among kin-groups affects the evolutionary origin and persistence of groups. Both can lead to the evolution of co-operation and group adaptations. But there is another level at which selection can act: selection can occur among groups, and the differential survival, 'reproduction', and extinction of groups may be important for the evolution of the properties found within them. It can be argued that, if some groups survive longer, multiply more, or become extinct less commonly than others because the individuals within them co-operate more or are more self-sacrificing, selection among groups will enhance the evolution of co-operative and altruistic behaviour.

For many years, it was assumed that selection between groups was unimportant in the evolution of what can be regarded as social adaptations. The theoretical possibility of group selection was not in doubt: since groups can multiply, vary and have properties that result in daughter groups resembling mother groups (i.e. they have heredity), natural selection at this level can occur.[70] The argument against the importance of group selection was that, when a behaviour benefits the group but detracts from the individual's fitness, individual selection is far more powerful, because the rate at which individuals reproduce and die is much greater than the rate of reproduction and death of groups. It was therefore assumed that selection favouring non-altruistic individuals within groups would override any beneficial effects of altruistic co-operation on the survival and reproduction of groups.

It is, of course, quite wrong to assume that individual selection and group selection will always pull in opposite directions. Often they will work together. In such cases, it is difficult to know whether or not group selection plays a role in the evolution of the trait. Take communication between individuals: a better ability to communicate may be selected because it benefits individuals within a group, and over time the frequency of groups with better communicators or means of communication will increase. Yet, a better ability to communicate may be selected at the group level too: groups with a higher proportion of good-communicators may survive longer or form more daughter groups than groups made up of individuals with poorer abilities to communicate, because communal activities like attack or defence are better co-ordinated. It is important to remember that the environment in which social interactions are selected is that of the social group. The fitness of an individual in a group depends on what other group members do. If selection of individuals as group members increases the relative efficiency and survival of the group as a whole, selection at both the individual and group levels will be going on at the same time and in the same direction.

How selection among groups actually takes place depends on the ontogeny of groups – on how new groups originate, grow and die out. There are two general types of 'group life cycles' that affect selection among groups. In the first, a group forms through the coming together of unrelated individuals that live as a group while producing progeny, then disperse, and later aggregate to form new groups with new members, and so on. A property like altruistic-cooperation, which may be to the benefit of the group as a whole but to the detriment of the individuals comprising it, can evolve in such a life cycle if, in spite of selection against altruists within the group, groups composed of a high proportion of altruists ('altruistic groups') do much better than those with fewer altruists. The balance between the selective advantage of co-operation for altruistic groups and the selective advantage for self-ishly acting individuals ('social parasites') within a group will determine whether altruism evolves through group selection. If, through group advantages, 'altruistic groups' flourish and consequently altruistic dis-persers are more abundant and more social groups containing altruists are founded, altruistic co-operation will spread up to the point where its advantages are balanced by the within-group advantage of being selfish. 'Group reproduction' in this case is not a matter of a group actually budding off daughter groups, but rather of a group producing more individuals that will be future group formers. Such a life cycle is typical of the family-based social groups found in most birds and some mammals, where mates come together to form a family (or kin-group) made up of offspring who eventually disperse and associate with unrelated individuals to form new kin-groups.[71] From this perspective, kin selection is a special case of group selection.[72] The outcome would be the same, however, if non-related individuals formed groups based on non-genetic but heritable similarity in behavioural traits.[73] The simplest case is selection acting on pairs of co-operative mates, which we described in chapter 5; here 'groups' each consist of two individuals. If a co-operator tends to pair with another co-operator, and if co-operative pairs have higher productivity than non-cooperative pairs, then, if the behaviours of mates towards each other are heritable (either genetically, or through social learning, or both), pair selection will lead to the evolution of co-operation between mates. The frequency of co-operative pairs will increase over time, even if in 'mixed' pairs con-sisting of a co-operator and a non-cooperator the co-operative mate is slightly exploited.

In the second type of a life cycle, a mother group buds off a daugh-ter group, so the 'reproduction' of groups is by fission. Assuming, first, that emigration into groups is rare, so daughter groups are more simi-lar to mother groups than to non-related groups, and, second, that selec-tion against altruists within the group is weak while the positive effects of co-operation on group survival or reproduction are strong, theoreti-cal studies show that altruism can evolve via group selection.[74] Group fission is known to occur, for example in many species of monkeys, in the Australian white-winged choughs described in the previous chapter, and in African subterranean naked mole rats. When a colony of naked mole rats grows to such an extent that the dominant female is unable to suppress the reproduction of other females, one of the females at the periphery of the group becomes reproductively active, founds a new group, and the two parts of the original group split and become independent.[75]

The theoretical models show that altruistic co-operation could evolve through group selection, even when co-operation involves actions that are detrimental to the individual's interests. But, as we have already stressed, not all co-operative interactions within groups involve altruism. By looking at non-altruistic but group-relevant behaviours, Boyd and Richerson have introduced a completely different way of thinking about group selection, a way that is particularly important when much of the behavioural variation is culturally acquired and transmitted.[76] With many social interactions, once a particular behaviour pattern is sufficiently common, individuals who display this behaviour have a higher fitness than individuals who display any other behaviour. For example, if most individuals in a group assist those who co-operate with them, but ignore non-cooperators who need help or even actively punish them, it is worthwhile for any newcomer, whether a youngster born in the group or an immigrant, to become such a tit-for-tat co-operator. Similarly, in a group where the majority of individuals act selfishly, it is worthwhile for a newcomer to act selfishly too, for a co-operator will be thoroughly exploited. In both groups, there is selection for individuals displaying the majority behaviour, and selection against 'deviant' individuals.

How a particular behaviour becomes established in a group might be either a matter of chance, or an effect of changing local conditions that transiently favours one behaviour over another. When an individual from a group that employs behaviour A moves to a group that employs behaviour B, it either conforms by learning the local strategy, or pays a high price in fitness terms. Boyd and Richerson's models show that if strategies A and B differ in their effect on group productivity, groups using the more productive group behaviour could become more common through group selection. Although selfish individuals in a group with a majority of selfish individuals have the same relative fitness as tit-for-tat co-operative individuals in a group of such co-operators, the latter group is likely to have higher productivity, and is therefore likely to be selected.

Unlike the conventional models of group selection, in this model immigration has little effect if the number of immigrants is small relative to the group size; as long as the accepting group has a large enough proportion of individuals who use the dominant type of behaviour, the immigrants must conform or suffer. They will not cause a change in the behaviour of the group they join unless they manage to take over the leadership of the whole group and dominate it. When selection against deviants is strong and the groups are large, immigration is unlikely to be disruptive. Furthermore, as Boyd and Richerson stress, group extinction need not involve the death of all individuals within the group, but rather the disintegration of the group and the dispersal of its members to other groups where they either learn or fail to learn the local behavioural strategy. The conditions they describe are fairly common, so this type of group selection may be quite common too. Moreover, they show how the chances of group selection are further enhanced when additional forces, such as acquiring the behaviours that are most common in the group through social learning, increase behavioural unity. Looking at the conditions promoting group selection from this perspective, Boyd and Richerson have shown how, when culturally driven, group selection can be an important and dominant force in evolution.

Boyd and Richerson applied their model to human groups, where traditional forces are extremely strong and are strengthened by sanctions and rewards. Their ideas can also apply to groups of higher animals, providing the alternative traditions are both common enough and robust enough. Selection among groups can often be equivalent to selection among group traditions. Frequently, local traditions are quite stable and persist for a long time, resisting changes introduced by new immigrants. The group-binding mechanisms that we described in the previous sections stabilise the local habits within a group, since they define simple rules of conduct that are easily learnt or imposed. A majority effect may commonly be involved – as more group members adopt a behaviour, the local tradition, new members have more 'models' from whom to learn it. A new member with a deviant behaviour may be punished, ignored as a potential mate or avoided in other ways. However, even if the majority effect increases uniformity only through conformist social learning, it still enhances the chances of selection among groups.[77]

Groups over time – learnt habits and traditions
The cohesive mechanisms that we described earlier ensure the

functioning of groups, but they do not specify which particular behaviours will be seen within a group. For example, the same fundamental mobbing behaviour may be directed at different predators in different groups, simply because the types of predators the groups have encountered and their members have learnt about are not the same. Similarly, although tutoring may be universal in meerkat bands, exactly what is demonstrated in a particular band depends on local ecological conditions and local habits. Even the extent to which a group-binding mechanism is expressed may vary, as when the expression of emotional contagion is influenced by the number of individuals in the group, or depends on the presence of a strong leader. This behavioural plasticity means that the same distribution of genotypes in two populations allows very different behaviours, habits and traditions. The important consequence of this is that variations in behaviours and in traditions do not necessarily stem from variations in genes, and therefore cannot be explained in these terms.

Over the last twenty years, knowledge of the many life styles that can be found within a single social species has been expanding rapidly, changing our perception of what constitutes a characteristic species-specific behaviour or social structure. Whereas in the past we thought we could characterise a species as one with a particular type of breeding pattern, or mode of parental care, or social structure, today we often have to accept that these attributes vary, and sometimes vary a great deal. The answers to questions such as 'Is the mating system of this species monogamous, polygamous, or promiscuous?', 'Is it a solitary or a communal breeder?', 'Does it have paternal as well as maternal care?' nowadays all seem to begin with 'It depends . . .' . The problem is that some individuals or groups of the same social species behave differently from others. Behaviours that ensure good access to food, territory and mates are often sensitive to both social and non-social environmental conditions. Consequently, social behaviour, social structure and even whole life styles may change when the environment changes.[78]

In the previous chapter, we described how social and ecological factors influence the decision made about whether a bee-eater or silver-backed jackal should disperse and try to start breeding, or stay on the natal territory and become a helper. The speed of these individual behavioural responses to environmental conditions shows that variations in genes are not responsible for the variations in behaviour. The plasticity of behaviour is large enough to allow several alternative behavioural responses, several different life styles. But the number of alternative life styles is far from infinite. In fact, there seem to be only a few stable alternatives. However, different life styles are not alternative 'programs', pre-existing in the animal's brain, ever ready to be switched on and off by the appropriate environmental cues. They result from learning based on simple and general rules, and on active and often original individual responses to a changed environment. Often there are many possible behavioural solutions to an ecological situation, but only a few of them will become self-sustaining and form a tradition. For example, imagine that a new terrestrial, fox-like predator is introduced into an area in which a group of the imaginary tarbutniks that we described in chapter 1 is living. At first there would be many different responses to this predator – several variant alarm calls might be used, some tarbutniks might try to mob the predator, others run to their burrows, and still others might change their foraging patterns and locations to avoid it. After several generations, we would see that the responses to the predator are much less varied and more standardised. Some of the initial responses were too costly or did not fit well with the well-established tarbutnik habits, and consequently were rejected. Others that were particularly easy to learn and to transmit to the young, and that could be interpreted unambiguously and fitted well with already established habits, were adopted. A train of behaviours subsumed under simple rules that clearly define the interactions with the new predator developed. Social learning and selection among different behaviours led to a new tradition, to new stable, socially transmissible patterns of behaviour.

Frequently the different behaviours that we see in different groups or populations of a species are adaptations to local ecological conditions, but sometimes they stem from local traditions. Tradition is usually inferred when present ecological factors cannot account for alternative patterns of behaviour. For example, some meerkat groups forage for termites, while others do not, although termites are present in abundance in their territory.[79] This difference is unlikely to be genetic. It is probably due to tradition – in some groups termite-eating was never established, whereas in others it became a learnt habit and is transmitted through tutoring from one generation to the next. Similarly, when one group of common chimpanzees uses stones to crack open nuts and another never tries to crack them open, the observed differences in behaviour are ascribed to differences in custom.[80] However, if we found nut-cracking in all populations of common chimpanzees, we certainly could not infer from this that nut-cracking is not a tradition. The proof that it is a tradition is in the way the behaviour is perpetuated – traditions are perpetuated through social learning and can be transmitted across generations.

Tradition may sometimes be responsible for delays in adapting to changes in the environment, making behavioural responses lag behind ecological change. For example, although newly introduced monkeys were preying on their eggs and nestlings, Mauritius kestrels almost became extinct before they changed their traditional habit of nesting in trees to the new habit of nesting on predator-safe rocks.[81] Similarly, in North America, long-term traditions of habitat preference and habits of grazing are probably preventing mountain sheep from extending their ranges into terrain which, because of ecological changes over recent decades, has become suitable and available for them.[82] Similar conservatism is seen in many other sheep, goats and ibexes. When reintroduced into a suitable area, they disperse very little, become loyal to the new area, and use just a fraction of the available habitat.[83]

The inertia of strongly established socially transmitted behaviours often leads to conservatism, and in extreme cases can lead to extinction. In fact, traditional patterns of behaviour can be so stable that at first sight they seem to be the result of genetical rather than social transmission. For example, once the tradition of washing soiled sweet potatoes became established in a group of Japanese macaques, the washing behaviour was generalised to other foods, and may persist for a long time, even when the original tradition-inducing items (sweet potatoes) are no longer around. The general custom of washing soiled food may well remain for many generations to come. If they had no knowledge of the history of the group, future observers might be tempted to attribute the behaviour to a genetically selected 'washing-food brain-module'.

Almost every aspect of an animal's life can show variations in the traditions associated with it. There are traditions of foraging, hunting methods and places; traditions of courtship; traditions of territorial songs and other vocalisations; traditions of criteria for mate preference; traditions of nest (and other) building behaviours; traditions of displaying homosexual behaviours; and traditions of parental caring style. The scientific literature on animal traditions, like that on life-style variations, is steadily growing. It is impossible and unnecessary to review it all here, particularly since excellent accounts have recently been published,[84] but we want to look at a few examples in order to high-light some general aspects of traditions and discuss their evolutionary effects. The first is the socially learnt differences in vocalisations found in neighbouring populations of birds, primates and whales.[85] We have referred to these 'dialects' in previous chapters. Dialect differences can-not be ascribed solely to differences in ecology – to acoustical adapta-tions to local conditions – because neighbouring populations often occupy similar habitats.[86] Studies of birds have shown that some dif-ferences stem from 'cultural drift' – from copying errors that have no selective significance; others result from 'cultural diffusion' – from vocalisations introduced from other groups; and still others are the results of vocal innovations that are adaptive. From studies of chaffinch-es in England and New Zealand, it has been estimated that, as the young learn the local song, their copying error is 15 per cent, which is suffi-cient to account for the documented differences in the songs between different chaffinch populations. In the South American chingolo spar-row, however, dialects seem to be adaptations to some features of the habitat. The final trill rate is high in open areas, and much lower in woodland and forest.[87]

Once established, dialects are perpetuated through social learning within the group and may become further stabilised by acquiring a new function. In some cases, especially in relatively small groups, local dialects may be a kind of 'badge' that enables individual recognition and allows contact between group members over large distances. This can be seen in striped-backed wrens, which are patrilineal birds that live in extended family groups. Their calls are inherited in a family-specific and sex-specific manner. Unrelated birds almost never share calls, whereas the males in a patrilineage, whether they are in the same or different breeding groups, share the same male-calls, and females share lineage-specific female-calls. Males learn their calls from their male relatives, and females learn theirs from female relatives. This results in birds being able to recognise the lineage and sex of those near-by. The benefit of this seems to be related to their group-life: the philopatric males of a family group fight neighbouring groups over ter-ritorial boundaries, and the teams of related, dispersing females fight others for vacant breeding positions. Their calls seem to function as a group-specific badge, facilitating group co-ordination during these group contests.[88]

Social groups of whales (known as 'pods') also have local dialects. In killer whale groups, vocal traditions are very stable, although the members of different pods often interact with each other. The pods are long-lived units, made up of members of several overlapping generations belonging to several related matrilineages. There is a lot of co-ordination of activities in pods, including co-ordination of movements and foraging activities, and there are also co-operative activities such as shifts of baby-sitting, which require communication. It seems likely that the pod-dialects facilitate communication between pod members, and help to maintain social cohesion by allowing individuals to be recognised as group members over large distances.[89]

Recently, molecular biology has revealed some intriguing facts about whales, which have led the marine zoologist Hal Whitehead to suggest that cultural differences such as song may have led to selection between matrilineages.[90] Whitehead studied sperm whales, which live in cohesive, matrilineal groups, which individuals hardly ever leave. Like killer whales, different matrilineages have different song-dialects. Matrilineages also have different patterns and locations of marks on their bodies, including marks acquired from the teeth of predators, which suggests that there are family-specific ways of facing predators. Whitehead noted that, in sperm whales, killer whales and two other species of whales with a closed matrilineal social organisation and lineage-specific dialects, the diversity of mitochondrial DNA is very low compared with that of whales that do not have this type of social organisation. Mitochondria are small semi-autonomous cellular bodies that are associated with energy production; they have their own DNA. What is relevant here is that mitochondria are transmitted from generation to generation largely through the egg, not the tiny sperm. In other words, just like culture in matrilineal whales, mitochondrial DNA is transmitted through the female line. Therefore, while acknowledging that other explanations are possible, Whitehead suggested that the reduced diversity of mitochondrial DNA might be related to the cultural difference. If there is strong cultural selection for an adaptive tradition, it will lead to decreased diversity in both traditions and mitochondrial DNA. The mitochondrial DNA can be thought of as hitchhiking on the selected cultural variation. What happens to the DNA, to the genes, reflects this cultural selection, but it is functionally unrelated to it.

Different vocal traditions are not the only kind of cultural variations seen in neighbouring populations. Jared Diamond has described a fascinating example of dramatic differences in building style among populations of the Vogelkop bower-birds in New Guinea.[91] As we mentioned in chapter 3, the drab males of this species build some of the most elaborate and decorated architectural edifices of the bird world. These structures, which serve for courtship display, consist of stick towers that can be nearly three metres high, or large huts with a diameter of several metres. A large moss mat is put together, and the bower and its surroundings are decorated with hundreds of different items – flowers, beetle elytra, stones, butterfly wings, mushrooms, leaves, feathers, snail shells, sticks and other natural or man-made objects. Diamond compared several populations of this bower-bird species, all living high up in the mountains, separated from each other by 8 to 200 kilometres. Their body morphology differed only slightly, or not at all. However, they did differ, and sometimes differed dramatically, in the way they built their bowers. While there was individual variation within each population, populations differed from each other in both colour preferences and ways of building, so it was possible to distinguish distinct population-specific styles. The greatest differences were between two styles – one that of a population living on the Kumawa mountain, and another, a more widespread style, represented by a population on Wandamen mountain.

The Kumawa males build their bowers on broad flat areas on the highest part of mountain ridges. The bowers are so oriented that they are well lit by the morning sun, when the birds are most active. The males construct a stick tower on a central tall sapling, the 'maypole'. The side branches and lower leaves of the sapling are removed, and hundreds of sticks are piled upon it, woven around and glued to the maypole up to a height of three metres, thus creating a stick tower. Around the stick tower the male constructs a mat, often over two metres in diameter, perfectly round and shiny black. It is woven from dry dead fibres of moss, and the bird paints it glossy black, presumably using its own oily black excrement. In the middle of the mat and around the maypole, he constructs a moss cone, also painted black. Judging by their style, the Kumawa males and females have a taste for the elegant and the dignified. The objects decorating their bowers are all black, brown and grey, often organised into piles of the same colour. Piles of light-brown leaves, brown and grey shells, dark-brown acorns, decorative sticks painted black, beetle elytra also painted black, and dark-brown stones, are popular decorations among these males.

The bowers constructed by the males living on the Wandamen

mountain reveal a completely different style and taste. The bowers are built on sloping ground, the maypole is much shorter (only 40 to 80 centimetres), and around it is built a circular hut, 2 metres in diameter, with an entrance facing down the slope. The Wandamen birds cover the hut's ground with a bright-green mat made of live moss, and the sticks of the maypole tower and of the hut are not glued, they are just woven together. Their taste in decoration is along more vivid lines than that of Kumawa males – they decorate the inside of the hut as well as the entrance with colourful objects of many kinds. Blue, black and orange are the most popular colours, but there is significant use of red, yellow, green and purple objects. The habit of painting is unknown among them.

What is the reason for these architectural and decorative differences? Are genetic differences responsible? Are there genes for elegant Kumawa-type bowers and Wandamen vivid bowers; genes for painting and not painting; for preferring ripe fruit to leaves as decoration; for building huts; for weaving mats from dead or live moss? As Diamond argues and common sense suggests, this is highly unlikely. The birds all belong to the same species, and are all very similar in morphology. Neighbouring bowers tend to be more similar in style than distant ones, indicating that neighbours probably learn from each other. Males seem to take many years to learn how to build good bowers, and social learning by both males and females seems to be very important.[92] The different styles of building and decoration are not related to ecological differences, seeming to reflect an arbitrary convention rather than a local adaptation. Cultural evolution is surely the simplest and best explanation for the observed differences in styles between populations. If so, it is of great interest, because the many differences between the bowerbird populations testify to ongoing, cumulative evolution in bower construction and decoration. Cumulative cultural evolution is also the best explanation for the differences between related species of bower-birds, although in this case genetic biases which reinforce and stabilise the initial differences in tradition may also have been selected. Long-term studies and cross-fostering experiments within and between species are obviously needed to sort out genetic and cultural differences in these birds.

As with birds, our understanding of cultural evolution in mammals is limited because there are so few species for which the life styles of different populations have been observed over a long period. The common chimpanzee and the Japanese macaque are some of the most intensely studied. Here we will look at the culture of Japanese macaques rather than chimpanzees, because very good accounts of chimpanzee 'cultures' have been published recently,[93] and also because we want to establish the generality of cultural evolution by focusing on mammals that are not infra-human, so are very different from ourselves. The Japanese macaques have been studied since the late 1940s, mainly by Japanese primatologists. This work has shown that groups and matri-lineages differ in the kinds of food eaten, in the ways foods are handled, in the style of parental care displayed by males and females, in their vocalisations, and in the way they play with stones.[94] Even the manner in which these monkeys remove louse eggs as they groom their companions has been scrutinised and found to differ between lineages.[95] All of these patterns of behaviour are transmitted through social learning among individuals and between generations. They are local customs, or traditions.

In chapter 3 we described traditions of food handling by Japanese macaques on Koshima island. A remarkable young female, Imo, started washing sweet potatoes to remove the soil from them, and also found a way of separating wheat grains from sand by taking handfuls of sand and wheat to the water, letting the wheat float and the sand sink. These two methods of food handling soon spread among members of her group. A related habit, seasoning sweet potatoes by biting them and then washing them in the saltwater of the sea, also spread. Moreover, bathing and swimming in the sea, where all this food handling took place, in time became a local and very popular habit. More recently, another new habit, eating raw fish (when nothing better is available) has begun to spread among the Koshima monkeys. A new life style, involving several interlocking, adaptive habits, some of them new and associated with the sea and the nearby sandy beach, became integrated with the original mountain life style of the Koshima monkeys.

Four major phases can be seen in the development of new life styles in Japanese macaques and other animals. First, a new behaviour is learnt or invented by one or a few individuals. This new behaviour can be learned by chance, by trial and error, or through insight learning by a particularly bright individual. The chance of learning new skills or acquiring a new preference are enhanced by any developmental or ecological conditions that lead to increased individual exploration and experimentation. Among the Japanese macaques, it is usually the young or the more peripheral, non-dominant individuals who are the most inventive. Well-established individuals, like dominant males, seem to have less incentive to explore their surroundings or try out newly discovered techniques.

At the second stage, the newly acquired behaviour spreads, usually and most obviously among the young, and then from the young to older kin and to the parents. In mammals it spreads mainly to mothers, since usually they alone care for the young. The transmission process at this stage is irregular, depending on group structure, on the amount of contact between experienced and naïve individuals, on their cognitive and emotional states, and on local conditions. Since transmission is not certain, the behaviour can easily be lost. At the third stage, the parents become the major transmitters. These are either parents who have learnt the new behaviour from their young offspring, or young parents who acquired the behaviour as infants and now, as they became parents themselves, transmit the newly acquired behaviour to their young. In addition, peers from different lineages who play together will learn from each other, further disseminating the information in the group at large. This is a stable phase of information transfer, for parental transmission is both constant and reliable. The young infants are constantly exposed to the behaviours of their parents and acquire the new behaviour without effort, since early learning is extremely efficient.

At the fourth stage, the new pattern of behaviour may be further stabilised because it becomes associated with related learnt behaviours. Washing sweet potatoes by the Japanese macaques is an example of a behaviour that has strong effects on other behaviours. It seems to have become generalised into washing other soiled food, and to have induced the habit of bathing in the sea because infants get used to the water as they are inadvertently dipped into it whenever their mothers wash food. This, in turn, has led to swimming and diving, and possibly even to foraging for fish. The behaviour patterns stabilise each other, since the use of one behaviour increases the likelihood that associated behaviours will also be performed. A network of related behaviours is thus established, creating a new life style. The new urban life styles of grey squirrels and other mammals that have successfully moved into towns were probably established through a similar network of effects.

A new behaviour may also increase the likelihood of its own reoccurrence through its effect on the habitat. For example, the choice of mountain ridges as sites for building bowers may initially have been related to the chance presence there of the fruit trees on which the bower-birds feed. However, since the males feed on fruit and defecate near the bower, new trees will grow there, thus recreating the bower-birds' feeding environment and reinforcing the choice of the same locations for generations to come. The fruits, seeds, and fungi that bower-birds use as decorations are often able to develop, so by their behaviour bower-birds also ensure the long-term supply of the materials they choose as decorations. Caching seeds is another example of a habit that may be reinforced through the effect it has on the local environment. By caching seeds, animals provide themselves with a source of food for harsh winters, but since some of the cached seeds germinate, caching also provides new plants that will form seeds and create future caching opportunities.

The close, almost perfect fit of some habits with morphological and physiological features may tempt us to think that habits are the products of direct genetic evolution – that genetic variations with direct effects on these harmonious aspects of behaviour have been selected. This would probably be an incorrect conclusion. Since traditions can evolve more quickly than genes, traditions will adjust to pre-existing genotypes, rather than the other way round. In the same way as the glove evolved culturally to fit the hand, rather than the hand to fit the glove, so the more rapidly evolving 'cultural' adaptations will tend to evolve to eventually fit existing genetic pre-dispositions, without the latter necessarily having to change at all. This is true not just in the trivial sense that bower-building behaviour must be in harmony with the beak morphology of bower-birds, or that potato-washing, playing with stones and removing nits must fit the hand morphology of Japanese monkeys. It applies to more specialised traits. When we discussed the co-evolution of culture and genes in chapter 1, we made this point with regard to the imaginary tarbutniks, who never change genetically. We imagined how eating an acidic and not easily digestible local food led to the adoption of a new learnt habit (adding mud to the diet), which improved the efficiency of digestion, and how this was followed by additional, learnt feeding-adaptations. The series of cultural inventions resulted in a very efficient set of foraging adaptations, beautifully tailored to the pre-existing genotype. However the 'genetic adaptation' to the tarbutniks' new life style came about through purely cultural evolution! The way behaviours and culture adapt to pre-existing genetic biases is probably the reason why different species have such very different socially learnt communication systems, and why we find it so difficult to communicate with whales and birds.

Adjustments of culture to existing genotypes, together with the stabilising effects of networks of related behaviours, may explain why, although the group-binding psychological mechanisms we have described are very general, the behavioural adaptations of a particular species or population are quite specific. Certainly, the standard assumption that specificity is the outcome of the selection of genes producing particular behavioural programs should not be accepted uncritically. The notion that the end-product of social evolution is a genetically and optimally designed 'behavioural program' stems from the emphasis evolutionary psychologists have put on 'domain-specific modules'. These behavioural programs are supposed to have been selected to deal with local problems, with the result that there is a one-to-one correspondence between a particular genetic program and a specific set of social behaviours. However, although specific behavioural responses obviously occur, they may not be associated with such genetically selected programs at all. We have argued in previous chapters that behaviours such as adopting and helping may not be the outcomes of separate, specifically selected 'modules'. They can result from the interactions between mechanisms of social learning, general cognitive rules of social categorisation, and the typical, reoccurring ecological and social circumstances in which young-raising behaviours take place.[96] The specificity of the behavioural response does not necessarily point to an underlying genetic specificity, but rather to adaptive interactions of current traditions, which are based on the fundamental rules that organise social behaviour, on ecological conditions, and on the relatively stable morphological and physiological characteristics of the species. Within the bounds of species-specific behaviour, the gene–culture combination allows a lot of freedom. When it comes to object handling in Japanese monkeys or bower building in bower-birds, the general constraints of cognition and morphology are not very formidable.

When traditions in social groups become stabilised, group selection is possible. The tradition that leads to the greatest group productivity, either because the group survives longer or because it is better at colonising and founding daughter groups, may spread and become dominant. As daughter groups are formed, new variations in traditions may be introduced through adaptations to changed local conditions or through new discoveries by particularly lucky or bright individuals. The process of selecting among traditions and among groups will continue, leading to the evolution of richer and more complex traditions. In this way, the role of group selection in the evolution of social birds and mammals may be much greater than has usually been assumed.

Final stabilisation of new socially learnt behaviours may come through genetic changes congruous with the behaviours. If genetic variations affecting morphology, physiology or behaviour facilitate socially learnt behavioural adaptations, which could occur if the speed of learning or the reliability of performing the new behaviours is important, they will be selected. As we shall see in the next chapter, one useful way of viewing habits and traditions is as selective regimes that lead to the enhanced spread of those genotypes that help to stabilise the selective regimes themselves. But, before we go on to discuss the effect of traditions on adaptive genetic evolution, we want to look at the effects of traditions on speciation.

Habits and the origin of species

The idea that behavioural differences between populations may initiate or facilitate processes leading to the formation of new species has been a recurring theme in evolutionary biology.[97] Ernst Mayr, one of the leading evolutionary biologists of the twentieth century, has repeatedly stressed that changes in behaviour, such as acquired preferences for a new habitat, new type of food, or new host, may be the first stage in the subdivision of a species. Morphological adaptations usually follow, rather than precede, learnt habits. Alister Hardy suggested in the 1960s that the increased ability of birds and mammals to adapt to the local environment through changes in learnt habits was the driving force behind their great adaptive radiations. In the 1970s, Klaus Immelmann discussed the effects of early experience and imprinting-like learning on ecological and evolutionary adaptations in higher animals, and on the process of speciation. More recently, A. C. Wilson suggested that enhanced learning ability, especially social learning, has led to what he called a 'behavioural drive', in which behavioural innovations that become traditions accelerate the rate of morphological evolution and the formation of new species. In order to better appreciate these ideas, we first need to look at the processes that lead to the formation of new species.

In sexually reproducing organisms, the origin of new species involves the formation and establishment of reproductive isolation between individuals from different populations.[98] Reproductive isolation is a biological phenomenon – individuals do not interbreed either because they fail to meet, or if they meet they do not mate, or if they mate they produce sterile or inferior offspring. Over time, reproductive isolation can lead to the gradual divergence of populations, so that they become even more different than before in habits, morphology and genes.

It is common to distinguish between two major types of reproductive isolation, one that precedes the formation of zygotes (pre-zygotic isolation) and one that follows it (post-zygotic isolation). Pre-zygotic isolation involves mechanisms and circumstances that prevent sperm and eggs from individuals belonging to different populations meeting, so it includes breeding at different times or in different habitats, having different courtship behaviours and having morphologically incompatible genitalia. Post-zygotic isolating mechanisms are consequences of the sterility or inferior viability of hybrids, and are usually the result of cytoplasmic, chromosomal or genic incompatibility between sperm and eggs. Here we are interested mainly in pre-zygotic isolating mechanisms, because differences in habits and traditions often prevent members of different populations from mating. However, behaviour is sometimes involved in post-zygotic isolation, since hybrids can have behaviours that are incompatible with those of either of the parent populations, and they therefore fail to find or attract mates; such hybrids are physiologically fertile, but their sexual behaviour leads nowhere. Functionally, they are sterile.

Behavioural incompatibility is thought to be the most important pre-zygotic isolating mechanism in animals. Usually, differences in ecological habits, or sexual preferences or both are involved. Before we discuss the role of differences in habits and traditions in the origin of species, we need to digress briefly to look at the general ecological circumstances that initiate reproductive isolation. Although there is an ongoing debate about which conditions are most likely to result in speciation, we need not go into details here, since we are concerned only with the way in which habits and traditions promote speciation. There is general agreement that geographic isolation, the separation of parts of a population by a geographic barrier, is the most common reason why an ancestral species splits into two. This type of speciation is referred to as 'allopatric' speciation. It can occur either because a geographic barrier like a river or glacier divides a large population into parts, or because a small peripheral population, perhaps even a single gravid female, becomes geographically isolated from the main population. For example, it could occur if one or a few mainland-inhabiting individuals accidentally reach an island. The difference between the two modes of allopatric speciation is in the population sizes and in the ecological and social circumstances in which the separated animals find themselves. In whatever way it occurs, as a result of geographic separation there is, over time, phenotypic and genetic divergence of the separated populations, and sometimes this divergence is sufficient to make the populations incapable of interbreeding if their members meet again. Reproductive isolation is an accidental, though probably inevitable, by-product of long-term geographic separation.

'Sympatric' speciation occurs without geographic isolation: a new species forms within the geographic range of the ancestral species, even though individuals of the original and the new incipient species have the possibility of meeting and mating. The development of reproductive isolation in this type of speciation is less straightforward than in allopatric populations, since, if members of the two incipient species can meet, some special circumstances must prevent their mating. However, such special circumstances may not be rare. For example, in an insect in which mating and egg-laying usually take place on the plant species on which its larvae feed, reproductive isolation could be initiated by larvae feeding on a new species of plant host.[99] This could easily happen if, either by mistake or because the usual host is rare or otherwise inaccessible, a female lays her eggs on the 'wrong' plant. If the adults that develop from the larvae that survived on the new host plant species then use plants of that species as their mating and egg-laying site, the association will be perpetuated, leading to the formation of a partially reproductively isolated subpopulation with different habits and preferences. The new host preference may not be the result of a genetic change; initially, it may well be an acquired preference learnt in the larval stages.

Klaus Immelmann discussed the way in which new learnt preferences and habits may lead to speciation in birds.[100] He argued that reproductive isolation could be initiated by altered ecological and sexual preferences that become imprinted in the young. If some birds become accustomed to a new habitat and their offspring become imprinted on it, this imprinting may promote mating between individuals originating from this type of habitat. Moreover, if through chance changes or for adaptive reasons birds in the new habitat also have a different song-dialect, the tendency to mate with an individual from the same habitat may be reinforced. Mating between individuals from different habitats will be restricted if females prefer the local song-dialects of the males from the population in which they were reared, and if young male birds are attracted to breed in areas where they often hear the familiar dialect. This type of reproductive isolation through imprinting could be initiated in either local populations that remain in touch with others, or populations that are geographically isolated.

It is easy to envisage how reproductive isolation based on different habits can develop when populations are geographically isolated. Take island populations, which ever since Darwin have been a source of inspiration for ideas about speciation. Although the divergence of an island population and the mainland population from which it originated will eventually be genetic, it is likely to be initiated by behavioural differences. If a few individuals from the mainland reach an island and manage to establish a new population there, they are very likely to develop new habits, both because the habitat is probably different from their original one, and because the small number of founders may affect normal social structures. New habits will persist through early learning, and over a relatively short period of time they may form a 'package' of traditions that would reduce the chances of mating with members of the original population if these became available.

In order to detect incipient speciation caused by socially learnt habit differences, it is necessary to look at newly founded island populations that have existed for insufficient time for there to have been significant genetic divergence, to see if there are established changes in habits that could lead to reproductive isolation. A bird population on Rottnest Island, 20 kilometres off the coast of Western Australia, meets these conditions.[101] The island has been isolated from the mainland for six thousand years, and an individual singing honeyeater that originated from the mainland was first detected there in 1911. The songs of island and mainland singing honeyeaters have been compared, and found to differ. The variety of syllables from which the song is constructed is much smaller in the island population, but some syllables are apparently unique to this population. Island individuals also have fewer song-types (11 on the island, 47 on the mainland). The young learn their songs from adults, so the differences in song-type are unlikely to be based on genetic differences. Significantly, singing honeyeaters from the mainland do not respond to the islanders' territorial song, and vice versa. The island population has become behaviourally differentiated from its mainland source population. Although it has not been shown that island and mainland singing honeyeaters are reproductively isolated, it would not be surprising to find a degree of isolation. Females from the mainland would probably tend to reject island males because of their impoverished songs.

There are other cases where birds that have recently colonised islands have been found to have impoverished songs.[102] Such impoverishment probably occurs because the small number of founders carry with them only part of the repertoire of syllable-types and song-types that were present in the original population, in the same way that they carry only part of the original population's gene pool. Mayr called the results seen when a new population is formed from a small number of individuals carrying a limited and biased sample of the genes in the parent population, the 'founder effect'.[103] He suggested that the initial random changes in gene frequencies would disrupt the harmonious interactions that were present among genes in the old-established gene combinations, and that this, together with different selection pressures, might produce a cascade of genetic changes. Since behaviours also have complicated networks of interactions, comparable knock-on effects might also follow the reduced song repertoire of founding populations. M. C. Baker, who has studied the singing honeyeaters, has pointed out that additional factors have to be taken into account when the founder effect is cultural rather than genetic. The age of the founders is very important: if the males are young and their song has not yet crystallised, then, because they do not hear the normal diversity of adult songs, their own adult songs are likely to have fewer syllables than those of males in the parent population. The vocalisations of the colonisers will also be affected by the acoustic environment of the island they have colonised, which includes both physical and social factors. Colonising an island is also likely to lead to other behavioural modifications: colonisers may develop new food and habitat preferences, new foraging skills and new ways of constructing nests. Together, these changes in habits and preferences may lower the chances of mating with the original population if individuals of the two separated populations ever come into contact again.

A newly separated population certainly need not be small for socially learnt habits to spread and evolve. New habits can spread rapidly in large populations, as is clear from the rapid diffusion of the milk-bottle opening habit of great tits in Great Britain. However, in large populations we do not expect to find the drastic reduction in the repertoire of learnt behaviour patterns that is usually observed in populations founded by very few individuals. Also, when a large population is divided by a natural or man-made barrier, the social and ecological conditions in the subpopulations will probably remain essentially intact, so there should be no additional pressures to acquire new habits. For these reasons, the initial rate of cultural evolution is likely to be more rapid when founding populations are small than when they are large.[104]

In recent years, some evolutionary biologists have favoured the idea that speciation often occurs 'parapatrically', that is, through the divergence of a subpopulation that is adjacent to the main population but in sufficient contact with it for individuals theoretically to be able to meet and mate. It is not difficult to see how, in birds or mammals, learning could contribute to the development of reproductive isolation between parapatric populations. Think what will happen if, because of increased competition with other conspecifics or because of a deteriorating environment, individuals settle in an area at the periphery of the population distribution, or choose a new, previously unexploited habitat within the original area. Since philopatry is very common and seems to be based on early experience, the offspring of individuals who settle in the new habitat may become imprinted on it, and consequently try to breed in such a habitat when they are adults.[105] In theory, this could initiate a separation of the new subpopulation from the original one.

Settling in new habitats at the margins of the original distribution is occurring regularly as birds and mammals colonise urban and suburban areas. Urban populations can become self-sustaining if, through being imprinted on the urban habitat, individuals breed with others from the same population. Growing urbanisation offers town populations increased opportunities for colonisation, so urban habits may spread. One of the best examples of a changed habitat preference leading to rapid colonisation of new areas is seen in the European mistle thrush. This species was originally forest dwelling, but a population invaded open parkland areas and spread from the north of France to northern Germany at a rate of between 5 and 10 kilometres a year. Habitat imprinting seems to underlie the newly acquired preference for open parklands.[106] We do not know how extensive the gene flow between the original forest-dwelling population and the new parkland populations is, but it seems that partial ecological isolation has taken place.

This case is not an isolated one. In Israel, where, since the founding of the state fifty years ago, urbanisation has been very rapid, there have been many cases of desert and Mediterranean scrub species of birds spreading into new, town garden habitats.[107]

Once a new habitat has been occupied and habitat-associated habits have become established, sexual imprinting can become very important in maintaining isolation between the source and the daughter populations. Usually, sexual imprinting is a conformity-promoting rather than innovation-promoting process. For example, if a new variant displaying an unusual colour pattern appears in a bird population where the young sexually imprint on the appearance of their parents, this new mutant would have difficulty attracting a mate, because all potential mates are imprinted on the common pattern. But, if there is a taste for mild novelty among the members of the group, and the new variant is not too extreme in its appearance, the greater attraction it enjoys could lead to the new trait spreading.[108] In general, however, sexual imprinting will promote the status quo. The more common a trait is, the more common it is likely to become. When populations have already become partially isolated, sexual imprinting will reduce the effect of immigration, because it will reduce the acceptance of individuals with atypical traits. It will thus preserve and enhance the already common local characteristics, and thereby increase reproductive isolation.

The importance of early learning and sexual imprinting in preserving reproductive barriers often becomes apparent when these barriers break down. The Grants have made a twenty-year study of four species of Darwin's finches in the Galapagos Island of Daphne Major.[109] They have found that matings between these species occur rarely (in less than 5% of matings), but, when they do, the hybrids produced are fit and completely fertile, sometimes surviving better than members of the parental species. There is obviously no genetic incompatibility between the different species. The rare cases of hybridisation occur as a result of mis-imprinting. If a young male loses contact with its father during the post-fledgling period when it learns its song, and is instead exposed to the song of a male of another species, it learns this foreign song and consequently is chosen as a mate by a female of the other species. If mis-imprinting underlies hybridisation, then imprinting must underlie the normal reproductive isolation of these species. In fact, whenever hybridisation is due to mis-imprinting it can be inferred that early learning has a role in reproductive isolation.

Some of the best evidence for the importance of imprinting in speciation comes from studies of the parasitic whydah birds we described in chapter 4.[110] Females of these species lay their eggs in the nests of host Estrilidine finch species. If the parasitic hatchling is to be accepted and reared by its foster-parents, it must mimic precisely the appearance, interior mouth-markings and begging behaviour of the offspring of its host species. When it becomes a breeding adult, the parasite must mate with an individual of its own species, who has been reared by the same species of finch. If it does not, the genetically determined mouth-markings of its offspring will fail to resemble those of its host's young, and the host will reject them. Nicolai has shown that the criteria for the parasites' choice of mate and host recognition are not genetically determined, but are learnt early by the young. It is the strong sex-related imprinting of the parasites on their foster-parents that ensures compatible mating. When a whydah female is of breeding age, she searches for individuals who look like her foster-parents, and she only becomes reproductively active and ovulates when she sees the reproductive activities of adult members of her particular host species. When she was young, she became visually imprinted on her foster-parents' images and on their nesting activities. Moreover, she also became imprinted on their vocalisations. Young male whydahs also become imprinted on the vocalisations of their hosts, and their own adult song is composed of two different parts – it has a whydah part and a foster-parent part. The latter includes all the foster-parents' vocalisations, including the song, begging calls, contact calls and so on. The whydah male's song therefore attracts a whydah female who was reared by members of the same species of host. The combination of morphological, genetically determined mimicry, and habitat and sexual imprinting has led to the co-evolution of hosts and parasites. However, the faster divergence of the parasitic species has led to non-overlapping phylogenetic trees. The role of early learning in initiating and consolidating reproductive isolation in whydah birds is therefore clear, and suggests that cultural evolution can be a significant factor in speciation.

Summary

To understand how group traditions form and how they affect the evolution of social behaviour, we have to understand the mechanisms that bind individuals into a functional group in which they co-operate and exchange information. Group-binding mechanisms include family-based behaviours that have become generalised into a wider social context, emotional contagion and reproductive suppression. These organise and channel social learning, and facilitate information-sharing. Information, unlike energy, is not subject to a law of conservation, so sharing information allows an individual in a sharing-collective to have more information and pay less for it. When the mechanisms that bind a group together are disrupted so that sharing information is impossible, or when socially learnt traditions can no longer be transmitted, the group dissolves or suffers social death. Since through its interactions with other group members the individual constructs the group, yet the group channels and organises individual behaviour, studying social cohesion and social death means constantly switching perspectives between the social behaviour of the individual and group dynamics.

The habits found in animal groups can be very stable and persist across generations. Several interacting factors contribute to this: first, living in a group promotes the acquisition, transfer and consolidation of habits. Second, different socially learnt and transmitted behaviours interact to produce a network of interdependent behaviours – a behavioural package. Third, feedback interactions between learnt behaviours and the physical and social environment enhance the stability of behavioural sets. Cultural evolution, which is relatively rapid, adjusts traditions to fit existing genotypes, and this leads to increasingly tighter interdependence between traditions and morphological and physiological traits. Finally, behavioural conformism, based on social learning and selection for adopting the dominant tradition within the group, creates long-term, transgenerational behavioural stability.

Selection can occur among groups with different traditions. Any heritable variation in group behaviour, whether genetic or cultural, which enhances group productivity can be selected. Selection among groups that differ in traditions may have been a potent force in the evolution of sociality in birds and mammals. New habits and traditions that become established in a subpopulation may initiate reproductive isolation between this subpopulation and the one from which it originated. Cultural speciation may therefore play an important role in the divergence of social vertebrates.

Notes

1 This description is based on some of the scenes in the BBC film *Meerkats United and Meerkats Divided* (1996, World of Wildlife), E. A.'s observations of a captive meerkat colony in the Tisch Zoo in Jerusalem, and Clutton-Brock *et al.*, 1998*a,b*; Doolan & Macdonald, 1996*a,b*, 1997*a,b*; Ducker, 1962; Estes, 1991; Ewer, 1963*a,b*; Moran, 1984.

2 Asch, 1952, p. 257.

3 Mech, 1970.

4 Schenkel, 1948, 1967.

5 See the entry on 'grooming' by Woolridge in McFarland, 1987, pp. 237–40.

6 Kummer, 1995.

7 Estes, 1991; Ewer, 1973.

8 Rasa, 1985.

9 Von Holst, 1998.

10 Brown, 1985.

11 Tajfel, 1978; Turner *et al.*, 1987.

12 Hatfield, Cacioppo & Rapson, 1994.

13 Aristotle, Problems Book VII, 886a 25, 887a 4–5, 887b 15–16.

14 See, for example, Teoh, Soewondo & Sidharta, 1975.

15 Provine, 1996.

16 Le Bon writes about the effect of emotional contagion in human crowds: 'the disappearance of the conscious personality, the predominance of the unconscious personality, the turning by means of suggestion and contagion of feeling and ideas in an identical direction, the tendency immediately to transform the suggested ideas into acts; these, we see, are the principal characteristics of the individual forming part of a crowd. He is no longer himself, but has become an automaton which has ceased being guided by his will.' (Le Bon, 1895, p. 12).

17 Miller, Murphy & Mirsky, 1959.

18 In his 1920 book *The Group Mind*, McDougall discusses organised groups rather than the crowds that Le Bon concentrates on. He talks about emotional contagion as a positive feedback mechanism, whereby certain signs are both an expression of a given emotional state as well as the inducers of this state in observers. In organised groups, historical continuity, division of labour, habits and traditions encouraging self-categorisation as a group member, and confrontations with other groups allow escape from the primitive crowd-state. Members of the organised group behave according to pre-fixed roles, and the group acts as an 'individual', even if it is a short-term one.

19 Smith, 1992; von Frisch, 1938.

20 Suboski *et al.*, 1990. The effect of the visual stimulus alone is not as strong as that of the same stimulus combined with the alarm stuff, and the alarm stuff can elicit a reaction without the stimulus. Alarm stuff also has secondary effects, which lead to the recognition of predators. For example, when the release of the alarm stuff is associated with the smell of the predatory pike,

the pike is recognised as dangerous and, later, the fear reaction can be elicited by the smell of pike alone.

[21] Cheney & Seyfarth, 1990; Conover, 1987; Curio, Ernst & Vieth, 1978.

[22] Baker, 1978.

[23] Doolan & Macdonald, 1997*a,b*; Jolly, 1967; Kummer, 1995.

[24] Ims, 1990.

[25] Knobil & Neill, 1994, pp. 328–30.

[26] In nearly 50% of the cases, dominant females kill pups that are born to subordinates. See Clutton-Brock *et al.*, 1998*a*.

[27] For Freud, crowd psychology was a consequence and augmentation of family psychology. He accepted that contagion plays an important role in crowd behaviour: 'There is no doubt that something exists in us which, when we become aware of signs of an emotion in someone else, tends to make us fall into the same emotion; but how often do we not successfully oppose it, resist the emotion and react in quite the opposite way? Why therefore do we invariably give in to this contagion when we are in a group?' (Freud, 1921, p. 672). Freud believed that the primary reason for the effectiveness of contagion in groups is the transference of father–child relationships into the group domain through the acceptance of the leader's authority. However, his theory cannot explain certain types of leader-indifferent contagion, such as flight and panic contagion, and many cases of mobbing and reproductive synchrony.

[28] Anderson, 1980; Schlicht, 1979.

[29] One species known to have several guards at the same time is the black-tailed prairie dog. See Hoogland, 1995.

[30] Lachmann, Sella & Jablonka, 2000.

[31] Clutton-Brock & Parker, 1995*a*.

[32] Ward & Zahavi, 1973.

[33] See Brown, 1986, 1988; De Groot, 1980; Galef, 1991; Galef & Wigmore, 1983; Greene, 1987; Gori, 1988; Wilkinson, 1992.

[34] Greene, 1987.

[35] Packer, Lewis & Pusey, 1992.

[36] Better protection from pathogens may lead to better learning, because healthy young are more attentive and alert. Kavaliers, Colwell & Galea (1995) have shown that when domestic mice are infected with a naturally occurring enteric protozoan parasite, their spatial learning is impaired; immunological protection may therefore have specific and direct effects on the learning ability of the young. The superior protection that communal suckling probably provides may enable communally fed young to learn more effectively.

[37] Heinrich, 1989; Heinrich & Marzluff, 1995; Marzluff, Heinrich & Marzluff, 1996.

[38] Packer, Tatar & Collins, 1998.

[39] Anecdotal evidence can be found in Kummer, 1995.

40 The modelling of this cycle may be similar to that suggested by Stark for an analogous situation in human societies. See Stark, 1995, chapter 3.

41 Aronson, 1982; Clutton-Brock, Guinness & Albon, 1982; Geist, 1971; Mendelssohn, 1974.

42 Poole, 1994.

43 For dolphins, see Evans, 1984, pp. 180–9.

44 Calhoun, 1962, 1963a,b, 1973.

45 Krebs & Myers, 1974; Tamarin et al., 1990.

46 Krebs, 1996; Tamarin et al., 1990; Von Holst, 1998.

47 Ginzburg (1998) interpreted and modelled these population cycles as the consequence of deleterious, environmentally induced maternal effects, with the stressed mothers producing physically and socially handicapped progeny who could not sustain a viable population.

48 Spurway, 1955.

49 Kruska, 1988.

50 Hemmer, 1990.

51 See Serpel, 1995. The existence of socially organised packs of dingoes in Australia and pharia dogs in South America shows that reversion to sustainable and independent social life is sometimes possible.

52 Boitani & Ciucci, 1995.

53 The development and behaviour of dog pups belonging to the Alaskan Malamute breed, which were adopted by a wolf in captive conditions, have been compared with those of adopted wolf pups. Important differences in the rate of development and rates of maturation were found (see Frank & Frank, 1982). However, the picture is not yet clear, because the behaviour of adopted dog pups in a wild wolf pack has not yet been followed.

54 Clemmons & Buchholz, 1997.

55 See Kleiman et al., 1996, especially chapter 31.

56 Beck et al. 1994; Sarrazin & Barbault, 1996.

57 See Stanley Price, 1989; Tear & Forester, 1992; Tear, Mosley & Ables, 1997.

58 Bertram, 1980; Hamilton, 1971; Kenward, 1978.

59 Hoogland & Sherman, 1976; Kruuk, 1964.

60 O'Connor, 1991, p. 250.

61 Curio, 1976.

62 Estes, 1974; Jarman, 1974.

63 Crook, 1965.

64 Dasmann & Taber, 1956.

65 Clutton-Brock, Guinness & Albon, 1982; Darling, 1937; Estes, 1974.

66 Stander, 1992.

67 For a review of the advantages of group-life for the individual, see Gosling & Petrie, 1981.

68 Among non-human animals, reciprocal altruism has usually been studied using pairs of individuals. Interactions between two individuals are a special case of reciprocal interactions, which may involve interactions between one

individual and a group, or interactions between groups. Such complex reciprocal interactions have been extensively studied in humans (Dawes, 1980), but not in higher animals.

[69] Contrary to what one might think, co-operation based on reciprocal altruism between two or more players is not learnt very readily. Psychologists experimenting with pairs of human subjects found that altruistic co-operative choices, which received greater rewards, increased only twofold (from 30% to 60%) after 300 trials. This is a rather disappointing result if one assumes humans behave rationally and seek to maximise their gains. Turner and his colleagues (1987) have reviewed the conditions that increase the effectiveness of altruistic co-operation, and concluded that the prior existence of group membership, however temporary and *ad hoc*, enhances the likelihood of co-operation. For suggestive evidence that such conditions may also enhance co-operation in animals, see Wilkinson, 1988.

[70] There can be little doubt that group selection has occurred during evolution. The existence of higher levels of organisation that are made of lower-level units (chromosomes made of genes, multicellular organisms made of cells) testifies to the fact that, at some stages of evolution, selection at the higher level, the group, must have been stronger than selection for the interests of the lower-level units that make up the group. See Jablonka, 1994, and Maynard Smith & Szathmáry, 1995, for discussions of these issues; for experimental evidence, see Goodnight & Stevens, 1997, and Wade, 1976; for a general critical evaluation of group selection, see Sober & Wilson, 1998, Williams, 1992 and Wilson, 1997.

[71] Wilson, 1983; Sober & Wilson, 1998.

[72] Some biologists do not accept that the sending off of future group-forming dispersers can be considered as genuine group reproduction. They insist that the 'groupiness' of a group has to be conserved at all stages of the life cycle. If this objection is accepted, then this life cycle is not relevant to group selection; it is an example of non-random assortment of genes or phenotypes. The position one adopts about group selection therefore depends on the definition of group reproduction. See Dawkins, 1979.

[73] In general, an altruistic phenotype reduces the fitness of the altruistic individual and enhances the fitness of those group members towards whom it directs its behaviour (some of whom are likely to be altruists). The altruistic behaviour will spread if the overall benefit of the behaviour on the reproductive success of altruists within the group is larger than its fitness cost to the 'actor'. The fitness of the phenotype therefore depends on the benefits its carrier confers on other individuals weighted by the probability that they, too, have altruistic phenotypes. These considerations are similar to those described by Hamilton for kin-groups, in which fitness depends on the benefit to individuals multiplied by their coefficient of relatedness. In the cultural case, we have to assume that altruistic behaviour is heritable through social learning.

[74] Wilson, 1983; Wilson & Sober, 1994. For the two life cycles described, the ratio between the number of groups and the number of individuals within each

group has a fundamental influence on the success of group selection. When the number of groups greatly exceeds the number of individuals within a group, selection among groups will overwhelm individual selection. See Szathmáry & Demeter, 1987.

[75] Sherman, Jarvis, & Alexander, 1991.

[76] Boyd & Richerson, 1990.

[77] Boyd & Richerson, 1985.

[78] For an early review of variations in social systems, see Lott, 1984; for variation between primate groups, see Normile, 1998; for different breeding patterns in domestic cats, see Feldman, 1993; for the effect of dispersal on social organisation and bonding among females in chimpanzee groups, see De Waal, 1994; for the effect of ecological conditions on social structure in Japanese macaques, see Nakagawa, 1998 and Yamagiwa & Hill, 1998; for the effect of ecological conditions on family organisation in dunnocks, see Davies, 1992, and chapter 5, pp. 145–6.

[79] Doolan & Macdonald, 1996*a*.

[80] McGrew, 1992.

[81] Collias & Collias, 1984.

[82] Geist, 1971.

[83] These studies are summarised by Geist, 1971.

[84] Bagemihl, 1999; Berman, 1997; Fairbanks, 1996; Heyes & Galef, 1996.

[85] Hauser, 1997.

[86] Mundinger (1980) distinguishes between geographic variations in vocalisations, which result from different ecological conditions in widely separated populations, and dialects, which are differences between neighbouring groups whose individuals often interact with each other.

[87] Catchpole & Slater, 1995.

[88] Price, 1998.

[89] Ford, 1990.

[90] Whitehead, 1998.

[91] Diamond, 1986, 1987, 1988; Gould & Gould, 1989; Pruett-Jones & Pruett-Jones, 1983.

[92] Direct evidence about the length and mode of learning in Vogelkop bower-birds is not available, but inferences are made on the basis of information on a closely related species, Macgregor's bower-bird, where behavioural development and social learning have been followed more closely.

[93] Whiten *et al.*, 1999; Wrangham *et al.*, 1994.

[94] Green, 1975; Huffman, 1996; Itani, 1958; Kawai, 1965; Tanaka, 1995; Watanabe, 1989.

[95] Tanaka, 1998.

[96] The combination of these factors during development may construct what the American psychologist Gilbert Gottlieb calls the 'developmental manifold'. This is the developmental network that leads to construction of stable, species-specific behaviour. Gottlieb, 2000.

[97] Hardy, 1965; Immelmann, 1975; Mayr, 1963; Wyles, Kunkel & Wilson, 1983.

[98] There are many problems associated with the species concept (see, for example, Tempelton, 1989). Essentially, the species concept we are using is the classical concept introduced by Muller, Dobzhansky and Mayr during the 'Modern Synthesis' of evolution in the late 1930s and 1940s. According to Mayr, species are 'groups of actually or potentially interbreeding natural populations, which are reproductively isolated from other such groups' (Mayr, 1942, p. 120).

[99] Tauber & Tauber, 1989.

[100] Immelmann, 1975.

[101] Baker, 1993, 1994, 1996.

[102] Baker & Jenkins, 1987; Baptista & Johnson, 1982.

[103] Mayr, 1963, chapter 17.

[104] In small populations there may be strong pressure for conformity, so, after the first period of cultural adaptation, they may show less diversity and slower cultural evolution than large populations.

[105] Such habitat imprinting also seems to be one of the factors ensuring host specificity in the European cuckoo. See Teuschl, Taborsky & Taborsky, 1998.

[106] Immelmann, 1975, reviews the information on the European mistle thrush.

[107] Yom-Tov & Tchernov, 1988.

[108] ten Cate & Bateson, 1988. See Laland, 1994a, for a mathematical model of this case.

[109] Grant & Grant, 1996, 1997.

[110] Nicolai, 1964.

9   Darwin meets Lamarck – the co-evolution of genes and learning

In this chapter we are going to look at tradition, genes and learning all at once, as they interact during evolution. We have shown in previous chapters how, irrespective of any genetic change, social learning can lead to independent cultural evolution and promote speciation. When the role of the transmission of learnt information is recognised, interpretations of the evolution of many important behaviours are altered. However, for a more complete picture of what happens during behavioural evolution, we need to look at the type of genetic changes that occur during the evolution of the mechanisms of learning and the various forms of memory. We need to know what drives the evolution of learning, and in what kinds of environments it is likely to evolve. Learning is not a monolithic process, of course. For example, the development of bird song involves imprinting-like learning, trial-and-error learning and several types of social learning, all entwined.[1] The same is true of the development of behaviours such as foraging, hunting, mobbing and even of nest building, the once classical illustration of an 'instinct'. So how do these different types of learning evolve, and how does behavioural transmission across generations affect learning and other processes and characters? In what follows, we are going to argue that learning is an important agent of its own evolution – that the evolution of learning is, to a large extent, self-propelled.

During the nineteenth century, answers to questions about the evolution of learning and behaviour were usually couched in Lamarckian terms. It was assumed that learnt habits are directly inherited. Through regular use and disuse, an animal's morphology, physiology and behaviour change, and these changes impinge on the hereditary material and alter it in an appropriate manner. Lamarck, Darwin and most other nineteenth-century biologists endorsed this position. However, a few people looked at the relationship between habits, instincts and morphology in a different way. They believed that learnt habits affect the evolution of morphology and instincts by shaping and guiding selection, not by having direct effects on hereditary variation. An early proponent of this view was Alfred Russel Wallace, one of the greatest biologists of the nineteenth century, and the co-discoverer of the principle of natural selection. Like many naturalists, Wallace was enchanted by the beauty, complexity and variety of birds' nests, and he attempted to explain these artefacts using the principle of evolution by natural selection. However, unlike most naturalists, Wallace did not believe that birds have an unfailing, species-specific, nest-building instinct. To him, experience seemed to have an important role. He believed that the great intricacy and stability of the nest forms of different species can be explained almost solely by learning guided both by ecological determinants such as the climate, vegetation, type of predators and so on, and by morphological limitations such as the size of the feet and beak. Wallace claimed:

> they [birds' nests] may be in a great measure explained by the general habits of the species, the nature of the tools that they have to work with, and the materials they can most easily obtain, with the very simplest adaptations of means to an end, quite within the mental capacities of birds. The delicacy and perfection of the nest will bear a direct relation to the size of the bird, its structure and habits.
>
> (Wallace, 1870, p. 216)

Once individuals had learnt to build a nest, the procedure was stabilised and refined: through the natural selection of transmitted nest-building habits, nest-building practices became better and better adjusted to conditions and to the pre-existing morphological peculiarities of the birds. If Wallace's arguments were couched in modern terms, we would say that he believed that there are no 'nest-building genes'; there are nesting practices which, through individual and social learning, have evolved to fit the bird's morphology and ecology, in the same way that a glove has evolved culturally to fit the hand it covers. Wallace wrote:

> consideration of the structure, the food, and the other specialities of a bird's existence, will give a clue, and sometimes a very complete one, to the reason why it builds its nest of certain materials, in a definite situation, and in a more or less elaborate manner ... Besides the causes above alluded to, there are two other factors whose effect in any particular case we can only vaguely guess at, but which must have had

an important influence in determining the existing details of nidification. These are – changed conditions of existence, whether internal or external, and the influence of hereditary or imitative habit; the first inducing alterations in accordance with changes of organic structures, of climate or of the surrounding fauna and flora; the other preserving the peculiarities so produced, even when changed conditions render them no longer necessary.

(Wallace, 1870, pp. 231–2)[2]

Wallace also advanced some very subtle arguments about the evolutionary relationship between habits and morphology. In addition to nesting traditions becoming adjusted to pre-existing morphology, Wallace believed that morphology could become adapted to culturally evolved nesting practices. He found a striking correlation between nest-building practices and the coloration of the female bird: almost all species in which the incubating female has bright colours have nests that are concealed, well covered or otherwise protected. According to Wallace, since the female needs more protection than the male during the egg-laying period, female birds that lacked a ready method of building a protected nest were under strong selection to lose their original gay colours. Hence, species that build unprotected nests often have females with drab, camouflaging coloration, whereas those that build well-protected nests often have colourful females. It was the stable, culturally transmitted nesting practices that led to selection of the easily modifiable hereditary factors affecting feather colour.[3] We shall return to Wallace's idea that cultural evolution in animals can affect morphological and physiological evolution in later sections of this chapter. But first we want to look a little more closely at the role of learning in nest building and at evolutionary changes in nesting practices, since these will help us to illustrate some general features in the evolution of behaviour.

Wallace was partly right about the importance of experience – changed circumstances and individual learning can certainly play an important role in nest building. There are many examples of birds adjusting their nest material, their building method and their nest location to local circumstances. For example, the strongest nest ever built was constructed by a pair of house martins in Britain who, instead of using the usual mud, built their nest under the eaves of a house using wet cement inadvertently put out by builders.[4] Wallace himself described a change in nesting practices among palm swifts in Jamaica. Before 1854, the birds nested exclusively in the palm trees on the island, but one colony established itself in two coconut trees in Spanish Town. When these trees were blown down in 1857, the swifts moved into the Piazza of the House of Assembly, drove away the local swallows, and built their nests on the tops of the walls and in the angles formed by the beams, where they continued to nest. Not only were the nests built in a very different place, they were also much less elaborate than those constructed in palm trees, because in the new location they were more protected from predators and climatic hazards. A third example of changed nesting practices is the case of the Mauritius kestrel, mentioned in chapter 8. This species was driven to near extinction because its nest, which was built in tree cavities, was vulnerable to the monkeys that were introduced onto the island. However, in 1974 one pair of Mauritius kestrels altered its nesting habits and nested on a cliff that was safe from the monkeys. The young raised by this pair apparently became imprinted on the cliff locality, since they, too, nested on cliffs. Within three years, the population tripled, thanks to the new and adaptive nesting tradition that had become established in these kestrels.

Learning clearly plays a role in some nest construction, but how important is it for normal nest building? For example, can birds that have been deprived of parental care and nest materials during early life build a normal, species-typical nest? The genetics of nest building are unknown, but there is little doubt that Wallace relied too heavily on learning from experience to explain variations in nest-building practices among different species. Some aspects of nest building do seem to be innate: given the appropriate nesting material, hand-reared domesticated canaries construct simple open nests that are as neat and perfect as those built by normally experienced birds. Even when deprived of normal nesting material, the canaries will go through all the motions of building their nest cup.[5] However, innate building behaviour may be enough for only relatively simple nests. Nicholas and Elsie Collias, who studied nest building in the village weaver-bird, found that, although this behaviour has been considered to be the classical case of 'instinct', it takes two years of practice in normal social conditions before a young male is able to weave an adequate nest.[6] To build his nest, the male has to select flexible materials, tearing long strips from leaves or grass stems; he then has to attach a strip to an appropriate branch, construct a ring and weave an egg chamber on one side and an antechamber on the other. When the ceiling of the nest has been woven, the male thatches it with wide leaf strips until the ceiling becomes opaque, and then adds an entrance to the side of the antechamber. In their first year, male village weavers build crude nests, with loose ends projecting from the outer surface. Females usually ignore these messy nests and their builders, choosing instead the neat and elegant nests of adult males. Males who have been deprived of nest material sometimes never learn how to tear the long strips necessary for the first stage of weaving. However, once supplied with ready-made strips, they usually learn to weave a nest, even though they require much more practice than birds that have been raised under normal conditions, and they are generally retarded in the development of their building techniques. In other bird species, the role of experience is sometimes even more pronounced: hand-reared American robins were found to be totally incapable of building nests when they were first presented with appropriate nest material at the age of one or two years.[7]

The interaction of innate and learnt nest-building behaviours can be seen in hybrid lovebirds.[8] Different species of these small parrots have different nesting practices. Peach-faced lovebirds carry nesting material to their nest site by tucking it into their feathers, whereas Fisher's lovebirds carry it in the more usual way, in their bills. Matings between these species are common in captivity, so hybrids have been studied. These studies show that when hybrids first begin to build their nests they act as if they are completely confused, and find it difficult to decide how to carry the strips they cut from wood and bark. They started by tucking the nesting material into their feathers, carrying it in their bills for only 6 per cent of the time. However, nesting material tucked into feathers fell off, so only material carried in the bill reached the nest site. Gradually the hybrids increased the bill-carrying practice, and after two years they behaved largely like Fisher's lovebirds, carrying material in their bills and only occasionally tucking some among their feathers. Although it was clear there was an innate tendency, inherited from the peach-faced parent, to tuck nesting material into the feathers, the effects of experience greatly modified the expression of this tendency, to the extent that it almost completely disappeared.

Surprisingly, the role of social learning in the building of complex nests has not been studied directly, but there are strong indications that it is important. For example, although observational learning is not necessary for the normal weaving behaviour of the male village weaver, and an isolated male eventually learns to weave the species-specific nest, the speed of learning among males who live with their parents and their peers is much higher than that of isolated males. Yearling males construct 'play-nests' and weave strips in each other's nests, experimenting together in their small 'play colonies'. It seems likely that both the early exposure to their parents' nest and nesting behaviour, and the communal social play of young fledglings, help males to acquire the skills needed to build a good nest.

The significance of social learning for the development of nest-building skills probably varies in different species. It is likely to be important in species or populations where helpers take part in the construction of nests, because social learning can then have a direct influence on the development of building practices. Earlier we described how white-fronted bee-eater helpers assist in digging the nest tunnel, and in grey-breasted jays the entire flock of 8 to 20 birds helps in nest construction. Some species of black and red forest weaver birds have helpers who participate in the building of their large and complex nests. It is believed that these helpers gain weaving skills and experience, while the breeding birds economise on energy.[9] Social learning is also very important for the bower construction of Vogelkop gardener bower-birds. In this case, social learning seems to have led to dramatically different local building traditions.

Clearly, nest building, like many other complex behaviours in birds and mammals, can involve elements that have to be learnt as well as innate behaviours that require little or no experience. What we want to know is how these patterns of behaviour have evolved. Obviously, if we are interested in nest building, we have to think about the possible past and present functions of nests – about whether they have or had a role in protection against predators, or as shelter from adverse weather conditions, or in courtship displays. This is the kind of thinking we used in chapter 1 when we created a hypothetical scenario for the evolution of nest building in tarbutniks. We also have to think about related changes in morphology and physiology – about whether selection for particular behavioural features led to changes in form and function, or whether a particular form and function were selected for other reasons and then influenced behavioural evolution. This is only the beginning, however, because we must also think about changes in the actual behaviour itself. To understand any complex pattern of behaviour, we have to consider the evolutionary mechanisms that affect what is learnt and how it is learnt, and how changes in the ability to learn occur; we also have to consider whether and how the ability to learn can affect selection processes. In other words, we have to think about *learning-evolution* – about the factors that mould the nature, speed and mode of learning so that, at one end of the spectrum, we find behaviours that are learnt only after many trials and errors, or through gaining information about how to behave from others, whereas, at the other end, there are behaviours that need so little experience they are effectively innate.

In the following sections, we are first going to consider what features of the environment influence learning-evolution. Our conclusion is that the most significant factor is the stability or otherwise of the environment over time. We shall then look at how evolution in constant, stable environments can lead to behaviours involving little or no learning, and how such innate or 'instinctive' behaviours can become established through the process known as genetic assimilation. From there we shall move on to look at behaviour in less constant environments, where individual and social learning are advantageous, and at how genetic assimilation is involved in the learning-evolution that takes place in these environments.

The ecology of learning

First then, what are the factors in the environment that are important in learning-evolution? Can we identify conditions that are likely to lead to the evolution of behaviour at the innate end of the spectrum, and others that lead to the evolution of behaviour at the end of the spectrum where there is a lot of individual and social learning? Although until now we have used the word 'environment' in a common-sense, intuitive way, we have repeatedly emphasised that 'environment' means different things for different animals. For a food specialist like the loveable koala, who feeds only on the leaves of a certain species of eucalyptus, the relevant feeding environment is much more limited than that of a generalist like the brown rat, who eats almost anything. Even when two species live in the same physical environment, what is relevant to each of them may be very different. So, when we talk about an animal's 'environment', what we have in mind is the ecological niche of the organism, the habitat as it is experienced and constructed by the animal itself. 'Environment' is a species-specific concept. In fact, it is also a lineage-specific and even an individual-specific concept. Think about food: a habitat may be complex and rich for the omnivore, uniform and simple for the extreme specialist; a pine forest may be a difficult feeding ground for most black rat lineages, but paradise for those that have learnt to get at pine seeds. The predictability of the environment is something else that is species-specific. What is experienced repeatedly and predictably by an African elephant who lives for seventy years may not be so predictable for the area's pygmy white-toothed shrews with their lifespan of only a few months. Whether or not an environment is experienced as stable depends on the lifespan of those living in it. It also depends on behaviour: with the hoarding birds who provide for winter by caching seeds, their hoarding behaviour and marvellous spatial memory create a relatively stable and predictable food supply in periods when the environment is 'objectively' insecure and hostile. The hoarding birds actively construct and stabilise their environment.

There are many other ways in which animals alter their environment and thus affect their own evolution. Nests, burrows, dams, foraging habits, are all ways in which organisms adjust the environment to themselves. Moreover, organisms often transmit the physical environment they construct, and hence the selection regime, to their descendants. Darwin discussed this in his book about earthworms.[10] The properties of the soil change considerably as the earthworms bore through it, mix it, pass it through their digestive systems, and leave their casts on its surface. The environment constructed by the earthworms' activities is the environment in which they and their descendants are selected. Many biologists, most notably, Hardy, Ewer, Waddington and Lewontin, have repeatedly emphasised that the organism is not a passive target of selection, but actively influences the selection that it experiences; it is an agent of its own evolution.[11] Activities such as hoarding or nest building affect the selective forces that impinge on the animals that carry them out. Hardy and Ewer argued that, in higher animals, new learnt habits are in fact the major 'engines' of evolution, influencing both its direction and its rate.

There is a problem in thinking about the environment in the way we have just been doing, however. If we think of the nature of a niche as being uniquely determined by the organism that is inhabiting, experiencing and constructing it, then maybe we are obliged to consider each case in isolation, and should not expect to find any environmental features that promote or restrain learning-evolution. If so, we are in much the same position as the seventeenth-century chemists, who had to discover what happened in every chemical reaction and remember each case individually, because they had no principles to make sense of what happened and use as a framework for ordering their knowledge. Fortunately we are not in this position. In our case, simple logic suggests that, because natural selection operates, learning-evolution must be associated with the degree of predictability and repeatability in animals' environments. If the environment an animal experiences is extremely chaotic and changes very rapidly, learning will be useless because, by the time the organism has learnt something, the knowledge is already obsolete. The best way of adapting to such an environment is to produce many, cheap, fast-reproducing offspring, a few of which will be lucky enough to survive and reproduce. On the other hand, if an animal's environment is fairly constant, or changes in a predictable way, it makes sense to invest in quality rather than quantity – to having fewer offspring, but offspring who can learn. Learning and remembering are worthwhile because the knowledge acquired is likely to be useful. A long lifespan, low rate of reproduction, and parental care are all likely to be favoured in such an environment.[12] Parental care helps learning in two ways: first, the young can obtain knowledge from their parents though social learning; second, the security that parental care imparts encourages the young to actively explore their environment, and so makes individual learning more likely.

Learning itself can increase the perceived stability of the environment, because the more an individual knows about its environment, the more predictable and reliable it becomes. As happens so often in biology, the cause-and-effect chain curls back on itself: the effect works back on the cause and strengthens it, leading to the amplification of both cause and effect. This positive feedback-loop between environmental predictability and learning is established because better learning capacities evolve when an environment is perceived as predictable, and the environment becomes more predictable for an individual who is 'better informed'. Something similar happens when animals help to construct their environment. The excellent memory of caching birds makes the food supply in their environment more predictable, which, in turn, strengthens selection for memory. A reinforcing, positive feedback-loop is established between the constructed aspects of the environment and the animal's learning ability. The regularities that animals find or actively construct in their environment enable learning, and learnt behaviours stabilise the environment.

If evolutionary logic tells us that learning is more likely in environments that are predictable in the sense that animals are likely to experience the same situation several or many times, can we classify predictable environments in a way that will enable us to gain more insight into the evolution of learning and memory? We believe that a classification based on the temporal nature of the environment is useful. Effectively predictable environments can be divided into three broad categories: those that are changing rapidly; those that are relatively constant; and those that are changing at an intermediate rate. In all cases, the rate of environmental change is relative to the lifespan of the animal. In the first class are environments in which changes occur on the time-scale relevant to individuals: re-occurring events happen during an individual's lifespan. In the second class, the changes in the environment occur so slowly that for many generations conditions remain practically unchanged; the time-scale is an evolutionary one. In the third class, the time-scale is an ecological one, since cycles of environmental change are longer than the lifespan of an individual, occurring over several generations.

We want to see how the rate at which the environment changes can affect different types of learning – individual learning, social learning and early learning with a strong innate component. There are some snags in the classification of environments we are using, because we shall talk about 'the environment' being constant or changing, when, in fact, different aspects of any animal's environment will vary in different ways. For example, the presence of a mother is a constant feature of the environment of a young mouse, but the food supply may not be. We also have to bear in mind that the time-scale of environmental changes is not always independent of the activities of the organism: the environment is often constructed as the animal interacts with it, so it may itself be an evolving feature of the organism–environment system. If a lineage of birds evolves a food-hoarding habit, then, as far as food supply is concerned, the environment becomes more constant and predictable. Nevertheless, in spite of the multifaceted and dynamic nature of the environment, when we consider the three time-scales of change, the types of learning and memory that we expect to find are somewhat different. We will start by looking at the constant environment, the one with long-term stability, and at what Darwin had to say about the evolution of behaviour.

The ecology and evolution of instinctive behaviour: evolution in constant environments

Darwin was fascinated by the evolution of behaviour. It is the major subject of two of his books, *The Descent of Man and Selection in Relation to Sex* and *The Expression of the Emotions in Man and Animals*. Yet today's evolutionists find many of Darwin's ideas about the evolution of behaviour strange, for Darwin often explained the evolution of partially or totally innate behaviours in straightforward Lamarckian terms. Darwin saw innate behaviour as the outcome of individuals learning the same behavioural actions in every generation until eventually, after many generations, the behaviour that was acquired or learnt became an inherited character, with learning no longer being required for its expression. There are many examples of this Lamarckian reasoning in all of Darwin's major books. We will give just two illustrations:

> some intelligent actions – as when birds on oceanic islands first learn to avoid man – after being performed during many generations, become converted into instincts and are inherited.
>
> (Darwin, 1871, pp. 37–8)

> I shook a pasteboard box close before the eyes of one of my infants, when 114 days old, and it did not in the least wink; but when I put a few comfits into the box, holding it in the same position as before, and rattled them, the child blinked its eyes violently every time, and started a little. It was obviously impossible that a carefully-guarded infant could have learnt by experience that a rattling sound near its eyes indicated danger to them. But such experience will have been slowly gained at a later age during a long series of generations; and from what we know of inheritance, there is nothing improbable in the transmission of a habit to the offspring at an earlier age than that at which it was first acquired by the parents.
>
> From the foregoing remarks it seems probable that some actions, which were at first performed consciously, have become through habit and association converted into reflex actions, and are now so firmly fixed and inherited, that they are performed, even when not of the least use, as often as the

same causes arise, which originally excited them in us
through the volition.

(Darwin, 1872, pp. 39–40)

Darwin thought that the effects of use and disuse and of learnt habits
are aided by natural selection. Some of the many differences in the
behaviour of individuals are heritable, and natural selection for the
most appropriate behaviours will increase their frequency in the population and gradually perfect them. In the case of the instincts of worker
ants and social bees, Darwin pointed out that they cannot have come
about through the inheritance of habits, because the workers do not
reproduce and therefore cannot pass on acquired habits. In other cases
of complex innate behaviours, however, Darwin believed that both natural selection and the heritable effects of acquired habits affected their
evolution. He summarised a discussion on inherited changes in habits
in domesticated animals with 'In most cases habit and selection have
probably concurred.'[13]

Darwin's Lamarckian ideas were in tune with his theory of heredity,
which he called the 'pangenesis' theory.[14] This was actually a more
detailed version of a theory suggested by the Hippocratic doctors in
Greece some 2400 years earlier, although Darwin apparently learnt of
the Greek theory only after he had developed his own. According to the
pangenesis theory, each part of the body, at each stage of development,
sends tiny representative particles of itself to the reproductive organs.
These particles, which Darwin called 'gemmules', form the germ of the
sex cells, and the union of male and female germs forms the offspring.
A part of the body that is modified by the environment, or by persistent use or disuse, liberates modified gemmules into the circulation, so
the consequences of new learnt habits are inherited.

Darwin's hypothesis that habits, through their effects on the organism's physiology, can *directly* alter hereditary factors in the reproductive
organs has proved to be wrong, of course. However, this does not mean
that there is no causal relationship between the acquisition of habits
and the evolution of instinctive behaviour that mimics these habits. As
Wallace appreciated, habits can shape the selective forces acting on the
animal. They can have *indirect* effects that can lead to the transition
from learnt to innate behaviours. In order to understand how this transition may occur, we need to think a little more about the conditions
in which innate behaviours can be beneficial.

When environmental conditions remain constant, with the same events re-occurring regularly for many generations, the faster and more automatic the responses to them, the better. Learning by individual trial and error, or even from the experience of others, can be a lengthy and sometimes dangerous business, so we would expect efficient, fast, early learning to evolve. Take filial imprinting in the young domestic chicken. The newly hatched chick typically follows its mother and becomes attached to her to the exclusion of all other individuals. This is crucial for its survival, because the mother is the source of security and information about the environment. Obeying the simple rule 'follow a conspicuous moving object when you hatch out of the egg' ensures, in the overwhelming majority of cases, that the chick will follow its own mother, since she is the first conspicuous moving object it is likely to see. The presence of a mother is a regular feature of the hatching chick's environment. Only in unusual circumstances, through the untimely death of its mother or the manipulations of an interested ethologist, will the young chick follow a passing red fox, a stuffed polecat, the ethologist Douglas Spalding, or a rotating red box. Some fail-safe measures are built into the system: if chicks are given a choice between a red fox and something resembling a mother hen, they prefer the latter. So, in the unlikely circumstances that a red fox competes with the mother, the mother will 'win' and the chick will become imprinted on her.

Given that behaviours with strong innate components can be beneficial, how can such predispositions actually evolve? It may be very advantageous for the chick to identify the first moving object as 'mother', but how do the genes in the germ cells become informed about this? Similarly, for a migratory bird a predisposition to learn the star patterns in the circumpolar area of the sky by using a fixed reference point is extremely beneficial, but how did the genotype become 'informed' about the fixed point of reference in the starry night sky? Did some chance genetic changes just happen to endow some individuals with these abilities, which were so advantageous that their owners were better able to survive and reproduce, and hence their behaviour and their genes were perpetuated. This is the kind of scenario that is mocked by the anti-evolutionists, and, of course, rightly so. Most biologists would argue that instincts did not appear out of the blue, but were honed by natural selection from similar behaviours that were initially very plastic and involved much more learning. One way in which the transition from a flexible learnt response to a fixed or 'instinctive' response could occur through Darwinian selection is through what has become known as the 'Baldwin effect'. The genetic process that might underlie it is 'genetic assimilation'. Because we believe this process to be of central importance in the evolution of all types of learning, we will explain it in some detail.

The Baldwin effect – genetic assimilation of learnt behaviour
In the late nineteenth century, J. M. Baldwin, Lloyd Morgan and Fairfield Osborne independently suggested how selection could bring about a transition from a learnt to an instinctive response. Their idea, which is now known as the Baldwin effect, was clearly expressed by Morgan:

> any hereditary variations which coincide in direction with modifications of behaviour due to acquired habit would be favoured and fostered; while such variations as occurred on other and divergent lines would tend to be weeded out . . . It may be urged therefore that if natural selection be accepted as a potent factor in organic evolution, and unless good cases can be adduced in which natural selection can play no part and yet habit has become instinctive, we may adopt some such view as the foregoing. While still believing that there is some connection between habit and instinct, we may regard the connection as indirect and permissive rather than direct and transmissive. *We may look upon some habits as the acquired modifications which foster those variations which are coincident in direction, and which go to the making of an instinct.*
>
> (Morgan, 1900, p. 115, our italics)

According to Morgan, if learnt habits enable an organism to survive, selection will favour hereditary changes that mimic these learnt habits. It is as if the learnt habits are part of the niche in which the animal is selected, and hereditary variants that make it better able to occupy that niche are favoured.

Further light on the relevance of flexible learnt or induced responses to fixed phenotypes and innate behaviours came from the work of the British geneticist and embryologist C. H. Waddington, nearly half a century later.[15] Waddington did not look at behaviour, although he knew the relevance of his work to it, but studied the way in which a developmental response that normally depends on induction by an external stimulus can become 'innate' and independent of this stimulus. He called this process 'genetic assimilation', and demonstrated its occurrence for several characters in fruit-flies. We will describe one of Waddington's experiments and his interpretation of the results, because we want to argue that the same kind of process is involved in the evolution of behaviour.

In one of his experiments, Waddington kept fruit-fly larvae on a food medium having a high concentration of salt. The larvae that survived on this medium had modified anal papillae, which are structures at the rear end of the larvae that are associated with salt regulation. The modification was an individual, adaptive, physiological response to the salty conditions. Waddington took the larvae that survived on the salt (a minority), bred from the adult flies, and repeated the procedure for several generations, gradually increasing the salt concentration. To see if the modified papillae appeared without exposure to salt, he took some larvae from his selected line and reared them on normal non-salty medium. During the first generations of selection, the modified papillae developed only if the larvae were exposed to the salt. However, after twenty-one generations of systematic selection, Waddington found modified papillae in larvae that were not exposed to a salty medium at all. The character whose development was originally dependent on exposure to high salt concentrations had become more fixed genetically, and less dependent on the salt concentration. It now appeared in the normal, non-salty environment too. In Waddington's jargon, the acquired character had been assimilated.

Waddington performed similar experiments with other inducible characters in fruit-flies. For example, he induced and assimilated a four-wing phenotype in flies. Flies normally have only two wings, but some can be induced to have four wings if they are exposed to ether vapour at an early stage of development. By selecting and breeding from these, Waddington obtained stocks in which a proportion of the flies had the four-wing phenotype without ether treatment. He also followed the genetic changes in his experimental populations. From his results, Waddington argued that the transition from a phenotype that had to be induced to one that did not was the result of the selection of the gene combinations that progressively produced more rapid and efficient responses to the inducer (i.e. to high concentrations of salt or to ether vapours).

A simple genetic model will make Waddington's reasoning clearer.

Imagine that $a^1a^1b^1b^1$ is the common genotype in a population of fruit-flies (where a and b are different genes, and the number is one particular type of allele). Individuals with this genotype cannot develop on the salty medium, because they cannot adapt physiologically to the salty environment. Let us assume that a^1 and b^1 alleles are predominant in the population, their frequency being $9/10$. There are other alleles in the population, a^2 and b^2, whose frequency is $1/10$. Now assume that individuals having genotypes with two or more of the rare type-2 alleles can develop the adaptive structure when grown on salt – they are responders, who can adapt physiologically to high salt concentrations. However, only individuals with four of the rarer alleles ($a^2a^2b^2b^2$) can develop the adaptive modified anal papillae on a normal, non-salty medium, and their frequency is, of course, very low, only $(1/10)^4$, one in ten thousand. The frequency of the genotypes with two or more type-2 alleles is over 5 per cent, so even in a population of a few hundred there are likely to be several responders.[16] By selecting these individuals that are *capable* of developing the adaptive structure, we select all the genotypes that have at least two of the rarer alleles. The frequency of a^2 and b^2 after selection has, of course, increased to over 50 per cent. To obtain the genotype homozygous for the rare alleles is now very easy: we breed from the selected individuals. They will have genotypes such as $a^1a^2b^1b^2$, $a^1a^1b^2b^2$, $a^1a^2b^2b^2$ and so on, and many of the matings between them will produce some individuals that are homozygous for the previously rare alleles. Such $a^2a^2b^2b^2$ individuals will have modified anal structures even on a non-salty medium. If we want to obtain a line of individuals that develop the anal structure without the need for the inducing stimulus, the high salt concentrations, we can select and breed from these individuals. As a result of the selection, the acquired response has become innate. It has been genetically assimilated. Notice that the genes underlying the innate response were present in the population before it was put on the salty medium. The genotype producing the innate response was formed by nothing more than the normal sexual shuffling of genes. With more genes and more alleles, it would be possible to show how, through selection, a variable capacity to respond adaptively would gradually improve over the generations until eventually it became a fixed response.

During the 1950s, Hardy, Ewer and Haldane used Waddington's type of reasoning to explain the evolution of innate behaviour.[17] Each gave examples or thought experiments to illustrate how genetic assimilation

could lead to the evolution of an instinct. Since there are still no experimental or comparative studies showing that innate behaviour can be
produced by selecting variations in an originally learnt response, we
will summarise some of their examples and arguments. We will then
look at other cases that show how readily and elegantly genetic assimilation of acquired habits can explain the evolution of various innate
behaviours.

Ewer suggested that the evolution of filial imprinting in the young
of domestic chickens is the result of genetic assimilation.[18] She argued
as follows: at first, the mother-following behaviour was learnt and error-
prone, with ancestral chicks wandering about and frequently being
eaten by predators. Since the young chicks were active immediately after
hatching and needed to be protected, a following response would be
very beneficial. In every generation, genetic variations in all the components of the perceptual, limbic and motor systems of chicks were shuf-
fled through the usual sexual processes. New combinations of genes
came together, and variations in the chick's responses to the mother's
presence were exposed to selection. Chicks with variations that led to
rapid attachment and a reliable following response survived better.
Gradually a response tailored to the chicks' imperfectly developed sensory ability was selected, a response that was unaffected by the mother's complex behaviour or her idiosyncratic features. Selection favoured
the simplest 'rule', one that involved accentuating the most stable, striking and easy-to-learn feature – the movement of a large object. Ewer
believed that imprinting is a case where, through genetic assimilation,
the motor response has been perfected, and the learning period has
become vestigial.

Thinking very much along the same lines, Haldane suggested the following scenario for the evolution of an imaginary instinct through
genetic assimilation:

> In an area A a particular volatile substance is produced by a
> nutritious plant, in area B by a poisonous plant. In area A,
> those insects of a certain species which learn most readily to
> recognize this odour and associate it with food are at an
> advantage. As the features of the nervous system which
> favour such learning are accentuated, a few insects appear to
> whom the odour is attractive without learning, as the odour
> of sheep appears to be attractive to sheep-dog puppies. They

are at a double advantage, and after some time, all members of the insect species are attracted by the odour without any learning. Similarly, in area B a race evolves which finds the odour repulsive. We know that there is in fact 'raw material' on which selection can act from a study of our own species, where there are considerable differences in the capacity for detecting smells and tastes, and in judgement as to whether they are attractive or repulsive. Some at least of these differences are genetically determined.

(Haldane, 1959, p. 146)

Haldane had already drawn attention to a delightful imaginary case of instinct-evolution involving both natural and sexual selection, which had been suggested almost a century earlier by Douglas Spalding:

Suppose a Robinson Crusoe to take, soon after his landing, a couple of parrots, and to teach them to say in very good English, 'How do you do, sir?' – that the young of these birds are also taught by Mr. Crusoe and their parents to say, 'How do you do, sir?', – and that Mr. Crusoe, having little else to do, sets to work to prove the doctrine of Inherited Association by direct experiment. He continues his teaching, and every year breeds from the birds of the last and previous years that say 'How do you do, sir?' most frequently and with the best accent. After a sufficient number of generations his young parrots, continually hearing their parents and a hundred other birds saying 'How do you do, sir?' begin to repeat these words so soon that an experiment is needed to decide whether it is by instinct or imitation; and perhaps it is part of both. Eventually, however, the instinct is established. And though now Mr. Crusoe dies, and leaves no record of his work, the instinct will not die, not for a long time at least; and if the parrots themselves have acquired a taste for good English the best speakers will be sexually selected, and the instinct will certainly endure to astonish and perplex mankind, though in truth, we may as well wonder at the crowing of the cock or the song of the skylark.

(Spalding, 1873, p. 11)

In Spalding's story, the learnt response became instinctive after intense selective breeding for the ability to learn to utter some words more quickly and more efficiently, combined with an incidental, sex-related function that the utterances gained. What we know of the way that bird-song develops lends credibility to this type of scenario. Singing is associated with learning in all songbirds; the European starling is a striking example, for this bird modifies and enriches its song throughout life, incorporating elements from other birds' songs as well as other sounds it finds attractive. Yet, even in birds like the European starling, there is a strong disposition to produce a song with the basic species-specific ingredients. Although each starling song is unique, the songs are ordered according to certain definite and simple rules: the song bout starts with whistles, is followed by variable and complex song phrases that include elements usually mimicked from other birds, then by rattling song, and finally ends with some high-frequency song-types.[19] It is likely that, in the evolutionary past of the species, these simple rules were initially learnt and then became assimilated. The selective advantage of a stable song structure was that it enabled reliable recognition and communication between members of the species, while allowing learnt elaboration of components of the song for more subtle social interactions.

The process of genetic assimilation explains the mystery of how the selection of effectively 'blind' genetic variations can, within a very short period of evolutionary time, produce an innate response which mimics one that was previously acquired through learning. However, for the process to work we have to assume that populations have abundant genetic variation, that this variation is expressed in new ways under new circumstances, and that, through selection and sexual reshuffling, it can be recruited and organised into new adaptive genotypes. What we know of the nervous system and of the abundance of genetic variation permits us to make these assumptions. In chapter 3 we described how the nervous system has an ability to produce adaptive behaviours that is almost always greater than that which is normally realised. The neural basis for this plasticity probably lies in the capacity of the nervous system to reorganise as behaviour patterns become habitual during ontogeny. It is the ability of animals to behave in a plastic and adaptive manner that enables them to adjust to changes in the external world. Behavioural flexibility also allows them to compensate for, neutralise or circumvent the effects of many genetic variations. For example, a mother goat who has given birth to non-identical twins may compensate for the weaker offspring's feebleness by suckling it more often.[20] Since behavioural plasticity can mask both environmental and genetic variations, many genetic variations are protected from selective elimination and can accumulate. The net effect is a large reservoir of genetic variation underlying the organisation of the nervous system. This variation is exposed and recruited when the environment changes. A previously unrealised genetic variation may become advantageous if new conditions require learning radically new things. The sexual process brings together genes affecting many different facets of a learning ability (memory, perception, associative ability, attention and so on) and 'constructs' new behaviour from many different elements. This 'construction' does not depend on a rare chance mutation, only on the already-available but as yet unorganised genetic variation in the population. The learning process thus plays a dual role: it exposes new variations in the capacity to learn, and it creates the environment in which these variations are selected. Through classical Darwinian selection, genetic assimilation can lead to the conversion of a learnt behaviour into one that no longer depends on extensive experience. This is how 'instincts' evolve.

The 'instinctive', experience-independent aspects of nest-building that we discussed earlier in this chapter could have evolved through genetic assimilation of learnt behaviours. It is likely that, in the evolutionary past, some nest-building behaviour that is now innate, such as that seen in hand-reared canaries, was learnt. Over time, the ability to learn how to build nests efficiently, to perform the appropriate motor actions rapidly and smoothly, to select the right type of material and the right location, were all subject to natural and sexual selection. Of course, the nervous system of the birds must have already been equipped to some degree for these tasks: it can be assumed that learning to find somewhere that was protected pre-existed, and that birds had a limited ability to make the breeding place more comfortable for incubation by manipulating locally found building material. But selection would favour heritable variations affecting the organisation of the nervous system that made learning depend on fewer learning trials. The conditions that promoted learning were, as with the example of the fruit-fly's salt-induced morphological phenotype, both 'inducing' and selecting: among the birds who were able to respond to their environment by learning, those with the best learning ability and the most fitting morphological correlates survived and produced young. Since building nests is advantageous under most conditions, and the more efficient the building behaviour the better, selection for effective and efficient nest building has probably been a constant pressure throughout the evolution of birds. But not all aspects of nest building are expected to become equally invariant. The general structure of the nest is likely to be more uniform and less experience-dependent than the choice of materials from which the nest is made, since the choice of material depends on factors that are more variable and over which the bird has less control. The main point is that persistent, steady selection for the reliable, relatively error-free building of sound nests would lead to the genetic assimilation of those nest-construction behaviours that are associated with the general features of the nest.

Genetic assimilation of a learnt response is the simplest and the most elegant way to interpret the evolution of many 'instincts'. The 'instinctive' avoidance of predators is a good example. Being able to avoid common and dangerous predators is an obvious advantage. The more effectively and rapidly an animal can learn to avoid a particular predator, the better its chance of surviving. Consequently, through selection, the ability to avoid a predator may improve, so that it becomes dependent on fewer and fewer experiences of the predator, and eventually avoidance becomes an automatic response to a single exposure. The characteristic visual, auditory or olfactory features of the predator elicit an immediate avoidance response in the nervous system of the animal – the response becomes innate. The avoidance response of hand-reared spotted hyenas to the smell of lions, or of many small mammals and birds to hissing snake-like noises, may be examples of innate behaviours that have arisen through genetic assimilation.[21] Since in some places lions and snakes are dangerous and common predators, individuals who learn quickly and remember the sound, sight or smell that should be avoided have more chances of surviving, and the genetic constitution of these fast-learning individuals will be passed on to the next generation. After generations of selection for fast association between certain sense impressions and the danger, the avoidance response will become innate, with its expression depending on a single exposure to the danger-indicating stimulus. An interesting by-product of such selection might be that, although it leads to a highly stereotyped response, it also leads, through stimulus generalisation, to avoidance responses to animals with similar features, and to a general increased readiness to learn and be emotionally aroused by similar patterns.

The transmission of behaviours through social learning may accelerate the process of genetic assimilation. Behaviours that become traditions are often more enduring than those acquired through individual
learning, because the young learn the behaviours from their elders and
do not need to reinvent them. A tradition is therefore part of the persistent selective environment created by the social structure of the
group. The tradition itself accelerates and channels the selection of
genetic variations that have effects that simulate the adaptive, consistent aspects of the behaviour that has become traditional. It can hasten
genetic assimilation, and lead to the former tradition becoming a genetically fixed rather than a flexible behaviour.[22]

*The ecology and evolution of social and individual learning: evolution in
rapidly varying environments*
In the last section, we showed how learning, habits and traditions can
all affect learning-evolution and how, in a constant environment, innate
behaviours, which require little or no learning, can evolve through
genetic assimilation. We want now to turn to the conditions that affect
the way things are learnt and whether or not they are learnt. The most
obvious circumstances in which learning is beneficial are when environmental changes during the lifespan of the individual are frequent
and either reoccur or persist for some time. It is then worthwhile to
learn and remember.[23] The changes, however, must be varied enough
to exclude the evolution of a completely genetically assimilated
response. A good example of relatively frequent and recurring changes
that require flexible responses are seasonal changes in food types and
food availability. For example, the range of insect species available to a
great tit is completely different during spring from that in autumn, so
for each season the bird has to learn how to recognise, catch and handle different edible insects, and avoid poisonous and obnoxious ones.
Moreover, in the European winter, when insects are scarce, the diet of
great tits changes to seeds. Great tits learn to find and handle whatever
food is available by social learning from their parents, by individual
trial-and-error learning and, in winter, when several local populations
flock together, by social learning from experienced neighbours.

As we have stressed before, most forms of social learning and individual asocial learning are fundamentally similar. Social and asocial
learning differ largely in the context in which learning takes place,
rather than in basic learning mechanisms, so which type of learning a

species uses depends on many different factors in its environment and its evolved life style. To be able to learn by individual trial and error is advantageous in many conditions. Even in very stable environments, some things vary, often as a result of the activity of the animal itself. Individual, asocial learning often fine-tunes a 'fixed' behaviour, or one acquired by social learning. Trial-and-error learning is also a vital part of the exploratory behaviour of higher animals, a behaviour that has probably been very important in their evolution. Through exploratory behaviour, birds and mammals find out about the positive and negative aspects of the environment before being exposed to them in an already critical situation.

As with individual learning, social learning is also beneficial in most environments, because it decreases the risk of making individual mistakes. This is particularly true when individual learning is very risky, as it is, for example, when learning to recognise a new predator. When they are not innate, alarm responses are usually acquired through early social learning rather than by individual trial and error. Social learning is similarly advantageous when individual learning requires a lot of time and effort. For example, for many vegetarian animals there are foods, such as leaves, which are edible only for a very specific and restricted period. The animals that feed on these foods have to recognise that they are edible only during that window of time. This window may not even be simply seasonal – it may depend on the history of each individual plant, since frequent browsing can induce plants to produce poisons that make normally edible leaves inedible. Clearly, for animals that feed on such plants, social learning has advantages over individual learning. The same is true for many omnivorous species: the breadth of their diet means that they can live in many different niches, but it also means that many mistakes would be made if they relied solely on individual, asocial learning. With so many potential types of food, it is most unlikely that a strong innate predisposition to recognise each food type will evolve. The evolutionary solutions for species with broad diets seem to be either they are poison-immune, like the European hedgehog, and can therefore afford extensive individual learning, or they learn a lot about food socially, from parents and other informed individuals.

How diverse the environment is will influence the type of learning that is advantageous and the length of the learning period. Post-fledgling parental care in songbirds lasts 2 to 4 weeks in temperate zone species such as European tits and finches, but, for tropical insect-eating species in the equatorial forest of Sarawak, it can be as long as 10 weeks.[24] In the equatorial forest, many of the numerous insect species have specialised anti-predator devices, which make them either cryptic and hard to find, or repulsive and toxic. Moreover, species that are palatable are often mimics of the toxic ones. Foraging in such diverse, perceptually complex and possibly dangerous environments requires a lot of minimal-risk learning, which is probably why the birds have such a long period of parental care. The skills learnt from parents are then fine-tuned by individual learning.

The complexity of what has to be learnt is another factor affecting the evolution of the type and length of learning. In the oyster-catcher, Norton-Griffiths found that, when populations live in conditions where easy-to-catch worms are abundant, the chicks stay with their parents for only 6 or 7 weeks. In contrast, when mussels are the major food source, the chicks stay with parents for 26 weeks.[25] In the first case, learning from parents is probably less important than in the second, where a complex mussel-opening technique has to be learnt; the lengths of the chicks' periods of association with their parents may reflect this. If the environments where the complex behaviour has to be acquired were to persist, more efficient social learning might evolve, for example through chicks paying more attention to their parents' behaviours and developing better voluntary control of their muscles.

When there is ongoing selection for increased individual and social learning, the advantage of some highly stereotyped responses may be diminished. For example, learning how to hunt from mother is often more efficient than having an innate hunting skill, because it is easier to adapt to the availability of different types of prey and to transmit appropriate techniques to offspring. If replacing innate responses by learnt ones is beneficial, the young may become more dependent on parental care and on learning from other adults. This may enhance the selective advantage of a more cohesive family or group structure and lead to longer offspring dependency. At the genetic level, selection would mould the genome so that the innate behaviour is suppressed, while social learning is improved. The innate behaviour may not be lost completely, however, but remain for some time as a 'backup system', which is expressed and exposed to selection only under very unusual conditions.

There are, of course, circumstances that do not allow much social learning. The pup of the hooded seal is weaned in just four days.[26]

During this short time, the mother pumps a remarkable amount of extremely energy-rich milk into her young, but very little direct behavioural information can be transferred in four days. The young seal relies on innate behaviours and individual learning. On the hazardous floating ice on which the seal is raised, the price to pay for a period of weaning that is long enough to enable a lot of social learning would be very high. Usually, however, birds and mammals living in complex environments rely to some extent on socially transmitted information. Whenever social learning consistently and significantly reduces the cost of asocial learning, social learning is likely to become prominent.[27] The interaction between social and asocial learning is usually complementary. Individual learning rarely erases or nullifies the effects of social learning, and, although it may modify what has previously been socially learnt, the modification is often a variation on the theme rather than a radical change. What is learnt is rarely completely unlearned.

The ecology and evolution of transgenerational transmission: evolution in environments with intermediate-length cycles
There are many environmental conditions in which social learning is likely to evolve, because there are many conditions in which it will decrease the costs of pure individual learning. However, social learning does not always lead to the establishment of traditions. For example, although young may learn from their parents which food is best and where it is, the availability of food may change during the youngsters' lives, so they will not pass on the same information to their own young. Nevertheless, it is easy to see why in some circumstances it is beneficial for socially learnt information to be transmitted to the next generation. When asocial learning involves some cost (in terms of time and mistakes), and the environment fluctuates slowly, so that the conditions in which the learnt information is beneficial last longer than the generation time of the animal, transgenerational transmission will be an advantage. The fluctuating conditions that favour transgenerational transmission of acquired information have been called ILC (Intermediate Length Cycle) environments.[28] Examples of ILC conditions are periods of drought and cold that last longer than the generation time of the animals experiencing them, or changes in the availability of different types of seeds to seed-eating rodents, which varies every several rodent generations. In such conditions, genetic fixation of appropriate behaviours through genetic assimilation will not occur, because the environment

switches between different selective regimes too often. Since individual asocial learning is costly, the best strategy is for the naïve individual to learn the appropriate response from an experienced one – for it to 'inherit' the adaptive behavioural phenotype. When the environment switches to different conditions, it is possible to alter the socially transmitted pattern of behaviour by individual learning, although this may incur a considerable cost.

Such changing, ILC environments are very common. Consider the almost century-long ecological succession from shrubbery to dense forest and back to shrubbery that is found in the Mediterranean type of ecosystem. Many of the species living in these areas are experiencing an ILC environment, in which the transmission of information through social learning is beneficial. The effect is more pronounced in the short-lived bird and mammal species, where a long series of generations will live in a similar environment. Because the young will probably experience an environment similar to that of their mother, the fitness of both will usually increase if mothers transmit to their offspring materials and information that lead to phenotypic similarity. In general, ILC conditions will lead to the evolution of all kinds of ways of transmitting information between generations, including, for example, maternal transmission through the egg and milk. The reliability of transmission and the number of complementary channels used are expected to increase in ILC conditions. The evolved stability of transmission will be related to the length of the cycle: the longer the environmental cycles, the more stable the transmission of the information. When environmental cycles are very regular, selection may lead to rather sophisticated information transmission. In insects that live for just a few weeks but have several generations a year, mothers may produce offspring that will live and breed in a different season from that in which they themselves lived and reproduced. In some species, evolution in these regularly fluctuating conditions has led to the mother being able to respond to predictive environmental cues by switching the pattern of development of her eggs in anticipation of the coming environmental change.[29] Thus, the information these species transmit varies with the season.

The receding horizon: the evolution of altered and enhanced learning

Identifying the environments in which individual learning, social learning and traditions are advantageous is only the first step towards understanding their evolution. We still need to know how the evolutionary processes actually proceed, and how learning from experience affects the genetical evolution of social and individual learning. We are going to argue that the evolution of asocial and social learning, just like the evolution of innate behaviours, is guided by the genetic assimilation of learnt habits. However, the types of environment that lead to evolutionary changes in learning ability are different from those leading to innate behaviours. When we discussed the evolution of innate behaviours, we assumed that, in constant environments, learning is 'internalised' through genetic assimilation, leading to quicker, more reliable and efficient performance of the entire, previously learnt, patterns of behaviour. The evolution of such learning-independent behaviours is a kind of specialisation: a plastic, learnt response became rigid; the range of possible responses is narrowed and channelled to become highly specific, stereotyped and automatic. In environmental conditions that vary, this cannot happen, but genetic assimilation can lead to new types of learning, and even to a better ability to learn. We are going to discuss three cases: assimilation that leads to a longer sequence of acts within the same behavioural scheme; assimilation that leads to a shift in what is learnt, from information received through one sensory modality (e.g. the visual system) to information received through another (e.g. the auditory system); and finally, assimilation that leads to a generalised, rule-dependent ability to learn. For simplicity, we shall call the first case sequence-lengthening, the second modality-shift, and the third rule making.

Lengthening the behavioural sequence – the assimilate–stretch principle
To see how a behavioural sequence can be lengthened without altering learning ability, we will start with an imaginary scenario. Imagine that a bird is able to learn a sequence of four consecutive actions, for example four actions that culminate in the building of a simple nest. Assume that there is some constraint on the learning capacity of this species, so its learning ability is unlikely to improve. However, there is consistent selection for efficient and reliable nest building, so slowly one of the steps in the sequence of behaviours becomes genetically assimilated – it becomes innate. The bird now needs to learn only three steps, and will construct its simple nest much more efficiently. However, since its learning ability has not been generally handicapped, it can add a

new adaptive behaviour, an additional learned action, to the remaining three; for example, it may learn to tie the nest to the branch with plant strips, so it is less likely to fall when the wind blows. We can now observe five consecutive actions, one of which is innate. If building nests rapidly and efficiently continues to be advantageous, another previously learnt action may become assimilated, and yet another learnt one can be added, so that the behavioural sequence is lengthened by yet another step. In this way, it is possible to gradually lengthen the sequence of actions without changing the capacity to learn: the genetic assimilation of previously learnt behaviours 'frees' the individual to learn additional actions, without extending the limits set by its learning capacity. Of course, it is not necessary to assume that a particular action is completely assimilated and is produced upon a single exposure to a stimulus; it is enough if the number of trials required for its effective performance is significantly reduced.

We call the evolutionary process just described the assimilate–stretch process, because genetic assimilation enables the lengthening of a behavioural sequence by making parts of it more automatic. This assimilate–stretch process may underlie the evolution of many of the long sequences of behaviours that have both innate and learnt components, such as nest building or birdsong. As formerly learnt behaviours are transformed into innate behaviours, and as new learnt behaviours are added to the sequence, the overall number of learnt elements in the sequence may be preserved.

It is possible to interpret the evolution of nest-building behaviours in some species of swallows in the light of the assimilate–stretch idea. Cliff swallows build mud nests.[30] After choosing an appropriate nesting site, which is usually a vertical surface beneath a ledge, the mates take turns to collect pellets of mud, one bird remaining at the chosen nest site while the other is away mud-collecting. The first pellet is placed about ten centimetres below the overhang of the nest site, and slowly the birds form a narrow line of mud, which is gradually built up into a crescent-shaped ledge. The birds perch with their feet in the centre of the prospective nest, and reach laterally or forward to place the pellets on the rim. They then extend the mud crescent into a rounded half-cup. If there is no cliff overhang or it is irregular, the lateral walls of the nest are extended until they meet to form a mud roof. Lining material is gathered and put into the nest, and at this point egg-laying starts. The nest is built out into a wide-mouthed retort, and sometimes an entrance tunnel is added. The opening of the nest is narrowed to a small circle, with the entrance directed away from the nearest neighbour's nest, which presumably minimises territorial disputes. The nest-building process takes an entire week; it has to proceed slowly, for each mud pellet has to dry and harden to form the basis for the next step. Nests that are built too rapidly or in wet weather often collapse before they are completed.

We know very little about the role of learning in this long and complex series of nest-construction activities. The variations found in cliff swallow nests and nest-building suggest that local conditions affect the way the nest is made: old nests in good condition are reused, and the birds may repair old nests; sometimes they roof their nests, but not if they are under an overhang; nests built in natural crevices are given shorter projections of mud at the entrance. Nevertheless, in spite of the variations, the behavioural sequence seems to be very stable. In fact, the nest-building practices of different species of swallows are so stable they have been used for the construction of phylogenies.[31] Although some learning may well take place, construction seems strongly channelled to produce a very stereotyped end-product.

So how did such a lengthy behavioural sequence evolve? It is very likely that genetic assimilation had a role in this strong channelling, but it could have happened only gradually. It is very improbable that the whole behavioural sequence was once totally learnt, because learning by trial and error at each stage would be extremely time-consuming and inefficient. It is also unlikely that the whole sequence became assimilated all at once, since many interacting sets of genes must be involved. It therefore makes much more sense to assume that evolution through genetic assimilation was piecemeal: as one part of the sequence became completely or partially assimilated, new learnt behaviours were added, more genetic assimilation occurred and so on. If 'additions' of learnt behaviour did indeed occur, then we expect to find that the phylogenetic series of swallow nests will recapitulate the ontogeny of nest building. In the mud builders in the swallow family we do see this: the simplest nest, built by the barn swallow, is an open mud cup; the next most complex one, a covered mud cup, is built by the house martin; the most elaborate mud nest, built by the cliff swallow, has an entrance tunnel as well. The ontogeny of the nest of the cliff swallow follows the same stages: open mud cup, covered mud cup, and then covered mud cup with a tunnel added to it. It is not unreasonable to assume that, during the evolution of the cliff swallow, each stage of nest construction was at one time learnt, and then became partially genetically assimilated, allowing the addition of further stages of construction (the stretch phase).

Similar processes of assimilate–stretch may underlie the evolution of singing in songbirds, where the mixture of learnt and innate behaviours is well characterised. Parts of the song that were formerly learnt may have become genetically assimilated and innate, thus allowing the elaboration and sophistication of the song by the addition of new learnt parts. In fact, many behaviours that consist of a long sequence of actions, some of which are highly stereotyped, probably evolved through such a process of genetic assimilation followed by stretching.

The evolution of new ways of learning: switching modalities
A second way in which genetic assimilation can affect learning, the modality-shift case, is through the most important sensory modality changing as evolution proceeds. Initially, learnt responses to stimuli received through one modality (e.g. auditory) are enhanced as a result of the disuse of another (e.g. visual). Think about the Palestine mole rat, which lives in subterranean tunnels and is deprived of visual stimuli. In this animal, the eyes are greatly reduced in size, and skin and fur grow over them. The animal is practically blind. However, the Palestine mole rat still has the brain regions that seeing mammals use for processing visual information. These regions have not become degenerate like the eyes, but have been recruited to process auditory information. Hearing is extremely important for Palestine mole rats, because it is through sound and vibrations that they become informed about their biotic environment and communicate with each other underground. Anatomical and physiological analyses of the mole rat's brain have shown that projections of auditory nerves have spread out and invaded the 'free' (non-used) visual centres. The visual cortex of the Palestine mole rat now processes auditory information.[32] This was probably possible because parts of the visual cortex in mammals are normally bimodal anyway, with some auditory neurons going into the visual cortex. The process of 'compensation' and recruitment of brain areas deprived of normal connections has also been simulated in the laboratory. Experiments involving complex surgical manipulation in newborn ferrets, hamsters and rats have shown that it is possible to re-route the nerve projections of one sensory modality into brain areas normally

processing a completely different one. During the evolution of the Palestine mole rat, natural compensatory rerouting and rewiring of the brain have accompanied the gradual degeneration of the eyes.

This process was probably initiated by a change in habits. It is most unlikely that a beneficial mutation made the Palestine mole rat blind and at the same time also adapted it to live and thrive in subterranean tunnels. It is much more plausible that initially individuals in a lineage of seeing Palestine mole rat ancestors changed their way of life, and for good adaptive reasons, such as protection from predators, began to spend more time in subterranean tunnels. This change in habits led to changes in the relative adaptive value of many aspects of their behaviour, anatomy and physiology. For example, during a life of vigorous digging underground, vision is not very important, while the presence of unprotected eyes exposed to soil rich in bacteria and fungi is a dangerous source of infection. Since the eyes were little used and were a health hazard, their degeneration was beneficial. At the same time, the significance of hearing grew, since this sense was now more important in enabling the animals to monitor the world around them and communicate with each other. We know that the final pattern of the wiring of a mammalian brain is affected by experience, mainly early experience: preventing input into a sense organ during early development sometimes results in the degeneration of the normal neural connections. Since the eyes of mole rats living in complete darkness were not used during early post-natal development, the neural connections from the retina to the visual areas of the brain probably degenerated. Individuals deprived of vision adapted through extensive use of their auditory system, and selection for ever more efficient learning from what is heard rather than what is seen must have been ongoing. Those individuals who were more able to recruit the unused visual region for auditory processing survived better, and, gradually, increased auditory learning became genetically assimilated. Selection operated on pre-existing genetic variations in brain plasticity and the mechanisms that allow 'unused' brain regions to be used by another modality, but, of course, these genetic variations only became exposed to selection through the changes in the Palestine mole rat's habits. The disuse of the visual sensory modality, and the increased use of the auditory modality, led to the degeneration of the eyes, to the rewiring of the brain and to the increased use of auditory information. The Lamarckian idea that use and disuse are important guiding forces in evolutionary adaptation is certainly correct, although, unlike Lamarck and Darwin, we believe that use and disuse affect the selection pressures rather than the generation of genetic variations.

The evolution of the rules of learning
Genetic assimilation can also lead to an increase in general learning capacity and to rule making. This is likely to happen when the environment changes over an intermediate (ecological) time-scale, or when selection gradually gets weaker over generations. In order to see how such circumstances can promote the evolution of a greater ability to learn, and lead to the evolution of general strategies of learning, we will look at what happens to a domestic mouse population that is exposed to a new predator. At first, individuals will learn about the predator through their own experiences, but later they may learn from others, particularly their mother. Slow-learning individuals are likely to get caught; it is those who learn more quickly, both through individual and social learning, who will survive. But what does 'those who learn more quickly' actually mean? The response to the predator is not yet stereotyped, so it may mean the mice who learn to associate the smell of the predator with danger more efficiently because they are good at generalising on the basis of experience; or it may mean those whose long-term memory is better, or are more attentive or begin learning at an earlier age. These are rather general features of learning capacity, and it is clear that selection for some such features must be going on. Now think of what would happen if the predator disappeared after several mouse-generations, or the selection that it exerted weakened because through better learning the mice had become better protected. The avoidance response to this particular predator would not have been assimilated, because there was not enough time, but the advantages accrued through selection as the mice adapted to this predator may well remain. The associative ability, the memory and the attentiveness of individuals will all have improved. Since the improvement involves general aspects of learning, individuals in the selected mouse population will now be able to generalise to similar predators more easily, memorise olfactory patterns and co-ordinate them with motor responses better, and generally be able to exploit more aspects of their environment.

It is easy to see how this can lead to a runaway process of selection for increasingly better learning ability. The more readily an animal learns, the more learning opportunities become available to it, because increased learning capacity makes the individual aware of more aspects of its environment. As more aspects of the habitat become information rather than 'noise', the perceived environment of the individual expands. As the niche expands, there is selection for a better ability to exploit it, often for an even better learning ability. The stronger the selection for improved general learning ability, the richer the perceived environment, and so on. At each point in time, there is selection for something well defined – for recognising a particular predator, exploiting a new food source, improving nest-building skills. However, since selection here is never long-term, the genetic variations promoting learning will not combine in ways that lead to a highly channelled development of the nervous system and to a highly specialised response. Thus, although selection is always highly specific, its consequences may be very general. The more the animal learns, the more the 'target' of learning moves. The selective 'horizon' constantly recedes into the distance. The learning ability and, hence, the phenotypic plasticity of the individual evolves continuously, often in unexpected directions; the 'target' moves forwards, sideways, in all the dimensions of the increasingly self-constructed niche.

One outcome of the type of selection just described is biased categorisation. The consistent association between some types of stimuli and some types of adaptive responses may lead to the construction of a new perceptual category. In chapter 3 we described some experiments by Garcia, who showed that rats can learn to avoid tastes and smells that are associated with later gastric problems more easily than they can learn to avoid clicking sounds. From an evolutionary point of view, this bias makes sense, since food-poisoning is associated with tastes and smells, not with sounds. If during evolution several different food types were associated with food-poisoning, each for a limited time, an association between a *particular* food and poisoning would not evolve to become innate. However, a very general bias for associating taste or smell with a subsequent gastric problem could become established through *partial* genetic assimilation. A general rule: 'avoid any food (whatever its specific taste and smell) if, after eating it once, you feel sick', could become established. The evolution of this rule would have entailed the genetic assimilation of the link between the avoidance response and the generalised experiential category.

Genetic assimilation can also result in the evolution of response generalisation, in which an existing behaviour becomes associated with new behaviours. For example, when group size increases beyond the narrow family group, behaviours that evolved in family conditions are often recruited and associated with other behaviours that are important in group-life. Behaviours such as the rituals of wolves before the hunt are similar to the behaviours of young pups towards their caregivers, but they have been modified and associated with new patterns of behaviour, so that they appear in a completely different situation. Courtship feeding in birds is another example – it is very similar to chick-feeding behaviour, and there is little doubt that it has been derived from it, but it has become associated with sex and mating, and now typically elicits copulatory behaviour in females. It is reasonable to assume that initially the use of established behaviours in a new situation was learnt. The link between the established behavioural act and the new social context would then have evolved to become more reliable through genetic assimilation.

The interplay of the evolution of social life and the evolution of learning is extremely complex, but important. Selection, both for better learning and for the ability to transmit what has been learnt, occurs in the same type of conditions – in ILC environments, where changes are slow relative to the lifespan of the organism, but fast relative to the time required for genetic fixation. It is therefore not surprising that increased sociality, as crudely reflected in group size, is well correlated with increased intelligence, as reflected behaviourally in the ability to solve ecological and social problems in better and more diverse ways, and anatomically in the relative size of neo-cortical areas.[33] However, there is probably more to this correlation than simply that sociality and intelligence are both adaptive results of selection in the same type of environment. Group living affects learning evolution, and learning affects group evolution. Suppose that there is selection for increased group size because intense predation encourages the formation of larger family groups, or success in foraging is improved when groups are larger. The enlarged group will present a new challenge to individuals, because their social environment has expanded – they now regularly encounter, and have to interact with, more individuals of their own species. For an individual, others of the same species are probably the most complex facet of its environment, because it is better equipped to perceive and respond to the behavioural acts of members of its own species than to anything else. The kind of intelligence likely to evolve in group-living birds and mammals is therefore mainly social

intelligence – intelligence that enables it to manage intricate social relations. Individual recognition of other group members and appreciation
of their social status and past behaviour become important. As we
described in chapter 8, animals manage relations with other group
members by using behaviour patterns and emotional responses that
evolved in the context of mate relations or parent–offspring relations.
Identification with the group, suppression of aggression and enhancement of co-operation are brought about by the generalisation of stimuli
and responses derived mainly from the family context. Once the group
size increases, a runaway selection process, leading to increasingly better social learning and to greater social intelligence, can begin. More
sophisticated social intelligence allows the formation of larger and more
complex groups, and large and complex groups select for better social
intelligence. The target of selection very obviously changes during social
evolution.

Selection through habits: the co-adaptation of genes and traditions

In the last section, we made the case that learning guides its own evolution. We now want to look at how learnt habits and traditions can
not only direct and shape their own evolution, but also guide the evolution of morphological traits. The role of persistent patterns of behaviour, customs and 'cultural' traditions in the evolution of morphology
has been discussed repeatedly for more than a century, ever since
Wallace advanced his ideas about the relationship between habits and
structural evolution. Eminent biologists and psychologists have argued
forcefully and convincingly for the central, directing role of learnt
behaviours in the evolution of morphology and physiology.[34] Yet, despite
their eloquent efforts, behaviour has not received the general recognition it deserves as a major factor in animal evolution. Perhaps the main
reasons for this is the tendency for twentieth-century scientific explanations to be based on bottom-up causative chains – events at higher
levels of organisation (the evolution of behaviour) are explained in terms
of the activity of units at a lower level (cells, genes). The opposite
approach, which starts as we did from learnt behavioural phenotypes
(the higher level) and explains how changes in genotypes (the lower
level) are guided by phenotypic behavioural adaptations, is pursued only
reluctantly. Yet, this bias in modern thinking involves a one-sidedness

that obscures important evolutionary mechanisms. At all times, genes and behaviour are acting simultaneously, and recognising that genes and behaviour interact *in both directions* allows a more balanced and better understanding of evolution than either a dogmatic bottom-up or an equally dogmatic top-down approach. Another obstacle to accepting that behaviour can play a significant role in guiding evolutionary change has been an uneasiness about the dreaded Lamarckian ghost, as well as a lack of recognition of the principles of genetic assimilation. Since we believe that the ghosts of the past are no longer so menacing, and the importance of genetic assimilation in evolution is increasingly being acknowledged, the situation may change.

It would be quite impossible to review here the legions of cases pointing to the effect of habits on the evolution of morphology, physiology and correlated patterns of behaviour.[35] We have already mentioned some, such as Wallace's explanation for the effect of learnt nesting practices on the evolution of feather coloration in female birds. A similar example is the evolution of drab colours in male bower-birds. As the males learnt to build more elaborate bowers, the direction of sexual selection was shifted from the beauty of the male's coloration to the splendour of his bower. This decreased the importance of male colour as an indicator of excellence, and enhanced the selective importance of protective coloration. Clearly, behavioural changes initiated this process, and the evidence that there are local styles of bower building suggests that evolving traditions may have been the driving force that led to the loss of conspicuous male coloration. In the previous chapter, we also discussed how imprinting and traditions could lead to speciation and have incidental effects on the evolution of morphological traits. However, rather than cataloguing examples, we will first outline the way in which habits and traditions may guide the evolution of morphology, physiology and other behaviours by speculating about an already familiar case. We will then go on to examine some evolutionary effects of social learning that have not been discussed extensively by other authors.

In order to see how a new tradition can affect the cultural and genetical evolution of a large set of characters, we are going back to the black rats in the Jerusalem-pine forests.[36] In chapter 4 we described how these rats, which have extended their range to include the Jerusalem-pine forests, completely changed their diet and life style and began feeding exclusively on the pine seeds that are enclosed within inedible pine-cones. The rats have developed an elaborate cone-stripping technique to reach the seeds, and this new technique is socially learnt. The cone strippers have spread in the northern Israel Jerusalem-pine forests in just a few years. As far as we know, at present the rats have no genetic adaptations that are specifically related to the new diet and foraging method, or to the many other changes in life style that the new diet has engendered. However, imagine that the Jerusalem-pine forests survive the next millennium of anticipated ecological catastrophes, that the black rats' new life style persists, and that we can come back to look at the rats a thousand years from now. The continuous use of pine seeds as a major source of food will have given adaptive value to any genetic variations in morphology or physiology that improve the finding, processing or digestion of pine seeds, or improve the ability of the rats to learn the appropriate behaviours for doing so.

Coming back to the Jerusalem-pine forests after a thousand years, we find that the black rats have become better adapted to their niche. A comparison of the digestive enzymes of past and present populations shows some subtle differences. Since the pine seeds became an exclusive source of food, enzymatic variants with different rates of synthesis or activity have been selected, so the gene frequencies have changed – an evolutionary process somewhat reminiscent of that which led to changes in the frequency of lactase-I enzyme variants following the domestication of cattle by humans. The morphology and musculature of the jaws, limbs and body have undergone some modifications to accommodate the stripping behaviour and the tree-bound life. Selection for greater dexterity, especially of mouth and front paws, has led to changes in musculature and in the motor areas of the brain. Since the black rats have encountered new types of parasites, we see a change in the frequency of histocompatibility alleles. We also see new adaptations to the prevalent predators, and some of these adaptations are very efficient, appearing very early in life and requiring only a little learning. We are not very surprised to discover that, in some of our futuristic black rat populations, acquiring the cone-stripping technique no longer depends on weeks of learning, but is achieved very much more rapidly. Clearly, those rats who learnt more quickly had been selected. New habits have also appeared: in some populations, the hoarding of cones is observed. In general, we see changes in behaviours and morphology that converge somewhat with the adaptations of squirrels, who live in similar habitats.

Alister Hardy believed that broader patterns of convergent evolution, such as that between marsupial and placental mammals, as well as specific cases, may result from processes of selection that are guided by similar habits.[37] He also interpreted the dramatic divergence of mammals following the extinction of the dinosaurs in terms of habit-influenced selection. The decline of the reptiles led to new opportunities for the small mammals, which multiplied and competed with each other for food. Their exploratory behaviour led them to try out many different styles of life, to adopt habitual behaviours such as burrowing, swimming, running, climbing, jumping and so on. The new habits selected for correlated behavioural, morphological and physiological traits. They moulded the bodies and the minds of the pioneering mammals, leading to the evolution of the major adaptive types. Further changes in life styles led to further selection and diversification.

Increased behavioural sophistication is, however, not the only evolutionary product of behavioural selection. The geneticist Helen Spurway looked at the other side of the coin – at selection for the breakdown of behavioural rules and traditions.[38] She noted that, during the domestication of animals, man has been selecting behaviours that are convenient for him without regard to established animal rules. To encourage increased reproduction, man selected for the slackening of mating criteria, a shorter period of parental care, reproduction at earlier ages, unresponsiveness to group hierarchy and group rules, and so on. Preexisting behavioural plasticity was exploited: man selected animals with behaviours that were indiscriminate and rule breaking. We have seen the devastating effects of this selection in groups of feral dogs who cannot rear young, and can maintain their group only by recruiting new adult members.

Hardy, Ewer and others looked at many broad patterns of behavioural and morphological evolution, and explained them as arising principally through habit-guided selection. Their arguments strongly suggest that learnt behaviours have been an important selective force, affecting many different aspects of animal evolution. We now want to extend the type of arguments they developed by presenting some conjectures of our own about the effects of maternally transmitted learnt information on the evolution of sex-specific behavioural characters and cognitive compatibility between mammalian mothers and their young.

Some evolutionary effects of maternally transmitted learnt behaviours

In previous chapters, we have noted that in most mammals the male's only contribution to the production of his offspring is his genes. In the absence of paternal care, the young receive most of their early acquired information from their 'single mother'; they receive nothing from their father. In effect, the mother clones her cultural phenotype, since through social learning her offspring become behaviourally similar to her. Furthermore, if as well as transmitting behaviours a mother provides developmental and ecological legacies for her offspring, this heritage will help mould the type of environment in which young live, and therefore the type of selection they will experience. The result of these maternally transmitted non-genetic legacies is that females affect evolutionary change more than males. In particular, females can affect the phenotypic evolution of behaviour, and through this affect the genetic evolution of behaviour and other traits. The influence of the 'single mother' may even go beyond this: exclusively female cultural transmission may lead to either more extreme sexual dimorphism or, paradoxically, increased similarity between males and females.

In order to see how uniparental transmission can lead to the natural selection of sexual dimorphism, let us assume that the mother alone socially transmits behavioural phenotypes, and there is no way that the male can do so. The mother transmits some patterns of behaviour – for example particular food preferences and matching foraging skills – to both male and female offspring, but only the daughters transmit it to the next generation. This means that any alteration in the behaviour that a male makes, however wonderful and adaptive, will not be passed on. Although male offspring may benefit from the adaptations of their mothers, they are not transmitters, so cultural evolution of the behaviour through males is not possible; it can occur only through females.

Now let us assume that there is a difference in male and female life styles that is relevant to this transmitted behaviour pattern – let us assume that females are philopatric, while the males disperse widely and occupy different habitats. The daughters will benefit greatly from the information they acquired from their mother, because what they find as adults in the natal territory will match the preferences and the skills that they acquired from their mother. Cultural evolution of these preferences and skills through the female line is therefore likely. However, it will not necessarily be to the benefit of the young males to acquire such strong preferences and become committed to the maternal ways, since a mother's knowledge and skills may not be of much use to them once they disperse. Indeed, too strong a preference and too ingrained a skill may be positively harmful, detracting from their ability to adapt to different habitats following dispersal. In such a situation, selection may proceed in opposite directions in males and females. The females, who are the transmitters of behaviour, will become better and better adapted behaviourally to their local environment, since adaptive traditions will evolve. No such cultural adaptation is possible in the males. Since they are not transmitters of behaviour, they cannot respond to the maternally transmitted behaviour by building up adaptive, male-specific, cultural traditions. The males can only respond genetically. Several evolutionary outcomes are now possible.

First, genetic variations that are congruous with adaptive female-traditions will be selected in females, resulting in a reinforcement of female-specific behaviours. In males, on the other hand, selection will favour genes that lead to the suppression of female-specific adaptive behaviours. Since the behavioural differences are sex-specific, one of the outcomes of selection is likely to be that the genes affecting the behaviour that is an advantage only to females come under the regulatory influence of the sex hormones. In the jargon of genetics, gene expression becomes 'sex-limited'. Sex-limited expression of genes affecting sex-adaptive behaviour would lead to the evolution of more stable, and possibly also more extreme, differences in the behaviour of males and females. It would be accompanied by physiological and morphological changes that complement the behavioural differences. In this way, the evolution of sexual dimorphism may be a consequence of natural selection, rather than sexual selection.

A second evolutionary option, when it is a disadvantage for males to inherit female-specific behavioural traditions, is for males to be weaned earlier than females, and spend the remaining pre-dispersal period either on their own or in male groups. The general information passed on from the mother to her sons in this case would be less than that passed on to her daughters. This would be true not only with respect to foraging skills and food preferences, but also for social information. Information about who is related to whom, who is dominant to whom, and so on would not be transmitted as comprehensively and reliably to the males, but, of course, it is not very relevant for them anyway if they soon disperse to new groups. Therefore, if males are weaned earlier than females, differences in social behaviour as well as differences in foraging may become emphasised, and once again greater behavioural dimorphism may be accompanied by evolutionary changes in any correlated morphological characters.

A third evolutionary route is for males to become more like females – to find habitats as similar as possible to the natal one, or even to alter their behaviour and remain in the natal territory. In this case, greater similarity in the life styles and behaviours of males and females will evolve, although the females will still be the only transmitters of traditions. A fourth way is for males to become transmitters too – to begin to care for the offspring and thereby help determine what is transmitted. Then they can transmit male-specific traditions and cultural adaptations to their sons. Both convergence and divergence of behavioural traits of males and females may thus evolve from an original state of exclusively maternal transmission.

There is another possible effect of exclusively maternal transmission. One of the things that is necessary for effective early learning is for the offspring and the parents to be tuned to each other. A young animal has to be attentive to the behaviour of the parent and be able to learn from it. The most obvious way to achieve this is by behavioural means, through increasingly early and more effective conditioning using many complementary routes. When the mother alone transmits information to the young, her behaviour may reinforce this. In many mammals, a pregnant female at some stage isolates herself from other individuals, either by finding a private, secluded place, or by staying where she is and driving everyone else away. This voluntary isolation usually ends when the young are weaned. There are several good adaptive reasons for this behaviour: it allows the establishment of effective recognition between the mother and her offspring, and provides both mother and young with safety from cannibalism, predation and disturbance.[39] But the behaviour also enhances the mother's impact as a 'teacher' at a time when the capacity of her offspring to learn is at its peak. We certainly do not want to argue that this is the original function of this behaviour, but the value of isolation is increased if it also leads to the better social transmission of an integrated and coherent package of learnt information.

When the mother is the major source of learnt information for the young, any mechanisms that increase the 'mental' understanding between them, keeping the youngsters tuned to their mothers, will be advantageous. The greater the similarity between the cognitive capabilities of mother and offspring, the greater the effectiveness of early learning. We therefore expect evolution to lead to the maximisation of cognitive similarity between mother and offspring. If genes that affect learning and attention are active when inherited from the mother, while the corresponding genes from the father are suppressed, the cognitive similarity between mother and young offspring will be increased. Studies of genomic imprinting in mammals suggest that something like this may have happened during evolution.[40] Generally, expression of maternally derived genes enhances the growth and development of the brain, while expression of paternally derived genes enhances the growth of non-neural tissues. Moreover, the areas of the brain in which maternally derived genes are expressed and the genes derived from the father are inactive are in the neo-cortex, especially the frontal cortex. These neo-cortical regions are the brain regions responsible for high cognitive functions – for learning and forward planning. In contrast, in the brain regions that affect feeding and sexual behaviour and are under strong hormonal influence, the paternal genes are preferentially expressed. This pattern of differential gene expression in the brain is that which would be expected if early maternally guided social learning has been a major force in brain evolution in mammals: strong selection for early maternally guided learning, and for cognitive similarity between mother and young, has led to the exclusive expression in offspring (both daughters and sons) of maternal genes in the regions of the brain that are involved in learning, attention and forward planning.[41]

But this is not the end of the story. The regions to which the maternal genome contributes relatively more, the so called 'executive' regions in the neo-cortex that are responsible for the high cognitive functions, become increasingly important in primate evolution: the relative size of the neo-cortex increases. The so-called 'emotional' regions, where the paternal genes are expressed, have become relatively smaller. It therefore seems that the role of intelligence in the primate lineage has grown, and control of sexual, feeding and parental behaviour has been shifted to the 'higher' cortical regions. We have already noted that the relative size of the neo-cortex is correlated with group size: as the size of the group increases, the size of the neo-cortex increases too. However, studies relating the size of the group to neo-cortex size show that it is the size of the *female* group, not the male group, which matters in primates! In most primates, the females are the stable nucleus of the group, the caregivers and those who engage in complex long-term social relationships, so it is the size of the female group that exerts the greatest selection on the evolution of intelligence.[42] It therefore seems that, in mammals, the evolution of learning and intelligence is driven mainly by females. First, selection for learning is driven by the early care and transmission of learnt information from mothers to offspring, and then it is driven by the cohesive and complex social structure of female-bonded groups. Which things are learnt, how they are learnt and from whom they are learnt, are important factors in the evolution of brains, intelligence and morphology.

Blaise Pascal, the great seventeenth-century French philosopher and mathematician, reflected about the relationship between habit and nature, and worried about the facile distinctions usually made between them. In a famous aphorism he said: 'Habit is a second nature that destroys the first. But what is nature? Why is habit not natural? I am very much afraid that nature itself is only a first habit, just as habit is a second nature.'[43] The student of the evolution of behaviour can safely endorse Pascal's idea: through genetic assimilation, habits become nature, and nature becomes a habit.

Summary

Learning guides genetic evolution in many different ways. It affects its own evolution, with the result that different types of learning are used in different environmental conditions. In constant environments, genetic assimilation of learnt behaviour may lead to behaviours becoming increasingly less dependent on experience, and finally culminate in the evolution of innate 'instinctive' behaviours. In frequently changing environments, where genetic changes are not fast enough to track recurring change, individual and social learning are more beneficial. Transmission of socially learnt behaviour across generations is a selective advantage in environmental conditions that fluctuate at an intermediate rate, which is slow relative to the lifespan of the organism, yet too fast to allow adaptation through genetic change. Such slowly changing environmental conditions may lead to the evolution of increasingly more effective ways of transmitting acquired information and behaviours to the young.

What animals learn, and the way they change their environments through their learnt behaviours, moulds their own evolution. We described three ways in which learning can shape the evolution of behaviour and subsequent learning. Learning can lead to: the evolution of a longer behavioural sequence; an evolutionary change in the use of sensory modalities; and the evolution of general rules of learning. In all three cases, evolutionary change occurs because learnt behaviours are partially genetically assimilated.

Many biologists have argued that learnt habits can drive morphological and physiological evolution. Some of the effects can be quite subtle, but very important. For example, when only the mother cares for the young, and most early learning is therefore through her, this can affect the evolution of behavioural and morphological difference between males and females. Maternal transmission of behaviours may have driven the evolution of intelligence in primates. In all higher animals, learnt behaviours have probably guided many aspects of genetic evolution.

Notes

[1] In the beginning of his review of the structure and function of the song of European starling, Eens (1997) writes: 'studies of the development of bird song have made clear that behaviour patterns cannot just be labeled 'innate' or 'learned' but arise through an intricate interplay between the two'.

[2] On page 236 of the same 1870 essay, Wallace makes it clear that by 'hereditary or imitative habit' he does not mean 'instinct', but 'persisted imitative habits'.

[3] Wallace's view would not be accepted by most evolutionary biologists today. First, his assumption that nest-building practices are largely learnt and culturally transmitted seems to be exaggerated, since there is evidence for an innate component in nest building. Second, his assumption that female birds lost rather than acquired their bright colours is the opposite of the opinion held today, which explains the bright colours of the male as a result of sexual selection, and drab protective coloration as the default state. Wallace believed that sexual selection also operates on females, with males preferring brightly coloured females. Therefore he thought both sexes should be brightly coloured, unless more pressing selection, such as differential predation pressure, changes the situation. Darwin, too, believed that features (like colour) would be transmitted to both male and female offspring, so females would become colourful if they liked colourful males, and their colour would be suppressed from this default state by natural selection. Although Wallace's opinion about the evolution of female coloration is not accepted as a general explanation, it is nevertheless applied to some species of birds (Burns, 1998; Martin & Badyaev, 1996).

[4] Collias & Collias, 1984, p. 164.

[5] Hinde, 1958.

[6] Collias & Collias, 1984, chapter 12.

[7] Scott, 1902.

[8] Dilger, 1962.

[9] Brosset, 1978.

[10] Darwin, 1881.

[11] See Ewer, 1960; Hardy, 1965; Lewontin, 1978; Waddington, 1975.

[12] There have been several attempts to find a principle or pattern relating different types of environments to the life strategies of the organisms living in them. The most general and influential was suggested by MacArthur and Wilson (1967) and advanced by Pianka (1970). Using the parameters in the logistic equation that describes population growth, they classified the type of selection found in different environments and the corresponding life strategies into two broad categories, r and K. The r strategy is found in unpredictable, ephemeral environments, where selection favours characters associated with rapid reproduction and dispersal, such as small size and a short lifespan. The K strategy is found in established stable habitats, where there is no advantage in rapid reproduction because the space is already crowded, but larger sized and more competitive animals are favoured. In spite of its attractions, this classification assumes that the relationship between organism and environment is more or less fixed, and is therefore not quite suitable for our purposes. For a more recent review of the concepts of r and K selection, see Begon, Harper & Townsend, 1990.

[13] Darwin, 1872, 6th edition of *The Origin*, p. 210.

[14] Darwin's pangenesis theory is fully described in chapter 27 of volume 2 of *The Variation of Animals and Plants under Domestication*, 2nd edn, 1883.

[15] Waddington summarised his work in the *Strategy of the Genes* (1957) and *The Evolution of an Evolutionist* (1975). The work on the assimilation of the physiological response of fruitflies to high salt concentrations has been extended by te Velde, Gordens & Scharloo, 1988.

[16] If p, the frequency of a^2 and b^2 alleles, is 0.1, and q, the frequency of a^1 and b^1 alleles, is 0.9, then the overall frequency of genotypes with two or more type-2 alleles will be $6p^2q^2 + 4p^3q + p^4 = 0.0523$.

[17] See Hardy, 1965, Lectures 6 and 7, which summarise many examples.

[18] Ewer, 1956.

[19] Eens, 1997.

[20] Klopfer & Klopfer, 1977.

[21] Edmunds, 1974; Kruuk, 1972.

[22] A recent theoretical model (Pál, 1998) shows that, when a population that is subject to persistent selection pressure is far from the optimal genotype (the genetically assimilated state), the ability to transmit phenotypes to the following generations has an advantage both over a fixed genetic strategy, and over a plastic strategy that is not inherited and has to be generated anew

every generation. However, as the population gets closer to the optimum, the generation of alternative heritable phenotypes becomes less advantageous, and a genetically fixed response becomes the optimal strategy. In Pál's model, the phenotypic variations are random, and learning is not involved in the process. Transgenerational transmission of learnt behaviours would be even more beneficial than inheritance of randomly generated phenotypes, because it would increase the fitness of those individuals that have the best learnt tradition.

[23] The conclusion that asocial learning will evolve in rapidly changing environments, whereas innate mechanisms evolve when the relevant environment is very stable, is supported not only by common sense, but also by precise and detailed theoretical models and computer simulations. For examples, see Boyd & Richerson, 1985; Cavalli-Sforza & Feldman, 1981. One common situation where classical genetic responses are inadequate and learning ability is favoured is when there is a conflict between organisms, as for example between predators and prey, or hosts and parasites, or, in many species, between potential mates. The party that changes first (by conventional genetic change) puts the other at a disadvantage; a genetic response in the latter is likely to be slow, because gene mutations are rare and even if sexual reshuffling can produce a solution, it is still rather slow. However, if some form of learning is possible, it may solve the response problem within the lifespan of the individual, and over evolutionary time better learning skills will probably evolve. Even when behavioural interactions between organisms start with stereotyped fixed responses, behaviours in the two protagonists may be selected to become more plastic and changeable through learning. The evolution of learning mechanisms may therefore be a general consequence of co-evolutionary interactions.

[24] O'Connor, 1984; 1991, pp. 246–53.

[25] Norton-Griffiths, 1969.

[26] Bowen, Oftedal & Boness, 1985.

[27] Boyd and Richerson (1985) have supported this conclusion with formal models that show that, in fairly constant but complex environments, social learning is beneficial.

[28] The importance of ILC environments for the evolution of transgenerational transmission of variations is discussed and modelled in Lachmann & Jablonka, 1996.

[29] Fox & Mousseau, 1998.

[30] See Emlen, 1954, for details of nest building in the cliff swallow.

[31] Winkler & Sheldon, 1993.

[32] Doron & Wollberg, 1994; Heil *et al.*, 1991.

[33] Dunbar, 1992.

[34] See Hardy 1965; Huxley, 1942; Piaget, 1978; Waddington, 1975. For more recent work showing that niche construction (of which a learnt habit is an instance) may significantly alter the dynamics of evolutionary changes with-

in a population, see Laland, Odling-Smee & Feldman, 1996; Odling-Smee, Laland & Feldman, 1996. We briefly discussed the effect of cultural traditions on genetic evolution in humans in chapter 1, when we described the effect of the domestication of cattle on the evolution of lactose absorption in adults. William Durham, who discussed the lactose case, has examined other important examples of co-evolutionary relationships between genes and cultures in human populations (Durham, 1991).

[35] Harvey (1998) has recently made a case for habit-guided genetic assimilation of morphological asymmetry in the hermit crab, *Clibanarius vittatus*. This crab inhabits empty gastropod shells, and is highly asymmetrical. The asymmetry is, to a large extent, environmentally regulated: there is an initial asymmetry, but it is almost completely lost if a young crab is deprived of a shell for several moults. The effect of inhabiting a dextrally coiled gastropod shell (most gastropod shells are dextrally coiled) is to accentuate the initial asymmetry, because the abdomen cannot grow on the right side. Harvey suggested that abdomen asymmetry was initially entirely determined by the interaction between growth and the structure of the shell, but, through genetic assimilation, this interaction has been partially taken over by an internal stimulus, resulting in an asymmetrical bias now appearing in the first juvenile stage. The role of genetic assimilation in the evolution of late-developing asymmetry in other animals has been thoroughly discussed by Palmer, 1996.

[36] Aisner & Terkel, 1992; Terkel, 1996.

[37] Hardy, 1965, chapter 7.

[38] Spurway, 1955.

[39] Maestripieri, 1992.

[40] For details of genomic imprinting in the brain, see Keverne *et al.*, 1996. The hypothesis that the differential expression of maternal genes affecting higher cognitive functions is related to the mother's role as a major transmitter of social learning arose in a discussion with Marion Lamb and Iddo Tavory. One prediction from the idea is that, in mammalian species that have practised a monogamous type of family organisation for a lengthy evolutionary period, the effect of the parental origin of genes will be less pronounced and more complex, because the behaviours of males and females are more similar.

[41] In chapter 6 (pp. 188–9) we discussed Haig's hypothesis that genomic imprinting may reflect male–female and parent–offspring evolutionary conflict, with expression of paternally derived genes leading to more extensive growth of the embryo at the expense of the mother and her present and future offspring, while expression of maternally derived genes counters the 'exploitative' effects of the paternal genes. The pattern of imprinting in the brain makes us look at the data in a new light. Maternal and paternal genes are both promoting growth, but in different and complementary ways. Maternal genes enhance the growth of energetically costly neural tissues, and paternal genes enhance the growth of non-neural tissues. The pattern of expression can be

seen as an effective 'division of labour' between maternally and paternally derived genes, resulting in a pattern of gene activity that is to the benefit of both parents and offspring. Both offspring and parents benefit from maternal gene expression in the brain leading to greater cognitive compatibility and better maternally guided learning in the offspring. They also benefit from paternal gene expression in the other somatic tissues, since larger offspring have a general advantage, especially for males. There is an allometric relationship between body size and brain size, with bigger bodies having larger brains.

[42] Keverne, Martel & Nevison, 1996.

[43] Pascal (translated by Krailsheimer, 1966, p. 61).

10   The free phenotype

In *The Descent of Man* and *The Expression of the Emotions in Man and Animals*, Darwin argues for evolutionary continuity between the minds of man and higher animals, stressing that higher animals share with us many complex mental capacities:

> the difference in mind between man and the higher animals, great as it is, is certainly one of degree and not of kind. We have seen that the senses and intuitions, the various emotions and faculties, such as love, memory, attention, curiosity, imitation, reason, &c., of which man boasts, may be found in an incipient, or even sometimes a well-developed condition, in lower animals.
>
> (Darwin, 1871, p. 105)

In one form or another, the continuity thesis is accepted by all evolutionary biologists. Even when a large mental gap between the minds of animals and man is recognised, the interpretation of this gap is based on the assumption that there is an underlying genetic and evolutionary continuity. However, notice how Darwin framed his statement: he did not claim that we are psychologically and cognitively simpler than we believe we are – that we are psychologically more like 'lower' animals. On the contrary, Darwin believed that 'lower' animals are more complex than is usually thought – that they are more similar to us, possessing more sophisticated capacities than we usually grant them.[1]

In this book, we have followed Darwin's approach, emphasising the learning capacities of higher animals, particularly their ability to learn from others. We have focused especially on one aspect of the similarity between higher animals and man that Darwin mentioned but did not develop, namely the transmission of habits. Darwin claimed that animals are capable of 'progressive improvement': over time they become more and more proficient in what they do. He believed this improvement occurs at two levels: first, animals improve during their individual lifetime, because they learn to be more adept and skilful through experience; second, animals become progressively more proficient as their habits are passed on and improved from one generation to the next. For example, he discussed how animals that have been exposed to poison and trapping for several generations become better at avoiding poisons and traps.[2]

Here we have tried to show how variations in transmissible habits have played a major role in the evolution of all higher animals, not just man, first because they are the raw material on which the selection of traditions, life styles and cultures are based, and second because they are one of the factors that construct the selective regime in which individuals live. Since alternative patterns of behaviour can be re-produced in successive generations through social learning and niche construction, it is wrong to assume a priori that differences in genes underlie all heritable differences in phenotypes. The extreme sociobiological view, which ignores the roles of heritable habits and traditions, and explains the specificity and the stability of all patterns of behaviour in terms of specific genetic 'programs', is of only limited use. Whatever behaviours are explored, whether they are familiar behaviours such as those of mice and tits, or the more exotic behaviours of bower-birds and meerkats, their evolutionary interpretation always requires a consideration of the role of both the genetic and behavioural inheritance systems.

We do not want to repeat here all the conclusions we have reached in previous chapters. The most important assessment of the ideas that we have presented will come from their empirical evaluation. Wherever possible we have presented empirical evidence that supports our views, and have suggested experiments and comparisons that can test them further. What we now want to do is first to sketch in what we see as some additional directions for research, since we believe our approach can be extended to invertebrate animals. We will then look at some of the more general philosophical and social implications of our view of behavioural evolution, and see whether animal cultures can offer us any insights into human culture, particularly into its most prominent and unique feature, symbolic language.

Social transmission of information in insects, and early learning in man

The importance of studying culture in higher animals is becoming more widely appreciated and is having some practical consequences. For example, encouraging social learning and the formation of social traditions in conservation practices is increasingly recognised as crucial for successful conservation.[3] The role and importance of culture for man, the most 'cultural' of all animals, has never been in doubt. This is not the case with invertebrates, however. Culture has never been assigned to invertebrates, whose prevalent image is that of 'soft automata'. But invertebrates such as the terrestrial insects, on which we will focus here, have a wide spectrum of social systems, including systems with elaborate forms of parental and alloparental care. They also have the most dramatic specialisations for social group living that can be found anywhere in the living world.[4] Moreover, studies of their learning capacities have shown for many species that they are frequently very good and fast learners.[5] Insects also have many routes for the transfer of behaviour-affecting substances, which are analogous to the placenta, milk, saliva and faeces routes of birds and mammals. In some species of cockroaches, substances are passed from mother to embryos through the placenta-like soft-skin of the internal brood sacs in which embryos are provisioned;[6] nourishing, milk-like substances are supplied by insect parents as diverse as termites and tsetse flies,[7] and trophallaxis, the mouth-to-mouth exchange of alimentary-originating liquid foods, is well researched in many social and some pre-social insects.[8] Finally, coprophagy, the consumption of faeces, particularly the consumption by young of faeces of adults of the same species, has long been known to be common in several groups of insects.[9] As with birds and mammals, most of these contributions to the next generation have been discussed in nutritional, not informational contexts. We know of no one who has looked to see whether, how and to what extent the transfer of variant substances affects the behavioural preferences of the young, and whether these preferences may lead to self-sustaining habits. It would be surprising if some of the many opportunities for information transfer were not realised, and if socially transmitted substances and behaviours did not sometimes induce new persistent habits.

Because of the preconceptions about insects' mental capacities, questions about their social learning and the possibility of behavioural inheritance have hardly ever been asked. There are scattered observations that indicate that observational learning may exist, and many observations that suggest that imprinting-like phenomena may be common. Long ago, Darwin noted that when a cabbage butterfly lays its eggs on cabbage and the caterpillars feed on this plant, the adult females produced will choose to lay their eggs on cabbage rather than on other related

plants.[10] More recently it has been found for several species of butter-fly that individual females from the same population prefer to lay their eggs on different host-plant species. These preferences are often trans-ferred from mothers to their daughters, although the precise mecha-nism and the role of genetic variations in the process are still unclear. It is not uncommon to find that, when caterpillars are fed on a new species of plant that is chemically similar to the typical host species, they later tend to prefer the new host plant. Moreover, the feeding pref-erences of young and egg-laying preferences of adults are often related, because females lay their eggs on the same species as that on which they fed as caterpillars.[11] Consequently, if a new preference is induced in a female, it can have self-sustaining effects which result in it being transmitted in her lineage.

Such food and host imprinting may be very common in insects. For example, embryo cockroaches may become imprinted on the food pref-erences of their mothers, who pass traces of food to them when they incubate them in the soft brood sac. Similarly, young tsetse flies whose mothers feed them with 'milk' may become imprinted on their mother's food. For the reasons discussed in the previous chapter, such transmit-ted food preferences may be important in environments in which there are fluctuations in the availability of hosts. Testing the hypothesis that food preferences are transmitted non-genetically should not be difficult, since the food consumed by the mothers can be manipulated experi-mentally, and the preferences of the young and the young's descendants can be tested.

Insects may also provide other ecological and social legacies for the next generation. In fire ants, some aspects of social organisation seem to be culturally transmitted. Whether a nest has one queen or multiple queens depends on the rearing conditions. The social environment in which a queen develops (whether in a nest with a single queen or one with several) affects her weight, which in turn biases her to join or found a colony like the one in which she herself was reared.[12] The tendency of some species of ants to form specific associations with certain plants may be transferred from one generation to the next by imprinting-like learning, since early experience can induce a later preference in colony-founding queens.[13] The raiding preferences of slave-making species of ants is also probably socially transferred from one generation of workers to another, because newly eclosed workers learn to recognise and prefer the kind of slave-ants already present in their nest.[14] The

transmission of such specific food and slave preferences across generations will probably be most pronounced in those species of ants in which
daughter colonies are formed by budding from mother colonies.

Although we believe that most persistent habits in insects are transferred via behaviour-inducing substances and through the transmission
of ecological legacies, there are some intriguing observations that suggest that observational learning may also occur. Foraging adult worker
ants of several species are known to carry newly hatched workers on
the underside of their abdomen, in a position in which the young
workers face the foraging route of the experienced ones.[15] Do the young
workers learn this foraging route, and later prefer to forage along this
route? We do not know, but in view of the excellent learning capacities
of ants, it would not be surprising if they did. Cockroaches, too, show
behaviour that suggests that observational social learning may occur.
Nymphs of some species are either taken to good feeding grounds by an
adult, or are given adult food. In one species, where the young are
known to be able to recognise siblings, they follow their mother as she
roams at night searching for food. They do this until they are almost
her size.[16] What is the function of this behaviour? Do the young acquire
parental preferences and knowledge as they follow their parents, in the
same way as the young of the chukar partridge do? Again, it would not
be difficult to find out by manipulating the environment of the parents
and their descendants. However, all such experiments require a basic
assumption – that social learning is not beyond the mental capacities
of insects, and therefore that such experiments are worth doing. A lot
of what we have learnt about insect behaviour in recent years suggests
that insects can learn, and that there are ample opportunities for the
social transmission of information.

Whether or not social learning, especially early social learning, can
have long-term effects on social insects' development and evolution
remains to be determined, but it is quite clear that early learning and
experience are enormously important for highly encultured human
beings. For example, early exposure to language is necessary for normal
language acquisition. However, although all educators agree on the
importance of early social learning, we still do not know enough about
which aspects of knowledge are best learnt early, and how to present
them to the young so that they will be learnt quickly and efficiently.
We do know, however, that early emotional deprivation can have devastating long-term effects, and may lead to cycles of deprivation, where

criminal or harmful behaviour is perpetuated across generations. We also know that in humans, as in other mammals, pre-natal conditions may have persistent effects on the health and the behavioural development of the young, and these effects may be transmitted for more than one generation.[17] Nevertheless, our understanding of the effects of early experience is still meagre, and our ability to control these effects and compensate for them is therefore limited. We know even less about what type of learning, and what methods of learning are most effective at later ages. For example, when is learning from an adult caregiver more effective than learning from and with peers? Is it more effective at certain 'sensitive' periods and for certain types of information? What is the optimal group size and group composition (with regard to age and sex) for learning? Evolutionary considerations like these might help educators when designing effective care and teaching schemes.

Between development and evolution: beyond the replicator

Our argument throughout this book has been that including behavioural inheritance in evolutionary thinking adds to, and sometimes alters, conventional evolutionary interpretations of many family and group behaviours in animals. But such learning-based interpretations of evolution do more than merely add yet another adaptive function to those that have already been suggested for a behaviour, or replace one possible adaptive function with another. As we claimed in chapter 1, recognising the role of behavioural inheritance challenges today's gene-centred view of heredity and evolution, because it alters the way we have to think about variation and selection.

When we consider the role of the behavioural inheritance system in the generation and selection of variations, the distinction between developmental and evolutionary processes becomes very fuzzy. With the genetic system, the distinction is relatively clear. The generation of gene variations is largely random with respect to the environment, and their transmissibility remains the same whatever the environmental conditions, because it depends on the distribution of chromosomes during cell division and sperm or egg production, which is usually very regular. Consequently, what happens during individual development usually does not affect the generation and transmission of genetic variations. This is not the case with cultural variations, where both the generation of variation and the transmissibility of variants depends on the environment in which the organism develops and to which it responds. A change in the environment often induces or promotes learning, so the generation of variations in learnt information is not random, and the transmissibility of what has already been learnt is altered when conditions change. Therefore, since the generation and the transmissibility of heritable variation depends on things that occur during an individual's lifetime, behavioural development must be seen as an integral part of behavioural inheritance. And, since learnt behaviours that are inherited are the raw material for selection and the evolution of traditions, development and evolution are continuous with each other. Furthermore, self-sustaining cultural variations have indirect effects on genetic variation, because they shape the selective environment that affects the frequencies of alleles. So even genetic evolution is more intimately associated with developmental changes than we have been accustomed to believe.

The view of evolution that we advocate is very different from that popular today. Most thinking about inheritance and evolution has been influenced by what has been learnt about the molecular nature of the gene – by what we know about its structural organisation, the conditions for its transmission, the way in which it is transmitted and the way it varies. In chapter 1 we discussed how this influence has gone beyond the strictly genetic realm and affected ideas about the evolution of culture. We mentioned briefly the meme concept, which was formed by analogy with the gene, and was intended to elucidate cultural evolution. The meme, like the gene, belongs to a class of entities that Dawkins called 'replicators'. A replicator was defined as 'anything in the universe of which copies are made' (Dawkins, 1982, p. 83). Although this definition may seem broad, the replicator concept entails a very special kind of copying. Its definition and application presuppose that only instructions, rather than the implementations of the instructions, can be meaningfully 'copied' or inherited. For example, the plan for building a house is a replicator, while the actual process of building the house and the final product, the house itself, are not. Dawkins therefore suggested a distinction between replicators (instructions) and 'vehicles'. He defined a vehicle as 'any unit, discrete enough to seem worth naming, which houses a collection of replicators and which works as a unit for the preservation and propagation of those replicators' (Dawkins, 1982, p. 114). Like genes, memes are replicators, entities whose 'copying' goes on independently of the development of their carriers.

Brains and individuals are the agents and vehicles of memes. It is a curious world that the memetic view draws – a world inhabited by selfish memes residing in and driving individuals and cultures. Feelings, ambitions and utopias are just ways in which memes manipulate their agents.

Our discussion of the behavioural inheritance systems in animals shows why this replicator-centred view of cultural inheritance and evolution is wrong. In chapter 4 we discussed the many channels through which substances that bias behavioural preferences can be transferred, usually from mothers to offspring. The information that is transferred by behaviour-influencing substances is not encoded like information in DNA or in symbolic language; it is always transmitted through reconstruction, through a developmental process. Transmissibility is often sensitive to local conditions, and variations are induced during development. The same is true of behaviours transmitted through various types of observational social learning. In non-human animals, social learning, including learning through imitation and even instruction, does not involve the transmission of encoded or symbolic information. That is why it is impossible to de-couple the transmission of information and its developmental function. Most transmission is not function-insensitive 'copying'. It is reconstruction – a function-sensitive developmental process. With this type of information transmission, *there is no unit of heritable variation that is not at the same time a unit of function that is constructed during development.* The replicator/vehicle dichotomy is meaningless in this case, and in all cases in which the transmission of information or the generation of new heritable information depends on development. And since it is the individual, and not the meme, which transmits and acquires information, the meme cannot function as an independent unit of cultural evolution. Even when culture is symbolically represented, as in many (but certainly not all) aspects of human culture, the generation of new symbolically encoded variation is non-random, so developmentally generated changes impinge on the course of cultural evolution. The replicator concept is associated with a very specialised type of information transmission, which does not cover all types of inheritance, and therefore cannot be the basis of all evolution. To understand evolution in all its richness and complexity, we have to go beyond the replicator. We have to consider different types of information transmission as aspects of developmental processes that lead to the re-generation and re-production of variant traits. Through this approach, the developing, active, interacting individual should regain its central role in evolution.

Cultural construction and cultural selection

A general conclusion that follows from our view of the relationship between development and evolution is that the way in which selection has structured the animal (and human) mind is often far more subtle and indirect than suggested by most evolutionary psychologists. We do not deny, of course, that there are species-specific built-in biases in attention, learning and memory. In the previous chapter we described how such biases can evolve through genetic assimilation of learnt behaviours. However, many species-specific behaviours probably emerge from a combination of general learning rules and mechanisms, self-sustaining ecological and social conditions, and the morphological and physiological adaptations that constrain and channel behaviour. The combination of these factors may result in a high level of specificity and transmissibility of patterns of behaviour, as Wallace suggested for species-specific nest-building behaviour in birds. It is therefore totally unwarranted to conclude that, when a behaviour is specific and found in every generation, it indicates that it has been selected genetically and is underlain by a corresponding, specially evolved, mental module in the brain.

The main attributes of a mental module are its relative autonomy, its speed of operation and inaccessibility to consciousness, and its dedication to the processing of specific incoming information. These features were originally intended to describe the neural circuits that underlie an innate pattern of perception or cognition.[18] However, they can also be seen in learnt habits, such as skilful bicycle riding by a circus chimpanzee or proficient piano playing. Such habits are specialised, and once automated they are performed quickly and largely unconsciously. So we could argue that a dedicated neural circuit or module must underlie every routinised habit. Yet the circuits we have just referred to are the result of learning and practice. Obviously, the particular neural circuit for cycle riding or piano playing cannot be found in the brain prior to learning – it is not sitting there waiting to be turned on. It has to be constructed from neural elements through structured learning. The difference between a module constructed by learning and an evolved module must be that the latter has the additional property that some of the neural elements from which it is constructed have a more constrained organisation prior to experience, so it needs less input (experience) to develop. This line of reasoning says that with innate behaviour there is a general 'neurological draft' in the brain prior to experience. A *species-specific evolved module* is one for which such a neurological draft, or a particular aspect of its organisation, would be found only in one particular species, and not in others. Behaviours like the temper tantrums of young baboons during weaning, or the helping behaviour of juvenile bee-eaters, are assumed to be the product of direct selection for the genes underlying these presumed adaptive behaviours, which are expressed as dedicated, partially 'pre-wired' neural modules. In contrast, since there was obviously no selection in chimpanzees for genes for performing circus tricks or for communicating with humans via symbols, the brain modules underlying them have to be constructed through learning.

The difference between a 'pre-wired', evolutionarily selected module and an experience-constructed module that is the product of development seems, theoretically, quite clear. But on closer scrutiny the distinctions between the two tend to get blurred. As we have repeatedly stressed, every complex behaviour in birds and mammals, whether it is the singing of a male songbird, or nest building or parental caring, has both innate and learnt aspects. It is not always easy to distinguish these, and it is even more difficult to identify what the direct target of past selection has been. It is too often assumed, without a shred of evidence, that an adaptive pattern of behaviour and the neural circuit underlying it have been selected as a whole – that they are the result of consistent selection of genes for this behaviour. Although this may sometimes be a valid assumption, there are many cases in which it probably is not. The behaviour seen in the young during weaning conflicts, which we discussed at length in chapter 6, may be a case in point. This suite of behaviours shows developmental specificity and is usually claimed to be the outcome of evolutionary conflict between parents and offspring. It is assumed that these juvenile behaviours were selected as a distinct psychological mechanism to outwit the parents and gain more from them. Parents are assumed to have a similar behavioural suite selected to counter their offspring's greed. We have suggested instead that weaning conflict is a consequence of youngsters gradually learning to become independent, with the squabbles being an inevitable consequence of the inherently frustrating process of learning. In most cases, there is no evidence that more parental indulgence would be beneficial to the offspring, and usually there are several good reasons to suppose that it would be counter-productive (in inclusive fitness terms) to all parties. According to our interpretation, weaning conflict is not a behavioural strategy that has been selected as a unit-character, and that is underlain by a distinct 'pre-wired' module. The behavioural and developmental specificity of weaning conflicts is due to the specificity of the recurring circumstances of young learning gradually in a family environment. This type of explanation may also be true for many other classes of behaviour, including, of course, human behaviour, which are seen by some to be indications of specifically selected modules in the brain

A human example will illustrate how the functional complexity and specificity of an adaptive suite of behaviours can lead to a mistaken inference about past genetic selection for these behaviours. In the case of the behavioural suite known as literate behaviour, i.e. reading and writing, we know from the archaeological record that it is a relatively recent cultural invention, the result of a process of cultural and not genetic evolution. But what would we think if we did not have this archaeological record, and if all humans lived in societies in which every normal child above a certain age showed literate behaviour? In order to see how easily such a situation could mislead us into concluding that selection of genes for reading and writing behaviours has taken place, let us imagine a literate world 500 years from now.[19] All the normal people in this world are literate. The environment in which every normal human child grows up is so structured that a child is exposed from birth to a flow of words and to linguistic, visual and tactile communicative symbols that stand for things, ideas and relations. These symbols are transferred not only via complicated machines (like computers), but also through other potent communicative devices. The result is that children acquire the ability to read without any formal instruction (as indeed do many children today, who learn to read from mere exposure to modern communications technology). They also easily learn how to write, since writing now requires simple motor actions such as pushing buttons.

Now imagine that a scientist from another planet is visiting this future world of ours, and her project is to understand the evolution of literacy. The visitor finds that all healthy individuals acquire the ability to read and write easily and almost without formal instruction, and there is no great variation between populations in the ability to read and write, although individuals may vary in the speed with which they learn. There is diversity in the kind of visual symbols used by different populations, but most systems are acquired with more or less the same ease, although there may be differences in the rate of acquisition of certain aspects of reading and writing. The alien also discovers that very young children, before they show any clear literate capacity, engage in apparently proto-literate behaviour, marking the environment with objects and assigning meaning to marks.

Looking at human neurology and genetics, the alien finds that there are specific defects, known as dyslexia, that primarily affect literacy. There are many different types of dyslexia, and they are to varying extents dissociated from other mental capacities and from general intelligence. The heritability of dyslexia is quite high, indicating that variations in genes affect literate behaviour. Moreover, hyperlexia, a neurological syndrome in which children acquire literate behaviour extremely early and efficiently, is associated with severe mental retardation, showing that literacy can be dissociated from general intelligence.[20] The alien also sees that there is a large learnt component in literate behaviour: socially deprived but neurologically normal children, and even adults, can learn to read and write at later ages, although not as easily as children exposed to literate behaviour from birth. When exposed to literacy at a more advanced age, lengthy formal instruction is necessary. Brain imaging techniques show fuzzy but non-random localisation of reading and writing, as for most other complex behaviours, such as speech comprehension and production. The complexity of literate behaviour is staggering. When humans read and write, information from several modalities is integrated within the framework of a symbolic system of rules that are not available to conscious scrutiny.

On the basis of the complexity of the behaviour, the relative facility of its acquisition and the genetic, neurological and developmental data, the alien comes to the conclusion that literate behaviour is a complex genetic adaptation, underlain by a distinct 'literacy' module, implying long genetic selection for the behaviour in the past. It is only after she consults the historical and archaeological literature that she realises that no direct selection for literacy (at least literacy as she has initially defined it) has occurred during human evolution! She now abandons her original hypothesis, and concludes that the 'literacy module' is, after all, constructed during the early development of each individual. In view of the complexity of the behaviour, she must consider a combination of various pre-existing cognitive adaptations, which came together to form literate behaviour through a long process (in historical terms) of *cultural evolution through cultural selection.*

We believe that the kind of problem that this thought experiment illustrates – the facile assumption that complex adaptations must have been *directly genetically* selected – is very common, with respect not only to man, but to animals too. We have argued that some cases of helping behaviour in birds and mammals may result from cultural selection leading to the co-ordination and stabilisation of parental behaviours to form a new coherent and robust behavioural adaptation. Bower building in different populations of the gardener bower-birds may be another case of cultural evolution through the selection of cultural variants. In humans, language may initially have evolved via cultural selection for elaborate social communication. Such cultural evolution brought together different cognitive adaptations into a coherent functional complex, initially without any genetic change. Of course we do not wish to argue that genetic assimilation did not occur subsequently. But we do want to stress the importance of behavioural and cultural selection, and its potential as a priming and guiding factor in evolution. We believe this is particularly important in the evolution of complex human behaviours, including the evolution of language.

Although we endorse Darwin's thesis that there is evolutionary continuity between the minds of animals and man, there is a break that Darwin too acknowledged. The introduction of symbolic language has altered the mental makeup of man. Cultural evolution in humans is also radically different from that found in animals, because information in humans became symbolically and linguistically represented. Although the same principles may initially have been at work in both human and animal behavioural inheritance and cultural evolution, we believe that, once the ability to communicate symbolically evolved, it created a run-away process that led to discontinuity between man and animals. We do not want to speculate here on the specific selection pressures that may have been involved in the evolutionarily shaping of human language, but we would like to discuss the emergence of this discontinuity, and examine the role of culture in the evolution of language.[21]

Language: a new inheritance system

Language is a unique and powerful system of representation and communication. What is transmitted from generation to generation through language are not overt patterns of behaviours, but symbolic representations – symbols, combinations of symbols, and narratives. The invention of discrete symbols allows for an almost indefinite increase in the aspects of the world that can be labelled, communicated and deliberately attended to. Grammar organises linguistic symbols into sentences by applying rules that allow hierarchical, recurring constructions of meaning-relations. This enables the sharing of complex experiences and expectations, and the construction and sharing of narratives describing possible situations that transcend any actual experience. This is a unique inheritance system: it creates an entirely new 'internal' social and cognitive environment for human beings, which alters their very experiences by making them interact with their linguistic representations of the world. There is a big jump here from the transmission of overt habits. What can be shared and transmitted is not only the actual, but also the possible, the imaginable, the fantastic. The acquisition of language led to the extension of memory, to the analysis as well as the better consolidation of habits, and to the ability to routinely share the consequences of this analysis.

In the previous chapters, we argued that tradition and cultural evolution can guide genetic evolution as well as adapt to it, so genes and cultures co-evolve to form complex adaptations. Since cultural evolution is faster than genetic evolution, learnt behaviours that are best adjusted to pre-existing physiological and morphological biases will be culturally selected. Culture therefore often determines what should be defined as a 'genetic pre-adaptation'. At a later stage, through genetic assimilation, variations in genes that are congruent with the evolving traditions may be selected too. Since culture has become increasingly important during the evolution of hominids, it is self-evident that cultural evolution has played a central role in the evolution of their cognitive and emotional traits. The increasing importance of cultural change over the ages may be the one uncontroversial issue in the heated discussions on the course of hominid evolution. It has probably played a central role in the evolution of language too.

Many evolutionary biologists and neurophysiologists argue for continuity between modern linguistic capacities and the pre-linguistic culture of the higher apes, suggesting various 'missing links'. The links were probably not strictly linguistic, although they must have increased the ability of hominids to represent the world and to communicate information to group members. A case for a bridging mimetic stage, during which hominids communicated via elaborate gestures, pantomime and song, has been persuasively suggested by the Canadian neurophysiologist Merlin Donald.[22] During this stage, information is assumed to have been represented symbolically, but not yet linguistically. The evolution of symbolical communication is thought to have involved all modalities. Donald suggested that it was associated with the increase in the size and organisation of areas in the brain that have predominantly general processing functions, such as co-ordinating information from several modalities, controlling memory, attention and voluntary motor control.

Cultural inventions, such as rituals, and their transmission through social learning, were probably the initiating events in the evolution of symbolic representation, and finally of the linguistic symbolic system. One cultural invention led to the selection of another, and to the selection of congruent gene variants. The evolution of language may consequently have proceeded in a boot-strapping fashion: a cultural invention, transmitted by social learning and elaborated by cultural evolution, led to new selection pressures for even more efficient behaviours; these selection pressures also led to the selection of genes affecting features of the nervous system that facilitated the behaviour. This may have involved, for example, genes affecting improved motor control, anatomical and physiological changes in the vocal apparatus, and changes in the organisation of the brain that led to more efficient transfer and decoding of symbolic information.[23] Such genetic changes increased the effectiveness of cultural transmission, and so on.

Assigning a leading role to cultural inheritance and cultural evolution in the evolution of language resolves the problem of how an improved linguistic ability that occurs in one individual could be an advantage for that individual if others in its group lacked it. In order to benefit from communication, one needs partners who understand.[24] But, as we have seen in the case of the spread of cultural innovations among Japanese macaques, and as we see in our own society, cultural innovations can spread within a group very quickly, especially among the young. A communicatory innovation can spread within one or two generations. As cultural innovations accumulate, they may stretch the learning capacity of the members of the group, so that cultural innovations that make the previous inventions more easily learnt and transmitted will be favoured. At the same time, this cultural selection pressure may lead to the genetic assimilation of part of the learning sequence that enables easier and earlier learning. The effect of language was to alter the evolutionary journey quite substantially by assigning a primary and directing role to cultural evolution. Genetic evolution became ever more dependent on culture, following, rather than leading, as it was channelled into the grooves drilled by cultural evolution. By enhancing the importance of cultural evolution, language increased the extent of individual and collective behavioural flexibility, and also enhanced the importance of the co-evolution of genes and culture. This led to a runaway process of language evolution, with positive feedback between cultural and genetic factors.

Focusing on culture highlights the behavioural plasticity of animals and man, and their practical liberation from rigid genetic determination. But culture also imposes harsh constraints on the behaviours of individuals and societies. Although culturally determined traits may be more easily changeable (in evolutionary time) than those determined by genetic variation, they are often very binding and constraining for individual life. Have we not substituted cultural determinism for genetic determinism?

Beyond cultural determinism: freedom and the hope of reason

Any kind of conservatism, whether genetic or cultural, has its price. The price of culture, of transmitting behaviours and symbols, is high. As with other inheritance systems, there is an in-built inertia in the cultural system. This cultural inertia may sometimes lead to the extinction of cultures and of cultural groups, and we are helplessly witnessing such extinctions in the modern world. Sticking to old habits in a changing world is hazardous. However, the cultural system can incorporate variation very much more readily than the genetic system. Since cultural change is both developmental and evolutionary, individuals can transform their cultures, up-date their information, and bridge the gap between the past and present. The urge to explore, prominent mainly in the juvenile period, allows individuals to experiment with their environment and improve on outdated cultural legacies. Moreover, cultures can die instead of their practitioners, who can 'convert' to another culture. With symbolic systems like the human one, there is also another possibility: the members of a cultural group may die, but some aspects of the group's culture may survive in the form of books and other

symbolically encoded artefacts, as did local Jewish–Yiddish cultural artefacts of eastern Europe whose creators and perpetuators perished during the holocaust. Nevertheless, we cannot ignore the powerful way in
which culture shapes our beliefs, preferences and behaviours. Is there
a way to overcome cultural norms other than following changing con-
tingencies by updating behaviour after the environment has already
changed? Is it possible to resist established cultural norms, and control
and construct the future in a more deliberate way?

Humans do have such an option, and it is our symbolic, linguistic system of thought and communication that provides us with it. Although
the co-evolution of culture and genes that has produced existing symbolic language may initially have been similar to other co-evolutionary
processes, it has led to amazing and paradoxical consequences. It allows
us, among numerous other things, to study geometry and enjoy its
beauty. It also allows us to go beyond our own particular culture and
to represent to ourselves the social group and the culture in which we
take part, to criticise and transcend cultural norms. First, the narrative,
the story (which initially could have been represented in ritual and pantomime), became a net that humans could cast on the future: it gives a
shared virtual reality that one can imagine, empathise with and therefore strive for. Second, linguistic abstraction and a systematic chain of
reasoning allow scrutiny of the story and the virtual reality it offers.
Language-based reasoning can even be taken beyond the realms of reason itself, to produce logical paradoxes. It is in these aspects of language
– the aspects that to a certain extent negate sensual reality (but not
passion!) – that we find our unique human freedom. They give us the
ability to view ourselves beyond the boundaries of sex, race, class,
nationality and even species. It is only through this kind of imaginative
abstraction that we can systematically study the world around us,
engage in self-exploration, think of justice embodied in law, demand
freedom or equality. Reason, imagination and shared narratives give us
the only way of popping out of both culture and biology. We can use
them to go beyond the norms of our own culture and reach something
that is more universally acceptable. Like the famous Baron Münchausen,
we can (though we rarely do) pull ourselves out of our own cultural
marshes with our own hair.[25]

Notes

[1] The American social psychologist Solomon Asch argued in much the same vein, challenging the behaviourist view of man. He wrote that Darwin 'attempted, so to speak, to bridge the gap by the argument that the capacities and tendencies of the mammalian organisms resembled those of men more closely than had been suspected. Subsequently, psychology simply reversed this emphasis; it directed its efforts to demonstrating that men are not as different psychologically from lower organisms as has been generally supposed' (Asch, 1952, p. 11). Today the proponents of cognitive ethology, such as Donald Griffin (1992), argue for a return to the Darwinian position, which recognises that animals feel, think and are conscious in ways that are very similar to those of humans.

[2] See Darwin, 1871, pp. 49–50.

[3] Clemmons & Buchholz, 1997.

[4] Choe & Crespi, 1997; Wilson, 1971.

[5] Papaj & Lewis, 1993.

[6] Nalepa & Bell, 1997.

[7] For information about termites, see Nalepa, 1994; for tsetse flies see Langley, 1977; Ma *et al.*, 1975.

[8] Hunt & Nalepa, 1994; Michener, 1974; Ross & Matthews, 1991.

[9] Hunt & Nalepa, 1994.

[10] Darwin, 1844, p. 28 (cited by Thompson & Pellmyr, 1991). Darwin did not say which of the two common species of cabbage butterfly he was observing.

[11] Bernays & Chapman, 1994; Tabashnik *et al.*, 1981; Thompson & Pellmyr, 1991.

[12] Ross & Keller, 1995.

[13] Jaisson, 1980.

[14] Goodloe, Sanwald & Topoff, 1987.

[15] See Hölldobler & Wilson, 1990, and references therein.

[16] Evans & Breed, 1984; Nalepa & Bell, 1997.

[17] Barker, 1994.

[18] Fodor, 1983; for a view focusing on social-intelligence modules see, Gigerenzer, 1997; see also chapter 1, p. 9.

[19] This thought experiment and the discussion are based on Jablonka & Rechav, 1996.

[20] Deacon, 1997.

[21] In the last ten years, there has been a veritable industry of speculations on the subject of language evolution. See, for example, Hurford, Studdert-Kennedy & Knight, 1998; Trabant, 1996, and references therein.

[22] Donald, 1991.

[23] Deacon, 1997; Jablonka & Rechav, 1996; Jablonka & Szathmáry, 1995.

[24] Of course, it is possible that features that led to an improved ability to communicate symbolically initially had an advantage that was not related to their function in communication. For example, the major advantage of better

control of the vocal apparatus could have been that it led to more pleasing articulation of sounds, and it was selected in a 'musical' context. However, once a communicative function is assumed to have been selected directly, the problem of parity between sender and receiver remains.

[25] Baron Münchausen, a German soldier who exaggerated his exploits and adventures, was the inspiration for R. E. Raspe's *Adventures of Baron Münchausen*, first published in 1785. In Raspe's story, Münchausen claimed that he once saved himself and his horse from perishing in a swamp by holding his horse firmly between his knees and using his strength to pull himself and his horse out with his own hair. (There is now a medical condition called Münchausen's syndrome, in which people feign illness or injury in order to get medical treatment.)

References

Adolphs, R., Tranel, D. & Damasio, A. R. (1998). The human amygdala in social judgement. *Nature*, **393**, 470–4.

Aisner, R. & Terkel, J. (1992). Ontogeny of pine cone opening behaviour in the black rat, *Rattus rattus*. *Animal Behaviour*, **44**, 327–36.

Alcock, J. (1993). *Animal Behaviour: An Evolutionary Approach*, 5th edn. Sunderland, Massachusetts: Sinauer.

Alexander, R. D. (1987). *The Biology of Moral Systems*. New York: Aldine de Gruyter.

Altmann, J. (1980). *Baboon Mothers and Infants*. Cambridge, Massachusetts: Harvard University Press.

Altringham, J. D. (1996). *Bats: Biology and Behaviour*. Oxford: Oxford University Press.

Anderson, J. R. (1980). *Cognitive Psychology and its Implications*. San Francisco: W. H. Freeman and Company.

Andersson, M. (1994). *Sexual Selection*. Princeton, New Jersey: Princeton University Press.

Apfelbach, R. (1973). Olfactory sign stimulus for prey selection in polecats (*Putorius putorius* L.). *Zeitschrift für Tierpsychologie*, **33**, 270–3.

Aristotle, *The Complete Works of Aristotle, The Revised Oxford Translation*. vol. 2, *Problems Book VII*, ed. J. Barnes. Bolleingen Series LXXI 2, 1985. New Jersey: Princeton University Press.

Armstrong, E. A. (1947). *Bird Display and Behaviour*. London: Lindsay Drummond.

Aronson, L. (1982). *The Nubian Ibex* (in Hebrew). Tel Aviv: Massada.

Arrowood, P. C. (1988). Duetting, pair bonding and agonistic display in parakeet pairs. *Behaviour*, **106**, 129–57.

Asch, S. E. (1952). *Social Psychology*. New York: Prentice-Hall.

Ashmole, N. P. & Tovar, S. H. (1968). Prolonged parental care in royal terns and other birds. *The Auk*, **85**, 90–100.

Atlan, H. & Koppel, M. (1990). The cellular computer DNA: program or data. *Bulletin of Mathematical Biology*, **52**, 335–48.

Avital, E. & Jablonka, E. (1994). Social learning and the evolution of behaviour. *Animal Behaviour*, **48**, 1195–9.

Avital, E. & Jablonka, E. (1996). Adoption, memes and the Oedipus complex: a reply to Hansen. *Animal Behaviour*, **51**, 476–7.

Avital, E., Jablonka, E. & Lachmann, M. (1998). Adopting adoption. *Animal Behaviour*, **55**, 1451–9.

Bagemihl, B. (1999). *Biological Exuberance: Animal Homosexuality and Natural Diversity*. London: Profile Books.

Baine, D. & Starr, E. (1991). Generalization of learning: an essential consideration in early childhood education. *International Journal of Early Childhood*, **23**, 58–67.

Baker, A. J. & Jenkins, P. F (1987). Founder effect and cultural evolution of songs in an isolated population of chaffinches, *Fringilla coelebs*, in the Chatham Islands. *Animal Behaviour*, **35**, 1793–803.

Baker, M. C. (1993). Evidence of intraspecific vocal imitation in singing honeyeaters (Meliphagidae) and golden whistlers (Pachycephalidae). *The Condor*, **95**, 1044–8.

Baker, M. C. (1994). Loss of function in territorial song: comparison of island and mainland populations of the singing honeyeater (*Meliphaga virescens*). *The Auk*, **111**, 178–84.

Baker, M. C. (1996). Depauperate meme pool of vocal signals in an island population of singing honeyeaters. *Animal Behaviour*, **51**, 853–8.

Baker, M. C. & Cunningham, M. A. (1985). The biology of bird-song dialects. *Behavioral and Brain Sciences*, **8**, 85–100.

Baker, R. R. (1978). *The Evolutionary Ecology of Animal Migration*. London: Hodder & Stoughton.

Baptista, L. F. & Johnson, R. B. (1982). Song variation in insular and mainland California brown creepers (*Certhia familiaris*). *Journal of Ornithology*, **123**, 131–44.

Barker, D. J. P. (1994). *Mothers, Babies, and Disease in Later Life*. London: BMJ Publishing Group.

Barrett, L., Dunbar R. I. M. & Dunbar, P. (1995). Mother–infant contact as contingent behaviour in gelada baboons. *Animal Behaviour*, **49**, 805–10.

Bateson, G. (1979). *Mind and Nature, a Necessary Unity*. London: Wildwood House.

Bateson, P. P. G. (1976). Rules and reciprocity in behavioural development. In *Growing Points in Ethology*, ed. P. P. G. Bateson & R. A. Hinde, pp. 401–21. Cambridge: Cambridge University Press.

Bateson, P. P. G. (1978). How does behavior develop? In *Perspectives in Ethology*, vol. 3, ed. P. P. G. Bateson & P. H. Klopfer, pp. 55–67. New York: Plenum.

Bateson, P. P. G. (1982). Preferences for cousins in Japanese quail. *Nature*, **295**, 236–7.

Bateson, P. P. G. (1988). The active role of behaviour in evolution. In *Evolutionary Processes and Metaphors*, ed. M.-W. Ho & S. W. Fox , pp.191–207. Chichester: Wiley.

Bateson, P. P. G. (1990). Is imprinting such a special case? *Philosophical Transactions of the Royal Society of London, Series B*, **329**, 125–31.

Bateson, P. P. G. (ed.) (1991). *The Development and Integration of Behaviour: Essays in Honour of Robert Hinde*. Cambridge: Cambridge University Press.

Bateson, P. P. G. (1994). The dynamics of parent–offspring relationships in mammals. *Trends in Ecology and Evolution*, **9**, 399–403.

Beauchamp, G. (1997). Determinants of intraspecific brood amalgamation in waterfowl. *The Auk*, **114**, 11–21.

Beck, B. B., Rapaport, L. G., Stanley Price, M. R. & Wilson, A. C. (1994). Reintroduction of captive-born animals. In *Creative Conservation: Interactive Management of Wild and Captive Animals*, ed. P. J. S. Olney, G. M. Mace & A. T. C. Feistner, pp. 265–86. London: Chapman & Hall.

Beecher, M. D. (1988). Kin recognition in birds. *Behavior Genetics*, **18**, 465–82.

Beer, A. E. & Billingham, R. E. (1976). Transmission of antibodies from mother to offspring before birth. In *The Immunobiology of Mammalian Reproduction*, ed. A. E. Beer & R. E. Billingham, pp. 190–216. New Jersey: Prentice Hall.

Beer, C. G. (1972). Individual recognition of voice and its development in birds. *Proceedings of the 15th International Ornithological Congress*, ed. K. H. Voous, pp. 340–56. Leiden: E. J. Brill.

Begon, M., Harper, J. L. & Townsend, C. R. (1990). *Ecology: Individuals, Populations and Communities*, 2nd edn. Boston: Blackwell Scientific Publications.

Bengtsson, H. & Rydén, O. (1981). Development of parent–young interaction in asynchronously hatched broods of altricial birds. *Zeitschrift für Tierpsychologie*, **56**, 255–72.

Bengtsson, H. & Rydén, O. (1983). Parental feeding rate in relation to begging behavior in asynchronously hatched broods of the great tit *Parus major*. *Behavioral Ecology and Sociobiology*, **12**, 243–51.

Benus, R. F. & Röndigs, M. (1996). Patterns of maternal effort in mouse lines bidirectionally selected for aggression. *Animal Behaviour*, **51**, 67–75.

Bergstrom, C. T. & Lachmann, M. (1998). Signalling among relatives. 3. Talk is cheap. *Proceedings of the National Academy of Sciences, USA*, **95**, 5100–5.

Berman, C. M. (1990). Intergenerational transmission of maternal rejection rates among free-ranging rhesus monkeys. *Animal Behaviour*, **39**, 329–37.

Berman, C. M. (1997). An animal model for the intergenerational transmission of maternal style? *Family Systems*, **3**, 125–40.

Bernays, E. A. & Chapman, R. F. (1994). *Host-Plant Selection by Phytophagous Insects*. New York: Chapman & Hall.

Berry, R. J. (ed.) (1981). *Biology of the House Mouse*. London: Academic Press.

Bertram, B. C. R. (1980). Vigilance and group size in ostriches. *Animal Behaviour*, **28**, 278–86.

Bertram, B. C. R. (1992). *The Ostrich Communal Nesting System*. Princeton, New Jersey: Princeton University Press.

Bhatia, M. S., Dhar, N. K., Singhal, P. K., Nigam, V. R., Malik, S. C. & Mullick, D. N. (1990). Temper tantrums: prevalence and etiology. *Clinical Pediatrics*, **29**, 311–15.

Birkhead, T. R. & Møller, A. P. (1992). *Sperm Competition in Birds: Evolutionary Causes and Consequences*. London: Academic Press.

Black, J. M. (ed.) (1996). *Partnerships in Birds: the Study of Monogamy*. Oxford: Oxford University Press.

Black, J. M. & Owen, M. (1988). Variations in pair bond and agonistic behaviors in barnacle geese on the wintering grounds. In *Wildfowl in Winter*, ed. M. W. Weller, pp. 39–57. Minneapolis: University of Minnesota Press.

Black, J. M. & Owen, M. (1995). Reproductive performance and assortative pairing in relation to age in barnacle geese. *Journal of Animal Ecology*, **64**, 234–44.

Black, J. M., Carbone, C., Wells, R. L. & Owen, M. (1992). Foraging dynamics in goose flocks: the cost of living on the edge. *Animal Behaviour*, **44**, 41–50.

Blackmore, S. (1999). *The Meme Machine*. Oxford: Oxford University Press.

Boakes, R. (1984). *From Darwin to Behaviourism: Psychology and the Minds of Animals*. Cambridge: Cambridge University Press.

Boitani, L. & Ciucci, P. (1995). Comparative social ecology of feral dogs and wolves. *Ethology, Ecology & Evolution*, **7**, 49–72.

Boness, D. J. (1990). Fostering behavior in Hawaiian monk seals: is there a reproductive cost? *Behavioral Ecology and Sociobiology*, **27**, 113–22.

Bonner, J. T. (1980). *The Evolution of Culture in Animals*. New Jersey: Princeton University Press.

Bost, C. A. & Clobert, J. (1992). Gentoo penguin *Pygoscelis papua*: factors affecting the process of laying a replacement clutch. *Acta Oecologica*, **13**, 593–605.

Bowen, W. D., Oftedal, O. T. & Boness, D. J. (1985). Birth to weaning in 4 days: remarkable growth in the hooded seal, *Cystophora cristata*. *Canadian Journal of Zoology*, **63**, 2841–6.

Box, H. O. & Gibson, K. R. (ed.) (1999). *Mammalian Social Learning. Comparative and Ecological Perspectives*. Cambridge: Cambridge University Press.

Boyd, R. & Richerson, P. J. (1985). *Culture and the Evolutionary Process*. Chicago: Chicago University Press.

Boyd, R. & Richerson, P. J. (1990). Group selection among alternative evolutionarily stable strategies. *Journal of Theoretical Biology*, **145**, 331–42.

Bradley, J. S., Wooller, R. D. & Skira, I. J. (1995). The relationship of pair-bond formation and duration to reproductive success in short-tailed shearwaters *Puffinus tenuirostris. Journal of Animal Ecology*, **64**, 31–8.

Breedlove, S. M. (1997). Sex on the brain. *Nature*, **389**, 801.

Breitwisch, R. (1989). Mortality patterns, sex ratios, and parental investment in monogamous birds. In *Currrent Ornithology*, vol. 6, ed. D. M. Power, pp. 1–50. New York: Plenum.

Brooke, M. & Birkhead, T. (ed.) (1991). *The Cambridge Encyclopedia of Ornithology*. Cambridge: Cambridge University Press.

Brosset, A. (1978). Social organization and nest-building in the forest weaver birds of the genus *Malimbus* (Ploceinae). *Ibis*, **120**, 27–37.

Brower, L. (1969). Ecological chemistry. *Scientific American*, **220**(2), 22–9.

Brown, C. R. (1986). Cliff swallow colonies as information centers. *Science*, **234**, 83–5.

Brown, C. R. (1988). Enhanced foraging efficiency through information centers: a benefit of coloniality in cliff swallows. *Ecology*, **69**, 602–13.

Brown, E. D. (1985). The role of song and vocal imitation among common crows (*Corvus brachyrhynchos*). *Zeitschrift für Tierpsychologie*, **68**, 115–36.

Brown, J. L. (1983). Cooperation: a biologist's dilemma. *Advances in the Study of Behavior*, **13**, 1–37.

Brown, J. L. (1987). *Helping and Communal Breeding in Birds: Ecology and Evolution*. Princeton, New Jersey: Princeton University Press.

Brush, A. H. & Power, D. M. (1976). House finch pigmentation: carotenoid metabolism and the effect of diet. *The Auk*, **93**, 725–39.

Burke, V. E. M. & Brown, L. H. (1970). Observations on the breeding of the pink-backed pelican *Pelecanus rufescens. Ibis*, **112**, 499–512.

Burley, N. (1981). Mate choice by multiple criteria in a monogamous species. *American Naturalist*, **117**, 515–28.

Burns, K. J. (1998) A phylogenetic perspective on the evolution of sexual diachromatism in tanagers (Thraupidae): the role of female versus male plumage. *Evolution*, **52**, 1219–24.

Burton, M. (1973). *The Hedgehog*. London: Corgi Books.

Byrne, R. (1995). *The Thinking Ape: Evolutionary Origins of Intelligence*. Oxford: Oxford University Press.

Caldji, C., Tannenbaum, B., Sharma, S., Francis, D., Plotsky, P. M. & Meaney, M. J. (1998). Maternal care during infancy regulates the development of neural systems mediating the expression of fearfulness in the rat. *Proceedings of the National Academy of Sciences, USA*, **95**, 5335–40.

Calhoun, J. B. (1962). Population density and social pathology. *Scientific American*, **206(2)**, 139–48.

Calhoun, J. B. (1963a). *The Ecology and Sociology of the Norway Rat*. Bethesda, Maryland: U.S. Public Health Service.

Calhoun, J. B. (1963b). The social use of space. In *Physiological Mammalogy*, vol. 1, *Mammalian Populations*, ed. W. V. Mayer & R. G. Van Gelder, pp. 1–187. New York: Academic Press.

Calhoun, J. B. (1973). Death squared: the explosive growth and demise of a mouse population. *Proceedings of the Royal Society of Medicine*, **66**, 80–8.

Carlstead, K. (1996). Effects of captivity on the behavior of wild mammals. In *Wild Mammals in Captivity: Principles and Techniques*, ed. D. G. Kleiman, M. E. Allen, K. V. Thompson & S. Lumpkin, pp. 317–33. Chicago: University of Chicago Press.

Caro, T. M. (1994). *Cheetahs of the Serengeti Plains*. Chicago: University of Chicago Press.

Caro, T. M. & Hauser, M. D. (1992). Is there teaching in nonhuman animals? *Quarterly Review of Biology*, **67**, 151–74.

Catchpole, C. K. & Slater, P. J. B. (1995). *Bird Song: Biological Themes and Variations*. Cambridge: Cambridge University Press.

Cavalli-Sforza, L. L. & Feldman, M. W. (1981). *Cultural Transmission and Evolution: A Quantitative Approach*. Princeton, New Jersey: Princeton University Press.

Cézilly, F. & Nager, R. G. (1996). Age and breeding performance in monogamous birds: the influence of pair stability. *Trends in Ecology and Evolution*, **11**, 27.

Chapais, B. (1992). The role of alliances in social inheritance of rank among female primates. In *Coalitions and Alliances in Humans and Other Animals*, ed. A. H. Harcourt & F. B. M. de Waal, pp. 29–59. Oxford: Oxford University Press.

Chapman, C. A. & Chapman, L. J. (1987). Social responses to the traumatic injury of a juvenile spider monkey (*Ateles geoffroyi*). *Primates*, **28**, 271–5.

Chardine, J. W. (1987). The influence of pair-status on the breeding behaviour of the kittiwake *Rissa tridactyla* before egg-laying. *Ibis*, **129**, 515–26.

Cheney, D. L. & Seyfarth, R. M. (1990). *How Monkeys See the World: Inside the Mind of Another Species*. Chicago: University of Chicago Press.

Choe, J. C. & Crespi, B. J. (ed.) (1997). *The Evolution of Social Behaviour in Insects and Arachnids*. Cambridge: Cambridge University Press.

Choudhury, S. & Black, J. M. (1993). Mate-selection behaviour and sampling strategies in geese. *Animal Behaviour*, **46**, 747–57.

Choudhury, S. & Black, J. M. (1994). Barnacle geese preferentially pair with familiar associates from early life. *Animal Behaviour*, **48**, 81–8.

Choudhury, S., Jones, C. S., Black, J. M. & Prop, J. (1993). Adoption of young and intraspecific nest parasitism in barnacle geese. *The Condor*, **95**, 860–8.

Clark, L. & Mason, J. R. (1985). Use of nest material as insecticidal and anti-pathogenic agents by the European starling. *Oecologia*, **67**, 169–76.

Clark, L. & Mason, J. R. (1988). Effect of biologically active plants used as nest material and the derived benefit to starling nestlings. *Oecologia*, **77**, 174–80.

Clark, M. M. & Galef, B. G., Jr. (1995). Parental influences on reproductive life history strategies. *Trends in Ecology and Evolution*, **10**, 151–3.

Clark, M. M., Karpiuk, P. & Galef, B. G., Jr. (1993). Hormonally mediated inheritance of acquired characteristics in Mongolian gerbils. *Nature*, **364**, 712.

Clayton, D. H. & Wolfe, N. D. (1993). The adaptive significance of self-medication. *Trends in Ecology and Evolution*, **8**, 60–3.

Clemmons, J. R. & Buchholz, R. (ed.) (1997). *Behavioral Approaches to Conservation in the Wild*. Cambridge: Cambridge University Press.

Clutton-Brock, T. H. (ed.) (1988). *Reproductive Success*. Chicago: University of Chicago Press.

Clutton-Brock, T. H. (1991). *The Evolution of Parental Care*. Princeton, New Jersey: Princeton University Press.

Clutton-Brock, T. H. & Parker, G. A. (1995a). Punishment in animal societies. *Nature*, **373**, 209–16.

Clutton-Brock, T. H. & Parker, G. A. (1995b). Sexual coercion in animal societies. *Animal Behaviour*, **49**, 1345–65.

Clutton-Brock, T. H., Guinness, F. E. & Albon, S. D. (1982). *Red Deer: Behavior and Ecology of Two Sexes*. Chicago: University of Chicago Press .

Clutton-Brock, T. H., Brotherton, P. N. M., Smith, R., McIlrath, G. M., Kansky, R., Gaynor, D., O'Riain, M. J. & Skinner, J. D. (1998a). Infanticide and expulsion of females in a cooperative mammal. *Proceedings of the Royal Society of London, Series B*, **265**, 291–5.

Clutton-Brock, T. H., Gaynor, D., Kansky, R., MacColl, A. D. C., McIlrath, G., Chadwick, P., Brotherton, P. N. M., O'Riain, J. M., Manser, M. & Skinner, J. D. (1998b). Costs of cooperative behaviour in suricates (*Suricata suricatta*). *Proceedings of the Royal Society of London, Series B*, **265**, 185–90.

Cockburn, A. (1998). Evolution of helping behaviour in cooperatively breeding birds. *Annual Review of Ecology and Systematics*, **29**, 141–77.

Cody, M. (1974). *Competition and the Structure of Bird Communities*. Princeton, New Jersey: Princeton University Press.

Collias, E. C. & Collias, N. E. (1964). The development of nest-building behavior in a weaverbird. *The Auk*, **81**, 42–52.

Collias, N. E. & Collias, E. C. (1984). *Nest Building and Bird Behavior*. Princeton, New Jersey: Princeton University Press.

Conover, M. R. (1987). Acquisition of predator information by active and passive mobbers in ring-billed gull colonies. *Behaviour*, **102**, 41–57.

Cooke, F. (1978). Early learning and its effect on population structure. Studies of a wild population of snow goose. *Zeitschrift für Tierpsychologie*, **46**, 344–58.

Cooke, F., Mirsky, P. J. & Seiger, M. B. (1972). Color preferences in the lesser snow goose and their possible role in mate selection. *Canadian Journal of Zoology*, **50**, 529–36.

Coombs, C. J. F. (1960). Observations on the rook *Corvus frugilegus* in southwest Cornwall. *Ibis*, **102**, 394–419.

Cowie, R. J., Krebs, J. R. & Sherry, D. F. (1981). Food storing by marsh tits. *Animal Behaviour*, **29**, 1252–9.

Cramp, S. (ed.) (1980). *Handbook of the Birds of Europe, the Middle East and North Africa; the Birds of the Western Palearctic.* vol. 2, *Hawks to Bustards*, Oxford: Oxford University Press.

Crook, J. H. (1965). The adaptive significance of avian social organizations. *Symposia of the Zoological Society of London*, **14**, 181–218.

Cross, H. A., Halcomb, C. G. & Matter, W. W. (1967). Imprinting or exposure learning in rats given early auditory stimulation. *Psychonomic Science*, **7**, 233–4.

Crowcroft, P. (1966). *Mice all Over*. London: Foulis.

Crowcroft, P. & Rowe, F. P. (1963). Social organization and territorial behaviour in the wild house mouse (*Mus musculus*). *Proceedings of the Zoological Society of London*, **140**, 517–31.

Curio, E. (1976). *The Ethology of Predation*. Berlin: Springer-Verlag.

Curio, E., Ernst, U. & Vieth, W. (1978). The adaptive significance of avian mobbing. II. Cultural transmission of enemy recognition in blackbirds: effectiveness and some constraints. *Zeitschrift für Tierpsychologie*, **48**, 184–202.

Daan, S., Dijkstra, C., Drent, R. & Meijer, T. (1988). Food supply and the annual timing of avian reproduction. *Proceedings of the 19th International Ornithological Congress*, pp. 392–407.

Dale, S., Amundsen, T., Lifjeld, J. T. & Slagsvold, T. (1990). Mate sampling behaviour of female pied flycatchers: evidence for active mate choice. *Behavioral Ecology and Sociobiology*, **27**, 87–91.

Darling, F. F. (1937). *A Herd of Red Deer*. Oxford: Oxford University Press.

Darwin, C. (1871). *The Descent of Man, and Selection in Relation to Sex*. London: Murray (Reprinted in 1981 by Princeton University Press as a photoproduction of the original edition.)

Darwin, C. (1872). *On the Origin of Species by Means of Natural Selection*, 6th edn. London: Murray.

Darwin, C. (1872). *The Expression of the Emotions in Man and Animals*, London: John Murray.

Darwin, C. (1881). *The Formation of Vegetable Mould, Through the Action of Worms, With Observations on their Habits*. London: Murray.

Darwin, C. (1883). *The Variation of Animals and Plants under Domestication*, 2nd edn. London: Murray. (Johns Hopkins University Press reprint edition, 1998.)

Darwin, C. (1909). Essay of 1844. In *The Foundations of the Origin of Species: Two Essays Written in 1842 and 1844 by Charles Darwin*, ed. F. Darwin. Cambridge: Cambridge University Press.

Dasmann, R. F. & Taber, R. D. (1956). Behavior of Columbian black-tailed deer with reference to population ecology. *Journal of Mammalogy*, **37**, 143–64.

Davey, G. (1989). *Ecological Learning Theory*. London and New York: Routledge.

Davies, N. B. (1976). Parental care and the transition to independent feeding in the young spotted flycatcher (*Muscicapa striata*). *Behaviour*, **59**, 280–95.

Davies, N. B. (1992). *Dunnock Behaviour and Social Evolution*. Oxford: Oxford University Press.

Davis, L. S. (1988). Coordination of incubation routines and mate choice in Adélie penguins (*Pygoscelis adeliae*). *The Auk*, **105**, 428–32.

Dawes, R. M. (1980). Social dilemmas. *Annual Review of Psychology*, **31**, 169–93.

Dawkins, R. (1976). *The Selfish Gene*. Oxford: Oxford University Press.

Dawkins, R. (1979). Twelve misunderstandings of kin selection. *Zeitschrift für Tierpsychologie*, **51**, 184–200.

Dawkins, R. (1982). *The Extended Phenotype*. Oxford: W. H. Freeman.

Dawkins, R. (1986). *The Blind Watchmaker*. Harlow, Essex: Longman.

De Groot, P. (1980). Information transfer in a socially roosting weaver bird (*Quelea quelea*; Ploceinae): an experimental study. *Animal Behaviour*, **28**, 1249–54.

De Waal, F. B. M. (1994). Chimpanzee's adaptive potential: a comparison of social life under captive and wild conditions. In *Chimpanzee Cultures*, ed. R. W. Wrangham, W. C. McGrew, F. B. M. de Waal & P. G. Heltne, pp. 243–60. Cambridge, Massachusetts: Harvard University Press.

Deacon, T. (1997). *The Symbolic Species: The Co-evolution of Language and the Brain*. New York: W. W. Norton.

Dennett, D. C. (1995). *Darwin's Dangerous Idea*. New York: Simon and Schuster.

Diamond, J. (1986). Biology of birds of paradise and bowerbirds. *Annual Review of Ecology and Systematics*, **17**, 17–37.

Diamond, J. (1987). Bower building and decoration by the bowerbird *Amblyornis inornatus*. *Ethology*, **74**, 177–204.

Diamond, J. (1988). Experimental study of bower decoration by the bowerbird *Amblyornis inornatus*, using colored poker chips. *American Naturalist*, **131**, 631–53.

Dilger, W. C. (1962). The behavior of lovebirds. *Scientific American*, **206**(1), 88–98.

Donald, M. (1991). *Origins of the Modern Mind*. Cambridge, Massachusetts: Harvard University Press.

Doolan, S. P. & Macdonald, D. W. (1996a). Diet and foraging behaviour of group-living meerkats, *Suricata suricatta*, in the southern Kalahari. *Journal of Zoology*, **239**, 697–716.

Doolan, S. P. & Macdonald, D. W. (1996b). Dispersal and extra-territorial prospecting by slender-tailed meerkats (*Suricata suricatta*) in the south-western Kalahari. *Journal of Zoology*, **240**, 59–73.

Doolan, S. P. & Macdonald, D. W. (1997a). Breeding and juvenile survival among slender-tailed meerkats (*Suricata suricatta*) in the south-western Kalahari: ecological and social influences. *Journal of Zoology*, **242**, 309–27.

Doolan, S. P. & Macdonald, D. W. (1997b). Band structure and failures of reproductive suppression in a cooperatively breeding carnivore, the slender-tailed meerkat (*Suricata suricatta*). *Behaviour*, **134**, 827–48.

Doron, N. & Wollberg, Z. (1994). Cross-modal neuroplasticity in the blind mole rat *Spalax Ehrenbergi*: a WGA–HRP tracing study. *NeuroReport*, **5**, 2697–701.

Ducker, G. (1962). Brutpflegeverhalten und Ontogenese des Verhaltens bei Surikaten (*Suricata suricatta* Schreb., Viverridae). *Behaviour*, **19**, 305–40.

Dudai, Y. (1989). *The Neurobiology of Memory*. Oxford: Oxford University Press.

Dugatkin, L. A (1996). Copying and mate choice. In *Social Learning in Animals: The Roots of Culture*, ed. C. M. Heyes and B. G. Galef, Jr., pp. 85–105. San Diego: Academic Press.

Dugatkin, L. A. (1997). *Cooperation Among Animals: An Evolutionary Perspective*. New York: Oxford University Press.

Dunbar, R. I. M. (1992). Neocortex size as a constraint on group size in primates. *Journal of Human Evolution*, **20**, 469–93.

Dunn, P. O., Cockburn, A. & Mulder, R. A. (1995). Fairy-wren helpers often care for young to which they are unrelated. *Proceedings of the Royal Society of London, Series B*, **259**, 339–43.

Durham, W. H. (1991). *Coevolution: Genes, Culture, and Human Diversity*. Stanford: Stanford University Press.

Eadie, J. McA., Kehoe, F. P & Nudds, T. D. (1988). Pre-hatch and post-hatch brood amalgamation in North American Anatidae: a review of hypotheses. *Canadian Journal of Zoology*, **66**, 1709–21.

Edmunds, M. (1974). *Defence in Animals: A Survey of Anti-Predator Defences*. Harlow, Essex: Longman.

Eens, M. (1997). Understanding the complex song of the European starling: an integrated ethological approach. *Advances in the Study of Behavior*, **26**, 355–434.

Ehrlich, P. R., Dobkin, D. S., Wheye, D. & Pimm, S. L. (1994). *The Birdwatcher's Handbook: A Guide to the Natural History of the Birds of Britain and Europe*. Oxford: Oxford University Press.

Emlen, J. T., Jr. (1954). Territory, nest building, and pair formation in the cliff swallow. *The Auk*, **71**,16–35.

Emlen, S. T. (1990). White-fronted bee-eaters: helping in a colonially nesting species. In: *Cooperative Breeding in Birds: Long-term Studies of Ecology and Behavior*, ed. P. B. Stacey & W. D. Koenig, pp. 489–526. Cambridge: Cambridge University Press.

Emlen, S. T. (1991). Evolution of cooperative breeding in birds and mammals. In *Behavioural Ecology: An Evolutionary Approach*, 3rd edn, ed. J. R. Krebs & N. B. Davies, pp. 301–37. Oxford: Blackwell.

Emlen, S. T. (1995). An evolutionary theory of the family. *Proceedings of the National Academy of Sciences, USA*, **92**, 8092–9.

Emlen, S. T. (1996). Living with relatives: lessons from avian family systems. *Ibis*, **138**, 87–100.

Emlen, S. T. (1997a). When mothers prefer daughters over sons. *Trends in Ecology and Evolution*, **12**, 291–2.

Emlen, S. T. (1997b). Predicting family dynamics in social vertebrates. In *Behavioural Ecology: An Evolutionary Approach*, 4th edn, ed. J. R. Krebs & N. B. Davies, pp. 228–53. Oxford: Blackwell Science.

Emlen, S. T. & Wrege, P. H. (1988). The role of kinship in helping decisions among white-fronted bee-eaters. *Behavioral Ecology and Sociobiology*, **23**, 305–15.

Emlen, S. T. & Wrege, P. H. (1989). A test of alternate hypotheses for helping behavior in white-fronted bee-eaters of Kenya. *Behavioral Ecology and Sociobiology*, **25**, 303–19.

Emlen, S. T. & Wrege, P. H. (1992). Parent–offspring conflict and the recruitment of helpers among bee-eaters. *Nature*, **356**, 331–3.

Emlen, S. T., Wrege, P. H. & Demong, N. J. (1995). Making decisions in the family: an evolutionary perspective. *American Scientist*, **83**, 148–57.

Enstrom, D. A. (1993). Female choice for age-specific plumage in the orchard oriole: implications for delayed plumage maturation. *Animal Behaviour*, **45**, 435–42.

Erickson, C. J. (1973). Mate familiarity and the reproductive behavior of ringed turtle doves. *The Auk*, **90**, 780–95.

Eshel, I. & Feldman, M. W. (1991). The handicap principle in parent–offspring conflict: comparison of optimality and population–genetic analysis. *American Naturalist*, **137**, 167–85.

Estes, R. D. (1974). Social organization of the African Bovidae. In *The Behaviour of Ungulates and its Relation to Management*, ed. V. Geist & F. Walther, pp. 165–205. Morges, Switzerland: IUCN.

Estes, R. D. (1991). *The Behavior Guide to African Mammals*. Berkeley: University of California Press.

Evans, L. D. & Breed, M. D. (1984). Segregation of cockroach nymphs into sibling groups. *Annals of the Entomological Society of America*, **77**, 574–7.

Evans, P. G. H. (1984). Dolphins. In *The Encyclopaedia of Mammals*, vol.1, ed. D. Macdonald, pp.180–9. London: Allen & Unwin.

Ewer, R. F. (1956). Imprinting in animal behaviour. *Nature*, **177**, 227–8.

Ewer, R. F. (1960). Natural selection and neoteny. *Acta Biotheoretica*, **13**, 161–84.

Ewer, R. F. (1963a). The behaviour of the meerkat, *Suricata suricatta* (Schreber). *Zeitschrift für Tierpsychologie*, **20**, 570–607.

Ewer, R. F. (1963b). A note on the suckling behaviour of the viverrid, *Suricata suricatta* (Schreber). *Animal Behaviour*, **11**, 599–601.

Ewer, R. F. (1968). *Ethology of Mammals*. London: Paul Elek.

Ewer, R. F. (1973). *The Carnivores*. London: Weidenfeld & Nicolson.

Fabre, J. H. (1913). *The Life of the Spider*. New York: Dodd, Mead & Company.

Fairbanks, L. A. (1996). Individual differences in maternal style: causes and consequences for mothers and offspring. *Advances in the Study of Behavior*, **25**, 579–611.

Farabaugh, S. M. (1982). The ecological and social significance of duetting. In *Acoustic Communication in Birds*, vol. 2, ed. D. E. Kroodsma & E. H. Miller, pp. 85–124. New York: Academic Press.

Fedigan, L. M. (1982). *Primate Paradigms: Sex Roles and Social Bonds*. Chicago: University of Chicago Press.

Feldman, H. N. (1993). Maternal care and differences in the use of nests in the domestic cat. *Animal Behaviour*, **45**, 13–23.

Ferguson, R. S. & Sealy, S. G. (1983). Breeding ecology of the horned grebe, *Podiceps auritus*, in southwest Manitoba. *Canadian Field Naturalist*, **97**, 401–8.

Ferrière, R. (1998). Help and you shall be helped. *Nature*, **393**, 517–9.

Fessl, B., Kleindorfer, S. & Hoi, H. (1996). Extra male parental behaviour: evidence for an alternative mating strategy in the moustached warbler *Acrocephalus melanopogon*. *Journal of Avian Biology*, **27**, 88–91.

Fisher, R. A. (1958). *The Genetical Theory of Natural Selection*, 2nd edn. New York: Dover.

Fleming, A. S., Morgan, H. D. & Walsh, C. (1996). Experiential factors in postpartum regulation of maternal care. *Advances in the Study of Behavior*, **25**, 295–332.

Fodor, J. A. (1983). *The Modularity of Mind*. Cambridge, Massachusetts: MIT Press.

Foelix, R. F. (1997). *Biology of Spiders*, 2nd edn. Oxford: Oxford University Press.

Ford, J. K. B. (1990). Vocal traditions among resident killer whales (*Orcinus orca*) in coastal waters of British Columbia. *Canadian Journal of Zoology*, **69**, 1454–83.

Forslund, P. & Pärt, T. (1995). Age and reproduction in birds – hypotheses and tests. *Trends in Ecology and Evolution*, **10**, 374–8.

Fox, C. W. & Mousseau, T. A. (1998). Maternal effects as adaptations for transgenerational phenotypic plasticity in insects. In *Maternal Effects as Adaptations*, ed. T. A. Mousseau & C. W. Fox, pp. 159–77. New York: Oxford University Press.

Fox Keller, E. (1992). *Secrets of Life, Secrets of Death*. New York: Routledge.

Fox Keller, E. (1996). Language and ideology in evolutionary theory: reading cultural norms into natural law. In *Feminism and Science*, ed. E. Fox Keller & H. E. Longino, pp. 154–72. New York: Oxford University Press.

Frank, H. & Frank, M. G. (1982). On the effects of domestication on canine social development and behavior. *Applied Animal Ethology*, **8**, 507–25.

Freeberg, T. M. (1998). The cultural transmission of courtship patterns in cowbirds, *Molothrus ater*. *Animal Behaviour*, **56**, 1063–73.

Freud, S. (1921). Group psychology and the analysis of the ego. In *Great Books of the Western World, 54. The Major Works of Sigmund Freud*, pp. 664–96, 1952. Chicago: Encyclopaedia Britannica.

Fry, C. H. (1984). *The Bee-eaters*. Calton, Staffordshire: Poyser.

Galef, B. G., Jr. (1988). Imitation in animals: history, definition, and interpretation of data from the psychological laboratory. In *Social Learning: Psychological and Biological Perspectives*, ed. T. R. Zentall & B. G. Galef, Jr., pp. 3–28. Hillsdale, New Jersey: Lawrence Erlbaum.

Galef, B. G., Jr. (1991). Information centres of Norway rats: sites for information exchange and information parasitism. *Animal Behaviour*, **41**, 295–301.

Galef, B. G., Jr. (1992). The question of animal culture. *Human Nature*, 3, 157–78.

Galef, B. G. Jr. (1996). Tradition in animals: field observations and laboratory analyses. In *Readings in Animal Cognition*, ed. M. Bekoff & D. Jamieson, pp. 91–105. Cambridge, Massachusetts: Bradford Book, MIT Press.

Galef, B. G., Jr. & Beck, M. (1990). Diet selection and poison avoidance by mammals individually and in social groups. In *Handbook of Behavioral Neurobiology*, vol. 10, *Neurobiology of Food and Fluid Intake*, ed. E. M. Stricker, pp. 329–49. New York: Plenum Press.

Galef, B. G., Jr. & Sherry, D. F. (1973). Mother's milk: a medium for transmission of cues reflecting the flavor of mother's diet. *Journal of Comparative and Physiological Psychology*, **83**, 374–8.

Galef, B. G., Jr. & Wigmore, S. W. (1983). Transfer of information concerning distant foods: a laboratory investigation of the 'information-centre' hypothesis. *Animal Behaviour*, **31**, 748–58.

Garcia, J., Hankins, W. G. & Rusiniak, K. W. (1974). Behavioral regulation of the milieu interne in man and rat. *Science*, **185**, 824–31.

Geertz, C. (1973). *The Interpretation of Cultures*. New York: Basic Books.

Geist, V. (1971). *Mountain Sheep: A Study in Behavior and Evolution*. Chicago: University of Chicago Press.

Gibson, R. M., Bradbury, J. W. & Vehrencamp, S. L. (1991). Mate choice in lekking sage grouse revisited: the roles of vocal display, female site fidelity, and copying. *Behavioral Ecology*, **2**, 165–80.

Gigerenzer, G. (1997). The modularity of social intelligence. In *Machiavellian Intelligence II: Extensions and Evaluations*, ed. A. Whiten & R. W. Byrne, pp. 264–88. Cambridge: Cambridge University Press.

Gilbert, A. N. (1995). Tenacious nipple attachment in rodents: the sibling competition hypothesis. *Animal Behaviour*, **50**, 881–91.

Gill, S. B. (1995). *Ornithology*, 2nd edn. New York: W. H. Freeman.

Gill, T. J. (1988). Immunological and genetical factors influencing pregnancy. In *The Physiology of Reproduction*, vol. 2, ed. E. Knobil & J. D. Neill, pp. 2023–42. New York: Raven Press.

Ginzburg, L. R. (1998). Inertial growth: population dynamics based on maternal effects. In *Maternal Effects as Adaptations*, ed. T. A. Mousseau & C. W. Fox, pp. 42–53. New York: Oxford University Press.

Giraldeau, L. A. (1997). The ecology of information use. In *Behavioural Ecology: An Evolutionary Approach*, 4th edn, ed. J. R. Krebs & N. B. Davies, pp. 42–68. London: Blackwell Science.

Godden, D. & Baddeley, A. D. (1975). Context-dependent memory in two natural environments: on land and under water. *British Journal of Psychology*, **66**, 325–31.

Godfray, H. C. J. (1995a). Evolutionary theory of parent–offspring conflict. *Nature*, **376**, 133–8.

Godfray, H. C. J. (1995b). Signaling of need between parents and young: parent–offspring conflict and sibling rivalry. *American Naturalist*, **146**, 1–24.

Godfray, H. C. J. & Parker, G. A. (1991). Clutch size, fecundity and parent–offspring conflict. *Philosophical Transactions of the Royal Society of London, Series B*, **332**, 67–79.

González-Mariscal, G. & Rosenblatt, J. S. (1996). Maternal behavior in rabbits: a historical and multidisciplinary perspective. *Advances in the Study of Behavior*, **25**, 333–60.

Goodloe, L., Sanwald, R. & Topoff, H. (1987). Host specificity in raiding behavior of the slave-making ant *Polyergus lucidus*. *Psyche*, **94**, 39–44.

Goodnight, C. J. & Stevens, L. (1997). Experimental studies of group selection: what do they tell us about group selection in nature? *American Naturalist*, **150**, S59–79.

Goodwin, T. W. (1950). Carotenoids and reproduction. *Biological Reviews*, **25**, 391–413.

Gori, D. F. (1988). Colony-facilitated foraging in yellow-headed blackbirds: experimental evidence for information transfer. *Ornis Scandinavica*, **19**, 224–30.

Gorman, M. L. (1974). The endocrine basis of pair-formation behaviour in the male eider *Somateria mollissima*. *Ibis*, **116**, 451–65.

Gosler, A. (1993). *The Great Tit*. London: Hamlyn.

Gosling, L. M. & Petrie, M. (1981). The economics of social organization. In *Physiological Ecology: An Evolutionary Approach to Resourse Use*, ed. C. R. Townsend & P. Calow, pp. 315–45. Oxford: Blackwell.

Gottlander, K. (1987). Parental feeding behavior and sibling competition in the pied flycatcher (*Ficedula hypoleuca*). *Ornis Scandinavica*, **18**, 269–76.

Gottlieb, G. (2000). A developmental psychobiological systems view: early formulation and current status. In *Cycles of Contingency*, ed. S. Oyama, P. Griffith & R. Gray. Cambridge Massachusetts: MIT Press.

Gottlieb, G. & Klopfer, P. H. (1962). The relation of developmental age to auditory and visual imprinting. *Journal of Comparative and Physiological Psychology*, **55**, 821–26.

Gould, E., Tanapat, P., McEwen, B. S., Flügge, G. & Fuchs, E. (1998). Proliferation of granule cell precursors in the dentate gyrus of adult monkeys is diminished by stress. *Proceedings of the National Academy of Sciences, USA*, **95**, 3168–71.

Gould, J. L. & Gould, C. G. (1989). *Sexual Selection*. New York: W. H. Freeman.

Gowaty, P. A. (1996). Field studies of parental care in birds: new data focus questions on variation among females. *Advances in the Study of Behavior*, **25**, 477–531.

Grant, B. R. & Grant, P. R. (1996). Cultural inheritance of song and its role in the evolution of Darwin's finches. *Evolution*, **50**, 2471–87.

Grant, P. R. & Grant, B. R. (1997). Hybridization, sexual imprinting, and mate choice. *American Naturalist*, **149**, 1–28.

Green, D. J. & Krebs, E. A. (1995). Courtship feeding in ospreys, *Pandion haliaetus*: a criterion for mate assessment? *Ibis*, **137**, 35–43.

Green, S. (1975). Dialects in Japanese monkeys: vocal learning and cultural transmission of locale-specific vocal behavior? *Zeitschrift für Tierpsychologie*, **38**, 304–14.

Greene, E. (1987). Individuals in an osprey colony discriminate between high and low quality information. *Nature*, **329**, 239–41.

Greenwood, P. J. (1980). Mating systems, philopatry and dispersal in birds and mammals. *Animal Behaviour*, **28**, 1140–62.

Griffin, D. R. (1992). *Animal Minds*. Chicago: Chicago Univeristy Press.

Griffiths, P. & Gray, R. D. (1994). Developmental systems and evolutionary explanations. *Journal of Philosophy*, **91**, 277–304.

Gyger, M., Marler, P. & Pickert, R. (1987). Semantics of an avian alarm call system: the male domestic fowl, *Gallus domesticus*. *Behaviour*, **102**, 15–40.

Haig, D. (1992). Genomic imprinting and the theory of parent–offspring conflict. *Seminars in Developmental Biology*, **3**, 153–60.

Haldane, J. B. S. (1954). Introducing Douglas Spalding. *British Journal of Animal Behaviour*, **2**, 1–11.

Haldane, J. B. S. (1959). Natural selection. In *Darwin's Biological Work*, ed. P. R. Bell, pp. 101–49. Cambridge: Cambridge University Press.

Hamilton, W. D. (1964). The genetical evolution of social behaviour. *Journal of Theoretical Biology*, **7**, 1–52.

Hamilton, W. D. (1971). Geometry for the selfish herd. *Journal of Theoretical Biology*, **31**, 295–311.

Hamilton, W. D. & Zuk, M. (1982). Heritable true fitness and bright birds: a role for parasites? *Science*, **218**, 384–7.

Hansen, T. F. (1996). Does adoption make evolutionary sense? *Animal Behaviour*, **51**, 474–5.

Harcus, J. L. (1977). The functions of vocal duetting in some African birds. *Zeitschrift für Tierpsychologie*, **43**, 23–45.

Hardy, A. (1965). *The Living Stream: A Restatement of Evolution Theory and its Relation to the Spirit of Man*. London: Collins.

Harris, M. A. & Murie, J. O. (1984). Inheritance of nest sites in female Columbian ground squirrels. *Behavioral Ecology and Sociobiology*, **15**, 97–102.

Harvey, A. W. (1998). Genes for asymmetry easily overruled. *Nature*, **392**, 345–6.

Hasler, A. D. & Scholz, A. T. (1983). *Olfactory Imprinting and Homing in Salmon: Investigations into the Mechanism of the Imprinting Process*. Berlin: Springer Verlag.

Hatfield, E., Cacioppo, J. T. & Rapson, R. L. (1994). *Emotional Contagion*. Paris: Cambridge University Press.

Hauser, M. D. (1992). Costs of deception: cheaters are punished in rhesus monkeys (*Macaca mulatta*). *Proceedings of the National Academy of Sciences, USA*, **89**, 12137–9.

Hauser, M .D. (1997). *The Evolution of Communication*. Cambridge, Massachusetts: MIT Press.

Hauser, M. D. & Marler, P. (1993). Food-associated calls in rhesus macaques (*Macaca mulatta*). II. Costs and benefits of call production and suppression. *Behavioral Ecology*, **4**, 206–12.

Heil, P., Bronchti, G., Wollberg, Z. & Scheich, H. (1991). Invasion of visual cortex by the auditory system in the naturally blind mole rat. *NeuroReport*, **2**, 735–8.

Heinrich, B. (1989). *Ravens in Winter*. New York: Summit Books.

Heinrich, B. & Marzluff, J. (1995). Why ravens share. *American Scientist*, **83**, 342–9.

Heinsohn, R. G. (1991a). Slow learning of foraging skills and extended parental care in cooperatively breeding white-winged choughs. *American Naturalist*, **137**, 864–81.

Heinsohn, R. G. (1991b). Kidnapping and reciprocity in cooperatively breeding white-winged choughs. *Animal Behaviour*, **41**, 1097–100.

Heinsohn, R. G. & Legge, S. (1999). The cost of helping. *Trends in Ecology and Evolution*, **14**, 53–7.

Heinsohn, R. G., Cockburn, A. & Cunningham, R. B. (1988). Foraging, delayed maturation and advantages of cooperative breeding in white-winged choughs, *Corcorax melanorhamphos*. *Ethology*, **77**, 177–86.

Hemmer, H. (1990). *Domestication: The Decline of Environmental Appreciation*. Cambridge: Cambridge University Press.

Hendrichs, H. (1996). The complexity of social and mental structure in nonhuman mammals. In *Evolution, Order and Complexity*, ed. E. L. Khalil & K. E. Boulding, pp. 104–21. London: Routledge.

Hepper, P. G. (1988). Adaptive fetal learning: prenatal exposure to garlic affects postnatal preferences. *Animal Behaviour*, **36**, 935–6.

Herrenstein, R. J. (1984). Objects, categories and discriminative stimuli. In *Animal Cognition*, ed. H. L. Roitblat, T. G. Bever & H. S. Terrace, pp. 233–61. Hillsdale, New Jersey: Lawrence Elrbaum.

Herter, K. (1965). *Hedgehogs*. London: Phoenix House.

Heyes, C. M. (1993). Imitation, culture and cognition. *Animal Behaviour*, **46**, 999–1010.

Heyes, C. M. (1994). Social learning in animals: categories and mechanisms. *Biological Reviews*, **69**, 207–31.

Heyes, C. M. & Galef, B. G., Jr. (ed.) (1996). *Social Learning in Animals: The Roots of Culture*. San Diego: Academic Press.

Hill, G. E. (1992). Proximate basis of variation in carotenoid pigmentation in male house finches. *The Auk*, **109**, 1–12.

Hill, G. E. (1993). Geographic variation in the carotenoid plumage pigmentation of male house finches (*Carpodacus mexicanus*). *Biological Journal of the Linnean Society*, **49**, 63–86.

Hill, G. E. (1994). House finches are what they eat: a reply to Hudon. *The Auk*, **111**, 221–5.

Hinde, R. A. (1952). The behaviour of the great tit (*Parus major*) and some other related species. *Behaviour*, Supplement 2, pp. 1–198.

Hinde, R. A. (1958). The nest-building behaviour of domesticated canaries. *Proceedings of the Zoological Society of London*, **131**, 1–48.

Hiraiwa, M. (1981). Maternal and alloparental care in a troop of free-ranging Japanese monkeys. *Primates*, **22**, 309–29.

Hölldobler, B. & Wilson, E. O. (1990). *The Ants*. Cambridge, Massachusetts: Belknap Press.

Holliday, R. (1990). Mechanisms for the control of gene activity during development. *Biological Reviews*, **65**, 431–71.

Hoogland, J. L. (1995). *The Black-Tailed Prairie Dog: Social Life of a Burrowing Mammal*. Chicago: University of Chicago Press.

Hoogland, J. L. & Sherman, P. W. (1976). Advantages and disadvantages of bank swallow (*Riparia riparia*) coloniality. *Ecological Monographs*, **46**, 33–58

Hooker, T. & Hooker, B. I. (1969). Duetting. In *Bird Vocalizations*, ed. R. A. Hinde, pp. 185–205. Cambridge: Cambridge University Press.

Huck, U. W., Labov, J. B. & Lisk, R. D. (1987). Food-restricting first generation juvenile female hamsters (*Mesocricetus auratus*) affects sex ratio and growth of third generation offspring. *Biology of Reproduction*, **37**, 612–17.

Huffman, M. A. (1996). Acquisition of innovative cultural behaviors in nonhuman primates: a case study of stone handling, a socially transmitted behavior in Japanese macaques. In *Social Learning in Animals: the Roots of Culture*, ed. C. M. Heyes & B. G. Galef, Jr., pp. 267–89. San Diego: Academic Press.

Hughes, R. N. (ed.) (1993). *Diet Selection*. Oxford: Blackwell Scientific Publications.

Hume, R. (1993). *The Common Tern*. London: Hamlyn.

Hunt, J. H. & Nalepa, C. A. (ed.) (1994). *Nourishment and Evolution in Insect Societies*. Boulder: Westview Press.

Hurford, J. R., Studdert-Kennedy, M. & Knight, C. (1998). *Approaches to the Evolution of Language*. Cambridge: Cambridge University Press.

Hurst, L. D. (1997). Evolutionary theories of genomic imprinting. In *Genomic Imprinting*, ed. W. Reik & A. Surani, pp. 211–37. Oxford: Oxford University Press.

Huxley, J. (1942). *Evolution, The Modern Synthesis*. London: Allen & Unwin.

Hyman, S. E. (1998). A new image for fear and emotion. *Nature*, **393**, 417–18.

Immelmann, K. (1975). The evolutionary significance of early experience. In *Function and Evolution in Behaviour: Essays in Honour of Professor Niko Tinbergen, FRS*, ed. G. Baerends, C. Beer & A. Manning, pp. 243–53. Oxford: Clarendon Press.

Ims, R.A. (1990). The ecology and evolution of reproductive synchrony. *Trends in Ecology and Evolution*, **5**, 135–40.

Itani, J. (1958). On the acquisition and propagation of a new food habit in the natural group of the Japanese monkey at Takasaki-Yama. *Primates*, **1**, 84–98.

Jablonka, E. (1994). Inheritance systems and the evolution of new levels of individuality. *Journal of Theoretical Biology*, **170**, 301–9.

Jablonka, E. & Lamb, M. J. (1995). *Epigenetic Inheritance and Evolution: The Lamarckian Dimension.* Oxford: Oxford University Press.

Jablonka, E. & Rechav, G. (1996). The evolution of language in light of the evolution of literacy. In *Origins of Language*, ed. J. Trabant, pp. 70–88. Budapest: Collegium Budapest.

Jablonka, E. & Szathmáry, E. (1995). The evolution of information storage and heredity. *Trends in Ecology and Evolution*, **10**, 206–11.

Jablonka, E., Lamb, M. J. & Avital, E. (1998). 'Lamarckian' mechanisms in darwinian evolution. *Trends in Ecology and Evolution*, **13**, 206–10.

Jaisson, P. (1980). Environmental preference induced experimentally in ants (Hymenoptera: Formicidae). *Nature*, **286**, 388–9.

Jamieson, I. G. (1986). The functional approach to behavior: is it useful? *American Naturalist*, **127**,195–208.

Jamieson, I. G. (1989). Behavioral heterochrony and the evolution of birds' helping at the nest: an unselected consequence of communal breeding? *American Naturalist*, **133**, 394–406.

Jamieson, I. G. (1991). The unselected hypothesis for the evolution of helping behavior: too much or too little emphasis on natural selection? *American Naturalist*, **138**, 271–82.

Jamieson, I. G. & Craig, J. L. (1987). Critique of helping behavior in birds: a departure from functional explanations. In *Perspectives in Ethology*, vol. 7, ed. P. P. G. Bateson & P. Klopfer, pp. 79–98. New York: Plenum Press.

Janetos, A. C. (1980). Strategies of female mate choice: a theoretical analysis. *Behavioral Ecology and Sociobiology*, **7**, 107–12.

Janeway, C. A., Jr. & Travers, P. (1994). *Immunobiology.* Oxford: Blackwell Scientific Publications.

Jarman, P. J. (1974). The social organisation of antelopes in relation to their ecology. *Behaviour*, **48**, 215–67.

Johannsen, W. (1911). The genotype conception of heredity. *American Naturalist*, **45**, 129–59.

Johns, T. & Duquette, M. (1991). Detoxification and mineral supplementation as functions of geophagy. *American Journal of Clinical Nutrition*, **53**, 448–56.

Johnsgard, P. A. (1997). *The Avian Brood Parasites: Deception at the Nest.* New York: Oxford University Press.

Johnston, T. D. (1982). Selective costs and benefits in the evolution of learning. *Advances in the Study of Behavior*, **12**, 65–106.

Johnstone, R. A. (1996). Begging signals and parent–offspring conflict: do parents always win? *Proceedings of the Royal Society of London, Series B*, **263**, 1677–81.

Jolly, A. (1967). Breeding synchrony in wild *Lemur catta*. In *Social Communication Among Primates*, ed. S. A. Altmann, pp. 3–14. Chicago: University of Chicago Press.

Jones, W. T. (1986). Survivorship in philopatric and dispersing kangaroo rats (*Dipodomys spectabilis*). *Ecology*, **67**, 202–7.

Källander, H. & Smith, H. G. (1990). Food storing in birds: An evolutionary perspective. In *Current Ornithology*, vol. 7, ed. D. M. Power, pp. 147–207. New York: Plenum Press.

Kavaliers, M., Colwell, D. D. & Galea, L. A. M. (1995). Parasitic infection impairs spatial learning in mice. *Animal Behaviour*, **50**, 223–9.

Kavanau, J. L. (1964). Compulsory regime and control of environment in animal behavior. 1. Wheel-running. *Behaviour*, **20**, 251–81.

Kavanau, J. L. (1987). *Lovebirds, Cockatiels, Budgerigars: Behavior and Evolution*. Los Angeles, California: Science Software Systems.

Kawai, M. (1965). Newly-acquired pre-cultural behavior of the natural troop of Japanese monkeys on Koshima Islet. *Primates*, **6**, 1–30.

Kawamura, S. (1959). The process of sub-culture propagation among Japanese macaques. *Primates*, **2**, 43–60.

Kendrick, K. M., Hinton, M. R., Atkins, K., Haupt, M. A. & Skinner, J. D. (1998). Mothers determine sexual preferences. *Nature*, **395**, 229–30.

Kenward, R. E. (1978). Hawks and doves: factors affecting success and selection in goshawk attacks on woodpigeons. *Journal of Animal Ecology*, **47**, 449–60.

Keverne, E. B., Martel, F. L. & Nevison, C. M. (1996). Primate brain evolution: genetic and funtional considerations. *Proceedings of the Royal Society of London, Series B*, **262**, 689–96.

Keverne, E. B., Fundele, R., Narasimha, M., Barton, S. C. & Surani, M. A. (1996). Genomic imprinting and the differential roles of parental genomes in brain development. *Developmental Brain Research*, **92**, 91–100.

Kilner, R. & Johnstone, R. A. (1997). Begging the question: are offspring solicitation behaviours signals of need? *Trends in Ecology and Evolution*, **12**, 11–15.

Kinsley, C. H. & Bridges, R. S. (1988). Prenatal stress and maternal behavior in intact virgin rats: response latencies are decreased in males and increased in females. *Hormones and Behavior*, **22**, 76–89.

Kinsley, C. H. & Svare, B. B. (1986). Sex-biased fecundity in prenatally-stressed female mice: alteration of sex ratio in favor of increased production of male offspring. *Biology of Reproduction*, **34**, suppl. 1, 67.

Kirkpatrick, M. & Lande, R. (1989). The evolution of maternal characters. *Evolution*, **43**, 485–503.

Kitchener, A. (1991). *The Natural History of the Wild Cats*. London: Christopher Helm.

Kleiman, D. G., Allen, M. E., Thompson, K. V. & Lumpkin, S. (ed.) (1996). *Wild Mammals in Captivity: Principles and Techniques*. Chicago: University of Chicago Press.

Klint, T. (1978). Significance of mother and sibling experience for mating preferences in the mallard. *Zeitschrift für Tierpsychologie*, **47**, 50–60.

Klopfer, P. & Klopfer, M. (1977). Compensatory responses of goat mothers to their impaired young. *Animal Behaviour*, **25**, 286–91.

Knobil, E. & Neill, J. D. (ed.) (1994). *The Physiology of Reproduction*, 2nd edn, vol. 2. New York: Raven Press.

Köhler, W. (1925). *The Mentality of Apes*, 2nd edn. London: Kegan Paul. (Reprinted by Penguin Books, 1957.)

Kokko, H. & Lindström, J. (1996). Evolution of female preference for old mates. *Proceedings of the Royal Society of London, Series B*, **263**, 1533–8.

Komdeur, J. (1996). Facultative sex ratio bias in the offspring of Seychelles warblers. *Proceedings of the Royal Society of London, Series B*, **263**, 661–6.

König, B. (1993). Maternal investment of communally nursing female house mice (*Mus musculus domesticus*). *Behavioural Processes*, **30**, 61–74.

König, B. (1994). Fitness effects of communal rearing in house mice: the role of relatedness versus familiarity. *Animal Behaviour*, **48**, 1449–57.

Krebs, C. J. (1996). Population cycles revisited. *Journal of Mammalogy*, **77**, 8–24.

Krebs, C. J. & Myers, J. H. (1974). Population cycles in small mammals. *Advances in Ecological Research*, **8**, 267–399.

Kreulen, D. A. (1985). Lick use by large herbivores: a review of benefits and banes of soil consumption. *Mammal Review*, **15**,107–23.

Kruijt, J. P., ten Cate, C. J. & Meeuwissen, G. B. (1983). The influence of siblings on the development of sexual preferences of male zebra finches. *Developmental Psychobiology*, **16**, 233–9.

Kruska, D. (1988). Mammalian domestication and its effects on brain structure and behavior. In *Intelligence and Evolutionary Biology*, ed. H. J. Jerison & I. Jerison, pp. 211–50. Berlin: Springer Verlag.

Kruuk, H. (1964). Predators and anti-predator behaviour of the black-headed gull (*Larus ridibundus L.*). *Behaviour Supplement*, **11**, 1–129.

Kruuk, H. (1972). *The Spotted Hyena: A Study of Predation and Social Behavior*. Chicago: University of Chicago Press.

Kummer, H. (1971). *Primate Societies: Group Techniques of Ecological Adaptation*. Chicago: Aldine-Atherton.

Kummer, H. (1995). *In Quest of the Sacred Baboon: A Scientist's Journey*. Princeton, New Jersey: Princeton University Press.

Kunkel, P. (1974). Mating systems of tropical birds: The effects of weakness or absence of external reproduction-timing factors, with special reference to prolonged pair bonds. *Zeitschrift für Tierpsychologie*, **34**, 265–307.

Kunz, T. H. (ed.) (1982). *Ecology of Bats*. New York: Plenum Press.

Lacey, E. P. (1998). What is an adaptive environmentally induced parental effect? In *Maternal Effects as Adaptations*, ed. T. A. Mousseau & C. W. Fox, pp. 54–66. New York: Oxford University Press.

Lachmann, M. & Jablonka, E. (1996). The inheritance of phenotypes: an adaptation to fluctuating environments. *Journal of Theoretical Biology*, **181**,1–9.

Lachmann, M., Sella, G. & Jablonka, E. (2000). On the advantage of information sharing. *Proceedings of the Royal Society of London, Series B*.

Lack, D. (1940). Courtship feeding in birds. *The Auk*, **57**, 169–78.

Lack, D. (1954). *The Natural Regulation of Animal Numbers*. Oxford: Clarendon Press.

Lack, D. (1966). *Population Studies of Birds*. Oxford: Clarendon Press.

Lack, M. (1998). Digging for memes: the role of material objects in cultural evolution. In *Cognition and Material Culture: the Archaeology of Symbolic Storage*, ed. C. Renfrew & C. Scarre, pp. 77–88. McDonald Institute for Archaeological Research.

Laland, K. N. (1994*a*). On the evolutionary consequences of sexual imprinting. *Evolution*, **48**, 477–89.

Laland, K. N. (1994*b*). Sexual selection with a culturally transmitted mating preference. *Theoretical Population Biology*, **45**, 1–15.

Laland, K. N. & Plotkin, H. C. (1993). Social transmission of food preferences among Norway rats by marking of food sites and by gustatory contact. *Animal Learning and Behavior*, **21**, 35–41.

Laland, K. N., Odling-Smee, F. J. & Feldman, M. W. (1996). The evolutionary consequences of niche construction: a theoretical investigation using two-locus theory. *Journal of Evolutionary Biology*, **9**, 293–316.

Langley, P. A. (1977). Physiology of tsetse flies (*Glossina* spp.) (Diptera: Glossinidae): a review. *Bulletin of Entomological Research*, **67**, 523–74.

Larsson, K. & Forslund, P. (1992). Genetic and social inheritance of body and egg size in the barnacle goose (*Branta leucopsis*). *Evolution*, **46**, 235–44.

Le Bon, G. (1895). *The Crowd. A Study of the Popular Mind*. 2nd edn. London: Fisher Unwin.

Leger, D. W., Owings, D. H. & Gelfand, D. L. (1980). Single-note vocalizations of California ground squirrels: graded signals and situation-specificity of predator and socially evoked calls. *Zeitschrift für Tierpsychologie*, **52**, 227–46.

Leonard, M. L., Horn, A. G. & Eden, S. F. (1988). Parent–offspring aggression in moorhens. *Behavioral Ecology and Sociobiology*, **23**, 265–70.

Lessells, C. M. (1990). Helping at the nest in European bee-eaters: who helps and why? In *Population Biology of Passerine Birds*, ed. J. Blondel, A. Gosler, J-D. Lebreton & R. McCleery, pp. 357–68. Berlin: Springer Verlag.

Lessells, C. M. & Krebs, J. R. (1989). Age and breeding performance of European bee-eaters. *The Auk*, **106**, 375–82.

Lessells, C. M., Avery, M. I. & Krebs, J. R. (1994). Nonrandom dispersal of kin: why do brothers nest together. *Behavioral Ecology*, **5**, 105–13.

Lessells, C. M., Rowe, C. L. & McGregor, P. K. (1995). Individual and sex differences in the provisioning calls of European bee-eaters. *Animal Behaviour*, **49**, 244–7.

Lessells, C. M., Coulthard, N. D., Hodgson, P. J. & Krebs, J. R. (1991). Chick recognition in European bee-eaters: acoustic playback experiments. *Animal Behaviour*, **42**, 1031–3.

Lévy, F., Porter, R. H., Kendrick, K. M., Keverne, E. B. & Romeyer, A. (1996). Physiological, sensory, and experiential factors of parental care in sheep. *Advances in the Study of Behavior*, **25**, 385–422.

Lewontin, R. (1978). Adaptation. *Scientific American*, **239**(3),157–69.

Lewontin, R. C. (1992). Genotype and phenotype. In *Keywords in Evolutionary Biology*, ed. E. Fox Keller & E. A. Lloyd, pp. 137–44. Cambridge, Massachusetts: Harvard University Press.

Lewontin, R. C. (1993). *The Doctrine of DNA : Biology as Ideology*. London: Penguin Books.

Leyhausen, P. (1979). *Cat Behaviour*. New York: Garland STPM Press.

Lieberman, A. F. (1993). *The Emotional Life of the Toddler*. New York: Free Press.

Ligon, J. D. (1997). *The Evolution of Avian Breeding Systems*. Oxford: Oxford University Press.

Ligon, J. D. & Ligon, S. H. (1978a). The communal social system of the green woodhoopoe in Kenya. *Living Bird*, **17**, 159–97.

Ligon, J. D. & Ligon, S. H. (1978b). Communal breeding in green woodhoopoes as a case for reciprocity. *Nature*, **276**, 496–8.

Ligon, J. D. & Ligon, S. H. (1982). The cooperative breeding behavior of the green woodhoopoe. *Scientific American*, **247**(1), 106–14.

Ligon, J. D. & Ligon, S. H. (1983). Reciprocity in the green woodhoopoe (*Phoeniculus purpureus*). *Animal Behaviour*, **31**, 480–9.

Löhrl, H. (1959). Zur Frage des Zeitpunktes einer Prägung auf die Heimatregion beim Halsbandschnäpper (*Ficedula albicollis*). *Journal für Ornithologie*, **100**, 132–40.

Lorenz, K. (1970). *Studies in Animal and Human Behaviour*, vol. 1. London: Methuen.

Lorenz, K. (1978). *The Year of the Greylag Goose*. New York: Harcourt Brace Jovanovich.

Lott, D. F. (1984). Intraspecific variation in the social systems of wild vertebrates. *Behaviour*, **88**, 266–325.

Lozano, G. A. (1998). Parasitic stress and self-medication in wild animals. *Advances in the Study of Behavior*, **27**, 291–317.

Lung, N. P., Thompson, J. P., Kollias, G. V., Olsen J. H., Zdiarski, J. A. & Klein, P. A. (1996). Maternal immunoglobulin G antibody transfer and development of immunoglobulin G antibody responses in blue and gold macaw (*Ara ararauna*) chicks. *American Journal of of Veterinary Research*, **57**, 1162–7.

Lyon, B. E., Eadie, J. M. & Hamilton, L. D. (1994). Parental choice selects for ornamental plumage in American coot chicks. *Nature*, **371**, 240–3.

Ma, W. C., Denlinger, D. L., Järlfors, U. & Smith, D. S. (1975). Structural modulations in the tsetse fly milk gland during a pregnancy cycle. *Tissue and Cell*, **7**, 319–30

MacArthur, R. H. & Wilson, E. O. (1967). *The Theory of Island Biogeography*. Princeton, New Jersey: Princeton University Press.

Macdonald, D. (1984). *The Encyclopaedia of Mammals*, vol. 1. London: Allen and Unwin.

Maestripieri, D. (1992). Functional aspects of maternal aggression in mammals. *Canadian Journal of Zoology*, **70**, 1069–77.

Maestripieri, D. & Call, J. (1996). Mother–infant communication in primates. *Advances in the Study of Behavior*, **25**, 613–42.

Maestripieri, D. & Wallen, K. (1997). Infant abuse runs in families of group-living pigtail macaques. *Child Abuse and Neglect*, **21**, 465–71.

Magrath, R. D. (1989). Hatching asynchrony and reproductive success in the blackbird. *Nature*, **339**, 536–8.

Malacarne, G., Cucco, M. & Camanni, S. (1991). Coordinated visual displays and vocal duetting in different ecological situations among Western Palearctic non-passerine birds. *Ethology, Ecology and Evolution*, **3**, 207–19.

Marinier, S. L. & Alexander, A. J. (1995). Coprophagy as an avenue for foals of the domestic horse to learn food preferences from their dams. *Journal of Theoretical Biology*, **173**, 121–4.

Marler, P. (1990). Song learning: the interface between behaviour and neuroethology. *Philosophical Transactions of the Royal Society of London, Series B*, **329**, 109–14.

Marler, P. & Evans, C. S. (1994). The dynamics of vocal communication in birds. In *Evolution and Neurology of Language*, ed. C. D. Gajdusek, G. M. McKhann & L. C. Bolis, pp. 81–9. Amsterdam: Elsevier.

Martin, K. (1995). Patterns and mechanisms for age-dependent reproduction and survival in birds. *American Zoologist*, **35**, 340–8.

Martin, T. E. & Badyaev, A. V. (1996). Sexual diachromatism in birds: importance of nest predation and nest location for females versus males. *Evolution*, **50**, 2454–60.

Marzluff, J. M. & Balda, R. P. (1992). *The Pinyon Jay: Behavioral Ecology of a Colonial and Cooperative Corvid*. London: Poyser.

Marzluff, J. M., Heinrich, B. & Marzluff, C. S. (1996). Raven roosts are mobile information centres. *Animal Behaviour*, **51**, 89–103.

Maynard Smith, J. & Szathmáry, E. (1995). *The Major Transitions in Evolution*. Oxford: W. H. Freeman.

Mayr, E. (1942). *Systematics and the Origin of Species*. New York: Columbia University Press.

Mayr, E. (1963). *Animal Species and Evolution*. Cambridge, Massachusetts: Harvard University Press.

McCarty, J. P. (1996). The energetic cost of begging in nestling passerines. *The Auk*, **113**, 178–88.

McDougall, W. (1920). *The Group Mind*. Cambridge: Cambridge University Press.

McFarland, D. (ed.) (1987). *The Oxford Companion to Animal Behaviour*. Oxford: Oxford University Press.

McGrew, W. C. (1992). *Chimpanzee Material Culture: Implications for Human Evolution*. Cambridge: Cambridge University Press.

McLeskey, J., Rieth, H. J. & Polsgrove, L. (1980). The implications of response generalization for improving the effectiveness of programs for learning disabled children. *Journal of Learning Disabilities*, **13**, 287–90.

Mech, L. D. (1970). *The Wolf: the Ecology and Behavior of an Endangered Species*. Minneapolis: University of Minnesota Press.

Meehan, A. P. (1984). *Rats and Mice: their Biology and Control*. East Grinstead: Rentokil.

Mendelssohn, H. (1974). The development of the populations of gazelles in Israel and their behavioural adaptations. In *The Behaviour of Ungulates and its Relation to Management*, ed. V. Geist & F. Walther, pp. 722–43. Morges, Switzerland: IUCN.

Michener, C. D. (1974). *The Social Behavior of the Bees: A Comparative Study*. Cambridge, Massachusetts: Belknap Press.

Miller, R. E., Murphy, J. V. & Mirsky, I. A. (1959). Non-verbal communication of affect. *Journal of Clinical Psychology*, **15**, 155–8.

Milner, B., Corkin, S. & Teuber, H.-L. (1968). Further analysis of the hippocampal amnesic syndrome: 14-year follow-up study of H.M. *Neuropsychologia*, **6**, 215–34.

Mock, D. W., Drummond, H. & Stinson, C. H. (1990). Avian Siblicide. *American Scientist*, **78**, 438–49.

Mock, D. W. & Forbes L. S. (1992). Parent–offspring conflict: a case of arrested development. *Trends in Ecology and Evolution*, **7**, 409–13.

Mock, D. W. & Forbes, L. S. (1995). The evolution of parental optimism. *Trends in Ecology and Evolution*, **10**, 130–4.

Mock, D. W. & Parker, G. A. (1997). *The Evolution of Sibling Rivalry*. Oxford: Oxford University Press.

Moehlman, P. D. (1979). Jackal helpers and pup survival. *Nature*, **277**, 382–3.

Moehlman, P. D. (1983). Socioecology of silver-backed and golden jackals (*Canis mesomelas, C. aureus*). In *Recent Advances in the Study of Mammalian Behavior*, ed. J. F. Eisenberg & D. G. Kleiman, pp. 423–53. American Society of Mammalogists.

Moehlman, P. D. (1986). Ecology of cooperation in canids, In *Ecological Aspects of Social Evolution*, ed. D. I. Rubenstein & R. W. Wrangham, pp. 64–86. Princeton, New Jersey: Princeton University Press.

Møller, A. P., Milinski, M. & Slater, P. J. B. (ed.) (1998). *Stress and Behavior. Advances in the Study of Behavior*, **27**. San Diego: Academic Press.

Moore, B. R. (1992). Avian movement imitation and a new form of mimicry: tracing the evolution of a complex form of learning. *Behaviour*, **122**, 231–63.

Moore, B. R. (1996). The evolution of imitative learning. In *Social Learning in Animals: The Roots of Culture*, ed. C. M. Heyes & B. G. Galef, Jr., pp. 245–65. San Diego: Academic Press.

Moore, C. L. (1982). Maternal behavior of rats is affected by hormonal condition of pups. *Journal of Comparative and Physiological Psychology*, **96**, 123–9.

Moore, C. L. (1995). Maternal contributions to mammalian reproductive development and the divergence of males and females. *Advances in the Study of Behavior*, **24**, 47–118.

Moran, G. (1984). Vigilance behaviour and alarm calls in a captive group of meerkats, *Suricata suricatta*. *Zeitschrift für Tierpsychologie*, **65**, 228–40.

Morgan, C. L. (1900). *Animal Behaviour*. London: Edward Arnold.

Morris, D. (1990). *Animal Watching*. London: Jonathan Cape.

Morris, J. S., Öhman, A. & Dolan, R. J. (1998). Conscious and unconscious emotional learning in the human amygdala. *Nature*, **393**, 467–70.

Morton, E. S. (1975). Ecological sources of selection on avian sounds. *American Naturalist*, **109**, 17–34.

Moss, C. J. (1988). *Elephant Memories*. New York: Morrow.

Moss, C. J. & Poole, J. H. (1983). Relationships and social structure of African elephants. In *Primate Social Relationships: An Integrated Approach*, ed. R. A. Hinde, pp. 315–25. Oxford: Blackwell Scientific Publications.

Mountjoy, D. J. & Lemon, R. E. (1996). Female choice for complex song in the European starling: a field experiment. *Behavioral Ecology and Sociobiology*, **38**, 65–71.

Mousseau, T. A. & Fox, C. W. (ed.) (1998). *Maternal Effects as Adaptations*. New York: Oxford University Press.

Mulder, R. A. & Langmore, N. E. (1993). Dominant males punish helpers for temporary defection in superb fairy-wrens. *Animal Behaviour*, **45**, 830–3.

Muller, R. E. & Smith, D. G.(1978). Parent-offspring interactions in zebra finches. *The Auk*, **95**, 485–95.

Mundinger, P. C. (1980). Animal cultures and a general theory of cultural evolution. *Ethology and Sociobiology*, **1**, 183–223.

Nadel, L. (1994). Multiple memory systems: what and why, an update. In *Memory Systems*, ed. D. L. Schacter & E. Tulving, pp. 39–63. Cambridge, Massachusetts: MIT Press.

Nakagawa, N. (1998). Ecological determinants of the behavior and social structure of Japanese monkeys: a synthesis. *Primates*, **39**, 375–83.

Nalepa, C. A. (1994). Nourishment and the evolution of termite eusociality. In *Nourishment and Evolution in Insect Societies*, ed. J. H. Hunt & C. A. Nalepa, pp. 57–104. Boulder: Westview Press.

Nalepa, C. A. & Bell, W. J. (1997). Postovulation parental investment and parental care in cockroaches. In *Social Behaviour in Insects and Arachnids*, ed. J. C. Choe & B. J. Crespi, pp. 26–51. Cambridge: Cambridge University Press.

Nelson, J. B. (1965). The behaviour of the gannet. *British Birds*, **58**, 233–88.

Nelson, J. B. (1972). Evolution of the pair bond in the Sulidae. *Proceedings of the 15th International Ornithological Congress*, 371–88.

Neville H. J. (1990). Intermodal competition and compensation in development. *Annals of the New York Academy of Sciences*, **608**, 71–91.

Newton, I. (1989). *Lifetime Reproduction in Birds*. London: Academic Press.

Newton, I. & Marquiss, M. (1983). Dispersal of sparrowhawks between birthplace and breeding place. *Journal of Animal Ecology*, **52**, 463–77.

Nicolai, J. (1964). Der Brutparasitismus der Viduinae als ethologisches Problem. *Zeitschrift für Tierpsychologie*, **21**, 129–204.

Nicolai, J. (1974). Mimicry in parasitic birds. *Scientific American*, **231**(4), 92–8.

Nisbett, I. C. T. (1973). Courtship-feeding, egg-size, and breeding success in common terns. *Nature*, **241**, 141–2.

Nisbett, I. C. T. (1977). Courtship-feeding and clutch size in common terns *Sterna hirundo*. In *Evolutionary Ecology*, ed. B. Stonehouse & C. M. Perrins, pp. 101–9. London: Macmillan.

Normile, D. (1998). Habitat seen playing larger role in shaping behavior. *Science*, **279**, 1454–5.

Norton-Griffiths, M. (1969). The organisation, control and development of parental feeding in the oystercatcher (*Haematopus ostralegus*). *Behaviour*, **34**, 55–114.

Nottebohm, F. (1972). The origins of vocal learning. *American Naturalist*, **106**, 116–40.

Nowak, M. A. & Sigmund, K. (1998). Evolution of indirect reciprocity by image scoring. *Nature*, **393**, 573–7.

O'Connor, R. J. (1984). *The Growth and Development of Birds*. Chichester: Wiley.

O'Connor, R. J. (1991). Parental care. In *The Cambridge Encyclopedia of Ornithology*, ed. M. Brooke & T. Birkhead, pp. 246–53. Cambridge: Cambridge University Press.

Odling-Smee, F. J. (1988). Niche-constructing phenotypes. In *The Role of Behavior in Evolution*, ed. H. C. Plotkin, pp. 73–132. Cambridge, Massachusetts: MIT Press.

Odling-Smee, F. J. (1995). Biological evolution and cultural change. In *Survival and Religion: Biological Evolution and Cultural Change*, ed. E. Jones & V. Reynolds, pp.1–43. Chichester: Wiley.

Odling-Smee, F. J., Laland, K. N. & Feldman, M. W. (1996). Niche construction.

American Naturalist, **147**, 641–8.

Owen, M., Black, J. M. & Liber, H. (1988). Pair bond duration and the timing of its formation in barnacle geese (*Branta leucopsis*). In *Wildfowl in Winter*, ed. M. W. Weller, pp. 23–38. Minneapolis: University of Minnesota Press.

Oyama, S. (1985). *The Ontogeny of Information: Developmental Systems and Evolution.* Cambridge: Cambridge University Press.

Packer, C., Lewis, S. & Pusey, A. (1992). A comparative analysis of non-offspring nursing. *Animal Behaviour*, **43**, 265–81.

Packer, C., Tatar, M. & Collins, A. (1998). Reproductive cessation in female mammals. *Nature*, **392**, 807–11.

Pagel, M. (1994). Parents prefer pretty plumage. *Nature*, **371**, 200.

Pál, C. (1998). Plasticity, memory and the adaptive landscape of the genotype. *Proceedings of the Royal Society of London, Series B*, **265**, 1319–23.

Palmer, A. R. (1996). From symmetry to asymmetry: phylogenetic patterns of asymmetry variation in animals and their evolutionary significance. *Proceedings of the National Academy of Sciences, USA*, **93**, 14279–86.

Papaj, D. R. & Lewis, A. C. (ed.) (1993). *Insect Learning: Ecology and Evolutionary Perspectives.* New York: Chapman and Hall.

Parker, G. A. & MacNair, M. R. (1978). Models of parent–offspring conflict. 1. Monogamy. *Animal Behaviour*, **26**, 97–110.

Pärt, T. (1991). Philopatry pays: a comparison between collared flycatcher sisters. *American Naturalist*, **138**, 790–6.

Pärt, T. (1994). Male philopatry confers a mating advantage in the migratory collared flycatcher, *Ficedula albicollis. Animal Behaviour*, **48**, 401–9.

Pascal, B., *Penseés*. Translated by A. J. Krailsheimer, 1966. Harmondsworth, Middlesex: Penguin Books.

Pennisi, E. (1998). A genomic battle of the sexes. *Science*, **281**, 1984–5.

Perrins, C. M. (1979). *British Tits.* London: Collins.

Perrins, C. M. & McCleery, R. H. (1985). The effect of age and pair bond on the breeding success of great tits *Parus major. Ibis*, **127**, 306–15.

Piaget, J. (1978). *Behavior and Evolution.* New York: Pantheon Books.

Pianka, E. R. (1970). On r- and K-selection. *American Naturalist*, **104**, 592–7.

Pianka, E. R. (1994). *Evolutionary Ecology*, 5th edn. New York: HarperCollins.

Pickering, S. P. C. (1989). Attendance patterns and behaviour in relation to experience and pair-bond formation in the wandering albatross *Diomedea exulans* at South Georgia. *Ibis*, **131**, 183–95.

Pierotti, R. (1991). Infanticide versus adoption: an intergenerational conflict. *American Naturalist*, **138**, 1140–58.

Poole, J. H. (1994). Sex differences in the behaviour of African elephants. In *The Differences Between the Sexes*, ed. R. V. Short & E. Balaban, pp. 331–46. Cambridge: Cambridge University Press.

Posner, M. I. & Raichle, M. E. (1994). *Images of Mind.* New York: Scientific American Library.

Pravosudov, V. V. (1985). Search for and storage of food by *Parus cinctus lapponicus* and *P. montanus borealis* (Paridae). *Zoologichesky Zhurnal*, **64**, 1036–43.

Price, J. J. (1998). Family- and sex-specific vocal traditions in a cooperatively breeding songbird. *Proceedings of the Royal Society of London, Series B*, **265**, 497–502.

Price, K. & Boutin, S. (1993). Territorial bequeathal by red squirrel mothers. *Behavioral Ecology*, **4**,144–50.

Prop, J. & de Vries, J. (1993). Impact of snow and food conditions on the reproductive performance of barnacle geese *Branta leucopsis*. *Ornis Scandinavica*, **24**, 110–21.

Provenza, F. D. & Balph, D. F. (1987). Diet learning by domestic ruminants: theory, evidence and practical implications. *Applied Animal Behaviour Science*, **18**, 211–32.

Provenza, F. D. & Cincotta, R. P. (1993). Foraging as a self-organizational learning process: accepting adaptability at the expense of predictability. In *Diet Selection*, ed. R. N. Hughes, pp. 78–101. Oxford: Blackwell Scientific Publications.

Provine, R. R. (1996). Contagious yawning and laughter: significance for sensory feature detection, motor pattern generation, imitation, and the evolution of social behavior. In *Social Learning in Animals: The Roots of Culture*, ed. C. M. Heyes & B. G. Galef, Jr., pp. 179–208. San Diego: Academic Press.

Pruett-Jones, M. A. & Pruett-Jones, S. G. (1983). The bowerbird's labor of love. *Natural History*, **9**, 49–55.

Pryce, C. R. (1996). Socialization, hormones, and the regulation of maternal behavior in nonhuman simian primates. *Advances in the Study of Behavior*, **25**, 423–73.

Pusey, A. E. & Packer, C. (1987). The evolution of sex-biased dispersal in lions. *Behaviour*, **101**, 275–310.

Pusey, A. E. & Packer, C. (1994). Non-offspring nursing in social carnivores: minimizing the costs. *Behavioral Ecology*, **5**, 362–74.

Racey, P. A. & Swift, S. M. (ed.) (1995). *Ecology, Evolution and Behaviour of Bats. Symposium of the Zoological Society of London*, **67**. Oxford: Clarendon Press.

Rasa, A. (1985). *Mongoose Watch: A Family Observed*. London: John Murray.

Redondo, T., Gomendio, M. & Medina, R. (1992). Sex-biased parent–offspring conflict. *Behaviour*, **123**, 261–89.

Reeve, N. (1994). *Hedgehogs*. London: Poyser.

Reyer, H.-U. (1990). Pied kingfishers: ecological causes and reproductive consequences of cooperative breeding. In *Cooperative Breeding in Birds: Long-term Studies of Ecology and Behavior*, ed. P. B. Stacey & W. D. Koenig, pp. 529–57. Cambridge: Cambridge University Press.

Ricklefs, R. E. (1993). Sibling competition, hatching asynchrony, incubation period, and lifespan in altricial birds. In *Current Ornithology*, vol.11, ed. D. M. Power, pp. 199–276. New York: Plenum Press.

Riedman, M. L. (1982). The evolution of alloparental care and adoption in mammals and birds. *Quarterly Review of Biology*, **57**, 405–35.

Robertson, F. W. (1960). The ecological genetics of growth in *Drosophila*. *Genetical Research*, **1**, 288–318.

Rohwer, S. (1986). Selection for adoption versus infanticide by replacement 'mates' in birds. In *Current Ornithology*, vol. 3, ed. R. F. Johnston, pp. 353–95. New York:

Plenum.

Roosenburg, W. M. & Niewiarowski, P. (1998). Maternal effects and the maintenance of environmental sex determination. In *Maternal Effects as Adaptations*, ed. T. A. Mousseau & C. W. Fox, pp. 307–22. New York: Oxford University Press.

Rose, S. (1992). *The Making of Memory.* Toronto: Bantam Press.

Rose, S., Kamin, L. J. & Lewontin, R. C. (1984). *Not in Our Genes.* New York: Pantheon.

Rosenblatt, J. (1987). Maternal Behaviour. In *The Oxford Companion to Animal Behaviour*, ed. D. McFarland, p. 362. Oxford: Oxford University Press.

Rosenblatt, J. S. & Snowdon, C. T. (ed.) (1996). *Parental Care: Evolution, Mechanisms, and Adaptive Significance.* San Diego: Academic Press.

Rosenfield, I. (1988). *The Invention of Memory.* New York: Basic Books.

Ross, K. G. & Keller, L. (1995). Ecology and evolution of social organization: insights from fire ants and other highly eusocial insects. *Annual Review of Ecology and Systematics*, **26**, 631–56.

Ross, K. G. & Matthews, R. W. (ed.) (1991). *The Social Biology of Wasps.* Ithaca: Comstock.

Rowell, T. E. (1988). What do male monkeys do besides competing? In *Evolution of Social Behavior and Integrative Levels*, ed. G. Greenberg & E. Tobach, pp. 205–12. Hillsdale, New Jersey: Lawrence Erlbaum.

Rowley, I. (1983). Re-mating in birds. In *Mate Choice*, ed. P. P. G. Bateson, pp. 331–60. Cambridge: Cambridge University Press.

Royama, T. (1970). Factors governing the hunting behaviour and selection of food by the great tit (*Parus major* L.). *Journal of Animal Ecology*, **39**, 619–68.

Sapolsky, R. M. (1990). Stress in the wild. *Scientific American*, **262**(1), 106–13.

Sapolsky, R. M. (1992). Neuroendocrinology of stress-response. In *Behavioral Endocrinology*, ed. J. B. Becker, S. M. Breedlove & D. Crews, pp. 287–324. Cambridge, Massachusetts: MIT Press.

Sarrazin, F. & Barbault, R. (1996). Reintroduction: challenges and lessons for basic ecology. *Trends in Ecology and Evolution*, **11**, 474–8.

Savard, J.-P. L. (1987). Causes and functions of brood amalgamation in Barrow's goldeneye and bufflehead. *Canadian Journal of Zoology*, **65**, 1548–53.

Schacter, D. L. & Tulving, E. (ed.) (1994). *Memory Systems.* Cambridge, Massachusetts: MIT Press.

Schaller, G. B. (1967). *The Deer and the Tiger.* Chicago: University of Chicago Press.

Schenkel, R. (1948). Ausdrucks-studien an Wölfen. *Behaviour*, **1**, 81–129.

Schenkel, R. (1967). Submission: its features and function in the wolf and dog. *American Zoologist*, **7**, 319–29.

Schlicht, E. (1979). The transition to labour management as a gestalt switch. *Gestalt Theory*, **1**, 54–67.

Schlicht, E. (1998). *On Custom in the Economy.* Oxford: Clarendon Press.

Schwabl, H. (1993). Yolk is a source of maternal testosterone for developing birds. *Proceedings of the National Academy of Sciences, USA*, **90**, 11446–50.

Scott, J. P. (1975). *Aggression*, 2nd edn. Chicago: University of Chicago Press.

Scott, W. E. D. (1902). Instinct in song birds. Methods of breeding in hand-reared robins (*Merula migratoria*). *Science*, **16**, 70–71.

Searcy, W. A. & Yasukawa, K. (1996). Song and female choice. In *Ecology and Evolution of Acoustic Communication in Birds*, ed. D. E. Kroodsma & E. H. Miller, pp. 454–73. Ithaca, New York: Cornell University Press.

Seibt, U. & Wickler, W. (1977). Duettieren als Revier-Anzeige bei Vogeln. *Zeitschrift für Tierpsychologie*, **43**,180–7.

Serpell, J. (1995). *The Domestic Dog: Its Evolution, Behaviour, and Interactions with People.* Cambridge: Cambridge University Press.

Sherley, G. H. (1990). Co-operative breeding in riflemen (*Acanthissitta chloris*): benefits to parents, offspring and helpers. *Behaviour*, **112**, 1–22.

Sherman, P. W., Jarvis, J. U. M. & Alexander, R. D. (1991). *The Biology of the Naked Mole Rat.* Princeton, New Jersey: Princeton University Press.

Sherry, D. F. (1985). Food storage by birds and mammals. *Advances in the Study of Behavior*, **15**, 153–88.

Sherry, D. F. & Galef, B. G., Jr. (1984). Cultural transmission without imitation: milk bottle opening by birds. *Animal Behaviour*, **32**, 937–8.

Sherry, D. F. & Schacter, D. L. (1987). The evolution of multiple memory systems. *Psychological Review*, **94**, 439–54.

Shettleworth, S. J. (1990). Spatial memory in food-storing birds. *Philosophical Transactions of the Royal Society of London, Series B*, **329**,143–151.

Shettleworth, S. J. (1993). Varieties of learning and memory in animals. *Journal of Experimental Psychology*, **19**, 5–14.

Shettleworth, S. J. (1998). *Cognition, Evolution, and Behavior.* New York: Oxford University Press.

Sing, C. F., Haviland, M. B. & Reilly, S. L. (1996). Genetic architecture of common multifactorial diseases. *CIBA Foundation Symposium*, **197**, 211–32.

Skolnick, N. J., Ackerman, S. H., Hofer, M. A. & Weiner, H. (1980). Vertical transmission of acquired ulcer susceptibility in the rat. *Science*, **208**,1161–3.

Skutch, A. F. (1935). Helpers at the nest. *The Auk*, **52**, 257–73.

Skutch, A. F. (1961). Helpers among birds. *The Condor*, **63**, 198–226.

Skutch, A. F. (1976). *Parent Birds and their Young.* Austin: University of Texas Press.

Slagsvold, T. & Lifjeld, J. T. (1985). Variation in plumage colour of the great tit *Parus major* in relation to habitat, season and food. *Journal of Zoology*, **206**, 321–8.

Slagsvold, T., Amundsen, T. & Dale, S. (1994). Selection by sexual conflict for evenly spaced offspring in blue tits. *Nature*, **370**, 136–8.

Slobodchikoff, C. N., Kiriazis, J., Fischer, C. & Creef, E. (1991). Semantic information distinguishing individual predators in the alarms calls of Gunnison's prairie dogs. *Animal Behaviour*, **42**, 713–19.

Smith, A. (1776). *An Inquiry into the Nature and Causes of the Wealth of Nations.* New York: Random House reprint edition,1937.

Smith, C. C. & Reichman, O. J. (1984). The evolution of food caching by birds and mammals. *Annual Review of Ecology and Systematics*, **15**, 329–51.

Smith, N. C., Wallach, M., Petracca, M., Braun, R. & Eckert, J. (1994). Maternal transfer of antibodies induced by infection with *Eimeria maxima* partially protects chicks against challenge with *Eimeria tenella*. *Parasitology*, **109**, 551–7.

Smith, R. J. F. (1992). Alarm signals in fishes. *Reviews in Fish Biology and Fisheries*, **2**, 33–63.

Smith, S. M. (1980). Demand behavior: a new interpretation of courtship feeding. *The Condor*, **82**, 291–5.

Smith, W. J. (1977a). *The Behavior of Communicating: an Ethological Approach.* Cambridge, Massachusetts: Harvard University Press.

Smith, W. J. (1977b). Communication in birds. In *How Animals Communicate*, ed. T. A. Sebeok, pp. 545–74. Bloomington: Indiana University Press.

Smith, W. J. (1994). Animal duets: forcing a mate to be attentive. *Journal of Theoretical Biology*, **166**, 221–3.

Smokler, R., Richards, A., Connor, R., Mann, J. & Berggren, P. (1997). Sponge carrying by dolphins (Delphinidae, *Tursiops* sp.): a foraging specialization involving tool use? *Ethology*, **103**, 454–65.

Smotheran, W. P. (1982). Odor aversion learning by the rat fetus. *Physiology and Behavior*, **29**, 769–71.

Sober, E. & Wilson, D. S. (1998). *Unto Others: The Evolution and Psychology of Unselfish Behavior.* Cambridge, Massachusetts: Harvard University Press.

Soukhanov, A. H. (ed.) (1992). *The American Heritage Dictionary of the English Language*, 3rd edn. Boston, Massachusetts: Houghton-Mifflin.

Southwick, C. H. (1968). Effect of maternal environment on aggressive behavior of inbred mice. *Communications in Behavioral Biology*, **1**, 129–32.

Spalding, D. (1873). Instinct with original observations on young animals. *Macmillan's Magazine*, **27**, 282–93. (Reprinted with an introduction by J. B. S. Haldane in 1954 in the *British Journal of Animal Behaviour*, **2**, 1–11.)

Spurway, H. (1955). The causes of domestication: an attempt to integrate some ideas of Konrad Lorenz with evolution theory. *Journal of Genetics*, **53**, 325–62.

Squire, L. R. (1987). *Memory and Brain.* New York: Oxford University Press.

Stacey, P. B. & Koenig, W. D. (ed.) (1990). *Cooperative Breeding in Birds: Long-term Studies of Ecology and Behavior.* Cambridge: Cambridge University Press.

Stamps, J., Clark, A., Arrowood, P. & Kus, B. (1985). Parent–offspring conflict in budgerigars. *Behaviour*, **94**, 1–40.

Stamps, J., Clark, A., Arrowood, P. & Kus, B. (1989). Begging behavior in budgerigars. *Ethology*, **81**, 177–92.

Stander, P. E. (1992). Cooperative hunting in lions: the role of the individual. *Behavioral Ecology and Sociobiology*, **29**, 445–54.

Stanley Price, M. R. (1989). *Animal Re-Introductions: The Arabian Oryx in Oman.* Cambridge: Cambridge University Press.

Stark, O. (1995). *Altruism and Beyond: An Economic Analysis of Transfers and Exchanges within Families and Groups.* Cambridge: Cambridge University Press.

Stenning, M. J (1996). Hatching asynchrony, brood reduction and other rapidly reproducing hypotheses. *Trends in Ecology and Evolution*, **11**, 243–6.

Stevens, T. A. & Krebs, J. R. (1986). Retrieval of stored seeds by marsh tits *Parus palustris* in the field. *Ibis*, **128**, 513–25.

Stoleson, S. H. & Beissinger, S. R. (1995). Hatching asynchrony and the onset of incubation in birds, revisited. In *Current Ornithology*, vol. 12, ed. D. M. Power, pp. 191–270. New York: Plenum Press.

Suboski, M. D., Bain, S., Carty, A. E., McQuoid, L. M., Seelen, M. I. & Seifert, M. (1990). Alarm reaction in acquisition and social transmission of simulated-predator recognition by zebra danio fish (*Brachydanio rerio*). *Journal of Comparative Psychology*, **104**, 101–12.

Sulloway, F. J. (1996). *Born to Rebel: Birth Order, Family Dynamics, and Creative Lives*. London: Little, Brown and Company.

Szathmáry, E. (1995). A classification of replicators and lambda–calculus models of biological organization. *Proceedings of the Royal Society of London, Series B*, **260**, 279–86.

Szathmáry, E. & Demeter, L. (1987). Group selection of early replicators and the origin of life. *Journal of Theoretical Biology*, **128**, 463–86.

Tabashnik, B. E., Wheelock, H., Rainbolt, J. D. & Watt, W. B. (1981). Individual variation in oviposition preference in the butterfly, *Colias eurytheme. Oecologia*, **50**, 225–30.

Tajfel, H. (1978). *Differentiation Between Social Groups: Studies in the Social Psychology of Intergroup Relations*. London: Academic Press.

Tamarin, R. H., Ostfeld, R. S., Pugh, S. R. & Bujalska, G. (ed.) (1990). *Social Systems and Population Cycles in Voles*. Basel: Birkhäuser Verlag.

Tanaka, I. (1995). Matrilineal distribution of louse egg-handling techniques during grooming in free-ranging Japanese macaques. *American Journal of Physical Anthropology*, **98**, 197–201.

Tanaka, I. (1998). Social diffusion of modified louse egg-handling techniques during grooming in free-ranging Japanese macaques. *Animal Behaviour*, **56**, 1229–36.

Tauber, C. A. & Tauber, M. J. (1989). Sympatric speciation in insects: perception and perspective. In *Speciation and its Consequences*, ed. D. Otte & J. A. Endler, pp. 307–44. Sunderland, Massachusetts: Sinauer.

te Velde, J. H., Gordens, H. & Scharloo, W. (1988). Genetic fixation of phenotypic response of an ultrastructural character in the anal papillae of *Drosophila melanogaster. Heredity*, **61**, 47–53.

Tear, T. H. & Forester, D. J. (1992). The role of social theory in reintroduction planning: a case study of the Arabian oryx in Oman. *Society and Natural Resources*, **5**, 359–74.

Tear, T. H., Mosley, J. C. & Ables, E. D. (1997). Landscape-scale foraging decisions by reintroduced Arabian oryx. *Journal of Wildlife Management*, **61**, 1142–54.

Tempelton, A. R. (1989). The meaning of species and speciation: a genetic perspective. In *Speciation and its Consequences*, ed. D. Otte & J. A. Endler, pp. 3–27. Sunderland, Massachusetts: Sinauer.

ten Cate, C. (1987). Sexual preferences in zebra finch males raised by two species. 2. The internal representation resulting from double imprinting. *Animal Behaviour*, **35**, 321–30.

ten Cate, C. & Bateson, P. (1988). Sexual selection: the evolution of conspicuous characteristics in birds by means of imprinting. *Evolution*, **42**, 1355–8.

ten Cate, C. & Vos, D. R. (1999). Sexual imprinting and evolutionary processes in birds: a reassessment. *Advances in the Study of Behavior*, **28**, 1–31.

ten Cate, C., Kruijt, J. P. & Meeuwissen, G. B. (1989). The influence of testing conditions on sexual preferences in double imprinted zebra finch males. *Animal Behaviour*, **37**, 694–6.

Teoh, J.-I., Soewondo, S. & Sidharta, M. (1975). Epidemic hysteria in Malaysian schools: an illustrative episode. *Psychiatry*, **38**, 258–69.

Terkel, J. (1996). Cultural transmission of feeding behavior in the black rat (*Rattus rattus*). In *Social Learning in Animals*, ed. C. M. Heyes & B. G. Galef, Jr., pp. 17–47. San Diego: Academic Press.

Teuschl, Y., Taborsky, B. & Taborsky, M. (1998). How do cuckoos find their hosts? The role of habitat imprinting. *Animal Behaviour*, **56**, 1425–33.

Thompson, J. N. & Pellmyr, O. (1991). Evolution of oviposition behavior and host preference in Lepidoptera. *Annual Review of Entomology*, **36**, 65–89.

Thorpe, W. H. (1972). Duetting and antiphonal song in birds: its extent and significance. *Behaviour*, suppl. 18, 1–197.

Tinbergen, N. (1951). *The Study of Instinct*. Oxford: Clarendon Press.

Todt, D. (1975). Effect of territorial conditions on the maintenance of pair contact in duetting birds. *Experientia*, **31**, 648–9.

Todt, D., Hultsch H. & Heike, D. (1979). Conditions affecting song acquisition in nightingales (*Luscinia megarhynchos* L.). *Zeitschrift für Tierpsychologie*, **51**, 23–35.

Tomasello, M. (1994). The question of chimpanzee culture. In *Chimpanzee Cultures*, ed. R. W. Wrangham, W. C. McGrew, F. B. M. de Waal & P. G. Heltne, pp. 301–17. Cambridge, Massachusetts: Harvard University Press.

Tomasello, M., Kruger, A. C. & Ratner, H. H. (1993). Cultural learning. *Behavioral and Brain Sciences*, **16**, 495–552.

Trabant, J. (ed.) (1996). *Origins of Language*. Budapest: Collegium Budapest.

Trivers, R. L. (1971). The evolution of reciprocal altruism. *Quarterly Review of Biology*, **46**, 35–57.

Trivers, R. L. (1985). *Social Evolution*. Menlo Park, California: Benjamin Cummings.

Turner, D. C. (1984). A myth exploded: hunting behavior of vampire bats. In *The Encyclopaedia of Mammals*, ed. D. Macdonald, vol. 2, pp. 812–13. London: Allen & Unwin.

Turner, J. C., Hogg, M. A., Oakes, P. J., Reicher, S. D. & Wetherell, M. S. (1987). *Rediscovering the Social Group: A Self-Categorization Theory*. Oxford: Blackwell.

van Rhijn, J. & Groothuis, T. (1987). On the mechanism of mate selection in black-headed gulls. *Behaviour*, **100**, 134–69.

Varela, F. J., Palacios, A. G. & Goldsmith, T. H. (1993). Color vision of birds. In *Vision, Brain and Behavior in Birds*, ed. H. P. Zeigler & H.-J. Bischof, pp. 77–98. Cambridge, Massachusetts: MIT Press.

Von Frisch, K. V. (1938). Zur Psychologie des Fisches-schwarmes. *Naturwissenschaften*, **26**, 610–6.

Von Holst, D. (1998). The concept of stress and its relevance for animal behavior. *Advances in the Study of Behavior*, **27**, 1–131.

Waddington C. H. (1957). *The Strategy of the Genes*, London: Allen & Unwin.

Waddington, C. H. (1959). Evolutionary systems – animal and human. *Nature*, **183**, 1634–8.

Waddington, C. H. (1975). *The Evolution of an Evolutionist*, Edinburgh: Edinburgh University Press.

Wade, M. J. (1976). Group selection among laboratory populations of *Tribolium*. *Proceedings of the National Academy of Sciences, USA*, **73**, 4604–7.

Wallace, A. R. (1870). *Contributions to the Theory of Natural Selection.* London: Macmillan & Co.

Wang, Z. & Insel, T. R. (1996). Parental behavior in voles. *Advances in the Study of Behavior,* **25**, 361–84.

Ward, P. & Zahavi, A. (1973). The importance of certain assemblages of birds as 'information-centres' for food-finding. *Ibis,* **115**, 517–34.

Wasserman, E. A. (1995). The conceptual abilities of pigeons. *American Scientist,* **83**, 246–55.

Watanabe, K. (1989). Fish: a new addition to the diet of Japanese macaques on Koshima island. *Folia Primatologica,* **52**, 124–31.

Watson, J. B. (1924). *Behaviorism.* New York: Norton.

Weller, M. W. (1965). Chronology of pair formation in some nearctic *Aythya* (Anatidae). *The Auk,* **82**, 227–35.

Welty, J. C. & Baptista, L. (1988). *The Life of Birds,* 4th edn. New York: Saunders.

Whitehead, H. (1998). Cultural selection and genetic diversity in matrilineal whales. *Science,* **282**, 1708–11.

Whiten, A. & Ham, R. (1992). On the nature and evolution of imitation in the animal kingdom: reappraisal of a century of research. *Advances in the Study of Behavior,* **21**, 239–83.

Whiten, A., Goodall, J., McGrew, W. C., Nishida, T., Reynolds, V., Sugiyama, Y., Tutin, C. E. G., Wrangham, R. W. & Boesch, C. (1999). Cultures in chimpanzees. *Nature,* **399**, 682–5.

Wickler, W. (1980). Vocal duetting and the pair bond: 1. Coyness and partner commitment. A hypothesis. *Zeitschrift für Tierpsychologie,* **52**, 201–9.

Wiesner, B. P. & Sheard, N. M. (1933). *Maternal Behaviour in the Rat.* Edinburgh: Oliver and Boyd.

Wilcox, R. S. & Jackson, R. R. (1998). Cognitive abilities of araneophagic jumping spiders. In *Animal Cognition in Nature,* ed. R. P. Balda, I. M. Pepperberg & A. C. Kamil, pp. 411–34. San Diego: Academic Press.

Wiley, R. H. (1991). Associations of song properties with habitats for territorial oscine birds of eastern North America. *American Naturalist,* **138**, 973–93.

Wilkinson, G. S. (1984). Reciprocal food sharing in the vampire bat. *Nature,* **308**, 181–4.

Wilkinson, G. S. (1988). Reciprocal altruism in bats and other mammals. *Ethology and Sociobiology,* **9**, 85–100.

Wilkinson, G. S. (1992). Information transfer at evening bat colonies. *Animal Behaviour,* **44**, 501–18.

Williams, G. C. (1992). *Natural Selection: Domains, Levels, and Challenges.* New York: Oxford University Press.

Wilson, D. S. (1983). The group selection controversy: history and current status. *Annual Review of Ecology and Systematics,* **14**, 159–87.

Wilson, D. S. (1997). Altruism and organism: disentangling the themes of multilevel selection theory. *American Naturalist,* **150**, s122–34.

Wilson D. S. & Sober, E. (1994). Reintroducing group selection to the human behavioral sciences. *Behavioral and Brain Sciences,* **17**, 585–608.

Wilson, E. O. (1971). *The Insect Societies.* Cambridge, Massachusetts: Belknap Press.

Winkler, D. W. & Sheldon, F. H. (1993). Evolution of nest construction in swallows (Hirundinidae): a molecular phylogenetic perspective. *Proceedings of the National Academy of Sciences, USA*, **90**, 5705–7.

Witt, P. N., Reed, C. F. & Peakall, D. B. (1968). *A Spider's Web: Problems in Regulatory Biology*. Berlin: Springer Verlag.

Wolf, J. B., Brodie, E. D. III, Cheverud, J. M., Moore, A. J. & Wade, M. J. (1998). Evolutionary consequences of indirect genetic effects. *Trends in Ecology and Evolution*, **13**, 64–9.

Wolf, U. (1997). Identical mutations and phenotypic variation. *Human Genetics*, **100**, 305–21.

Woodroffe, R. & Vincent, A. (1994). Mother's little helpers: patterns of male care in mammals. *Trends in Ecology and Evolution*, **9**, 294–7.

Woolfenden, G. E. & Fitzpatrick, J. W. (1978). The inheritance of territory in group-breeding birds. *BioScience*, **28**, 104–8.

Wrangham, R. W. & Peterson, D. (1997). *Demonic Males*. London: Bloomsbury.

Wrangham, R. W., McGrew, W. C., de Waal, F. B. M. & Heltne, P. G. (ed.) (1994). *Chimpanzee Cultures*. Cambridge, Massachusetts: Harvard University Press.

Wright, J. (1997). Helping-at-the-nest in Arabian babblers: signalling social status or sensible investment in chicks? *Animal Behaviour*, **54**, 1439–48.

Wright, J. (1998a). Helping-at-the-nest and group size in the Arabian babbler *Turdoides squamiceps*. *Journal of Avian Biology*, **29**, 105–12.

Wright, J. (1998b). Helpers-at-the-nest have the same provisioning rule as parents: experimental evidence from play-backs of chick begging. *Behavioral Ecology and Sociobiology*, **42**, 423–9.

Wright, J. & Cuthill, I. (1990). Biparental care: short term manipulation of partner contribution and brood size in the starling, *Sturnus vulgaris*. *Behavioral Ecology*, **1**, 116–24.

Wright, J. & Cuthill, I. (1992). Monogamy in the European starling. *Behaviour*, **120**, 262–85.

Wright, R. (1994). *The Moral Animal*. London: Abacus.

Wunderle, J. M., Jr. (1991). Age-specific foraging proficiency in birds. In *Current Ornithology*, vol. 8, ed. D. M. Power, pp. 273–324. New York: Plenum Press.

Wyles, J. S., Kunkel, J. G. & Wilson, A. C. (1983). Birds, behavior, and anatomical evolution. *Proceedings of the National Academy of Sciences, USA*, **80**, 4394–7.

Yamagiwa, J. & Hill, D. A. (1998). Intraspecific variation in the social organization of Japanese macaques: past and present scope of field studies in natural habitats. *Primates*, **39**, 257–73.

Yom-Tov, Y. & Tchernov, E. (ed.) (1988). *The Zoogeography of Israel*. Dordrecht: Junk.

Young, J. Z. (1981). *The Life of Vertebrates*, 3rd edn. Oxford: Clarendon Press.

Zahavi, A. (1975). Mate selection – a selection for a handicap. *Journal of Theoretical Biology*, **53**, 205–14.

Zahavi, A. (1990). Arabian babblers: the quest for social status in a cooperative breeder. In *Cooperative Breeding in Birds: Long-term Studies of Ecology and Behavior*, ed. P. B. Stacey & W. D. Koenig, pp. 103–30. Cambridge: Cambridge University Press.

Zahavi, A. (1999). Babbler altruism and kin selection – a reply to Jon. *Journal of Avian Biology*, **30**, 115.

Zahavi, A. & Zahavi, A. (1997). *The Handicap Principle: a Missing Piece of Darwin's Puzzle.* New York: Oxford University Press.

Zentall, T. R. & Galef, B. G., Jr. (ed.) (1988). *Social Learning: Psychological and Biological Perspectives.* Hillsdale, New Jersey: Lawrence Erlbaum.

Zuk, M. (1992). The role of parasites in sexual selection: current evidence and future directions. *Advances in the Study of Behavior*, **21**, 39–68.

Zuk, M. (1994). Immunology and the evolution of behavior. In *Behavioral Mechanisms in Evolutionary Ecology*, ed. L. A. Real, pp. 354–68. Chicago: University of Chicago Press.

Index of species

Bird names are according to Howard, R. & Moore, A. (1991) *A Complete Checklist of the Birds of the World*, 2nd edn, London: Academic Press; mammal names are according to Nowak, R. M. (ed.) (1991) *Walker's Mammals of the World*, 5th edn, Baltimore: Johns Hopkins University Press.

Adélie penguin, *Pygoscelis adeliae* (Spheniscidae, Sphenisciformes) 169(n67)

African elephant, *Loxodonta africana* (Elephantidae, Proboscidea) 87, 196, 262

African ground squirrel, *Xerus inauris* (Sciuridae, Rodentia) 244, 245–6

Alaskan malamute, *see* domestic dog

albatross, species of Diomedeidae (Procellariiformes) 145

American coot, *Fulica americana* (Rallidae, Gruiformes) 193

American crow, sometimes called common crow, *Corvus brachyrhynchos* (Corvidae, Passeriformes) 160, 253

American robin, *Turdus migratorius* (Turdidae, Passeriformes) 308

ants, species of Formicidae (Hymenoptera) 63, 96, 176–7, 206(n27), 256, 315, 355–6; *see also* fire-ant, harvesting ant, slave-making ant

Arabian babbler, *Turdoides squamiceps* (Timaliidae, Passeriformes) 216–17, 240(n22), 256

Arabian oryx, *Oryx leucoryx* (Bovidae, Artiodactyla) 268

Atlantic salmon, *Salmo salar* (Salmonidae, Salmoniformes) 79, 90, 122

baboons, *Papio* spp. (Cercopithecidae, Primates) 178, 361; *see also* gelada baboon; hamadryas baboon; yellow baboon

barnacle goose, *Branta leucopsis* (Anserinae, Anatidae, Anseriformes) 124, 154–5, 158, 168(n47), 169(n66), 242(n64)

barn swallow, *see* swallow

Bateleur eagle, *Terathopius ecaudatus* (Accipitridae, Falconiformes) 244

bee-eaters, species of Meropidae (Coraciiformes, Aves) 278, 361; *see also* European bee-eater; white-fronted bee-eater

bees, species of Apidae (Hymenoptera); when 'social bees' is used, we refer to species of *Apis* and in particular *A. mellifera* 35, 96, 176, 206(n27), 315

Bengalese finch, *Lonchura striata* (Estrildidae, Passeriformes) 230

blackbird (European), *Turdus merula* (Turdidae, Passeriformes) 82, 83, 102(n42), 206(n29)

black grouse, *Tetrao tetrix* (Tetraonidae, Galliformes) 167(n6)

black rat, *Rattus rattus* (Muridae, Rodentia) 133–6, 137, 339–40

black-tailed deer, *Odocoileus hemionus columbianus* (Cervidae, Artiodactyla) 271

black-tailed prairie dog, *Cynomys ludovicianus* (Sciuridae, Rodentia) 299(n29)

Index of subjects